The Secret Influence
of the
MOON

"This is a book that I have been waiting to read. For untold centuries the great orb in the sky has been causing large numbers of our species to become transformed into lovers, looters, or lunatics. In addition to providing interesting scientific data about the Moon, Proud also explores the many occult and esoteric traditions that have grown up around Moon lore and traces our satellite's influences on psychic seers, sensitives, oracles, and prophets."

BRAD STEIGER, AUTHOR OF *REAL ENCOUNTERS,*
DIFFERENT DIMENSIONS, AND OTHERWORLDLY BEINGS

"Louis Proud's *The Secret Influence of the Moon* is intriguing and most enjoyable. It is the kind of thing that sparks the imagination and the sense of wonder."

JOHN SHIRLEY, AUTHOR OF *GURDJIEFF:*
AN INTRODUCTION TO HIS LIFE AND IDEAS

"Louis Proud takes us on a fascinating journey of the Moon's mysteries in ancient lore and what we may actually have found there. As one who has long suspected that not only did we go but also that we may have found something that caused a tight lid of security to be clamped down, I recommend this book as a valuable addition to the library of anyone interested in space anomalies."

JOSEPH P. FARRELL, PH.D., AUTHOR OF
FINANCIAL VIPERS OF VENICE AND
COVERT WARS AND BREAKAWAY CIVILIZATIONS

The Secret Influence
of the
MOON
Alien Origins
and Occult Powers

LOUIS PROUD

Destiny Books
Rochester, Vermont • Toronto, Canada

Destiny Books
One Park Street
Rochester, Vermont 05767
www.DestinyBooks.com

Text stock is SFI certified

Destiny Books is a division of Inner Traditions International

Library of Congress Cataloging-in-Publication Data
Proud, Louis.
  The secret influence of the Moon : alien origins and occult powers / Louis
Proud.
      pages cm
  Summary: "The Moon is not a cold, dead rock but a rich, fascinating world just
as alive as Earth"—Provided by publisher.
  Includes bibliographical references and index.
  ISBN 978-1-59477-494-2 (paperback) — ISBN 978-1-62055-165-3 (e-book)
  1. Moon. 2. Human beings—Effect of the moon on. 3. Moon—Origin. 4.
Extraterrestrial beings. 5. Occultism. I. Title.
  BF1723.P76 2013
  133.5'32—dc23
                                                                    2013018263

Printed and bound in the United States by Lake Book Manufacturing, Inc.
The text stock is SFI certified. The Sustainable Forestry Initiative• program
promotes sustainable forest management.

10 9 8 7 6 5 4 3 2 1

Text design and layout by Brian Boynton
This book was typeset in Garamond Premier Pro with V used as the display
typeface

To send correspondence to the author of this book, mail a first-class letter to the
author c/o Inner Traditions • Bear & Company, One Park Street, Rochester, VT
05767, and we will forward the communication, or contact the author directly at
http://louisproud.net.

CONTENTS

ACKNOWLEDGMENTS

I'd like to thank everyone who encouraged and supported me, either directly or indirectly, with the writing and publication of this book. Those I'd like to thank especially are:

The staff at Inner Traditions, in particular Jon Graham, for deciding this project was worthy of publication, as well as my courteous project editor Chanc VanWinkle Orzell; my dear wife, Tseada Zekarias, for coping with the challenges that come from being married to an author of weird books; Geoff Henshall, for providing me with part-time employment while I wrote the first half of the book; Tess Warrener, also for providing me with employment; my parents Suellen Fuller and Phillip Proud (1947–2010); John Shirley; William Patrick Patterson; Nick Redfern; Brad Steiger; Farouk El-Baz; Clive R. Neal; Joseph P. Farrell; Mike Bara; Jason Horsley (aka Aeolus Kephas); David and Robert of *New Dawn* magazine; Rich Lilly; G. Jeffrey Taylor; Duncan Roads; Paul McLean; Neil Hague; Linda Atherton; Patrick Huyghe; and last but not least the man in the Moon.

INTRODUCTION

*There is something haunting in the light of the Moon; it
has all the dispassionateness of a disembodied soul, and
something of its inconceivable mystery.*

JOSEPH CONRAD

Even though it isn't always visible, the Moon consistently appears in the
sky, orbiting the Earth as it has done for billions of years. If you go outside tonight and gaze up at the heavens, you're likely to see it. Its glowing presence, both beautiful and haunting, has fascinated and inspired
humankind since the very earliest times.

Most of us occasionally glance at the Moon, but few of us—with
the exception of astrologers and astronomers, perhaps—actually pay it
considerable attention. This is not surprising, since we've been led to
believe the Moon is little more than a cold, dead "rock" that orbits the
Earth—a pretty object, but one deserving of little excitement.

The Moon has been orbited and photographed by satellites, visited
by probes, and walked on—even driven on—by astronauts. Its entire
surface has been mapped, so we're told. It has, in the physical sense at
least, been quite extensively explored. It has now been about forty years
since *Apollo 17* made its journey to the Moon and returned. Completed
in December 1972, it remains the last manned mission of its kind and

the last time human feet (wearing boots!) made contact with lunar soil.

Why hasn't the Moon been colonized, you ask? Where are the lunar bases? And why haven't scientific instruments, telescopes especially, been placed on the Moon's surface? The far side of the Moon—the side turned away from Earth—is completely devoid of radio interference, making it an ideal location for a radio telescope. Yet, no such facility exists. All is quiet on the Moon. Or is it?

What was discovered on the Moon that has kept humans away for so long? A hostile alien presence, perhaps?

The journey on which you are about to embark is, in part, a personal one. It begins in my childhood. Like all curious children, I was fascinated by the Moon—the largest, brightest, and most impressive object in the night sky. I kept a close eye on the Moon, as I do today, paying particular attention to its current phase. I remember being affected in an obvious way by the occurrence of a full moon; I would become twitchy, angry, and hyperactive, as would my brothers and sisters. That the full moon influences human (and animal) behavior, augmenting our irrational tendencies, is a persistent belief that deserves to be investigated. We will attempt to get to the bottom of this later.

At the age of five I began to think of the Moon as an artificial and inhabited object—a huge, cavern-filled "nest" constructed by giant insects. Like all children, I was intrigued by insects and their dwellings. Of particular interest to me were the wasps' nests that appeared frequently on the ceiling of my father's shed. The most common of which were the urn-shaped ones made from gray-colored clay. I asked myself, "Could the Moon have been constructed by wasplike creatures?" The idea was not an implausible one.

I remember describing my "insect moon theory" to one of my teachers in primary school. She found it cute and amusing but, to my great disappointment, failed to take it seriously. The Moon, she tenderly explained, formed as a result of "complex scientific processes," and had nothing to do with insects, giant or otherwise. Despite this, I remained convinced that the Moon was not what it seemed; it was simply too strange and artificial looking to have been built solely by Nature's hand.

I still have in my possession a story I wrote in my early childhood that reflects how I feel about the Moon to this day. Consisting of just one sentence, it reads: "Once there was an astronaut that [*sic*] hit the Moon and cracked the Moon." The story (if it deserves to be called one) includes a drawing of the Moon with a prominent crack down its side with an astronaut on a flying chair adjacent to it.

For a celestial object such as the Moon to be "cracked," it would have to be hollow. In order to be hollow, it would have to be artificial. To quote the late Carl Sagan in relation to the moons of Mars, "a natural satellite cannot be a hollow object."[1] What makes my story unusual is I don't remember hearing about any alternative theories regarding the Moon. How, then, did the idea of an artificial Moon form in my mind?

Believe it or not, the idea of a hollow, artificial Moon is not as far-fetched as it seems; there is plenty of evidence to indicate that such is indeed the case. Even the "insect moon theory" of my childhood is not without some merit—especially if you consider how little knowledge science provides for the Moon's origin. Today's leading theory of lunar origin, known as the big whack (or giant impact) hypothesis, is accurately described by Christopher Knight and Alan Butler in their book *Who Built the Moon?* as containing "more holes than a rusty colander."[2] We will explore this topic later.

I wish to make it clear to the reader that although this book is replete with scientific information, it is by no means a conventional scientific book. A more suitable classification, perhaps, would be para-scientific—*para* meaning "beyond." Rather than limit myself to a scientific perspective of the Moon, I have chosen to employ a number of different perspectives, including the occult and the spiritual, thus expanding the scope of my investigation beyond what science can tell us about the Moon.

Science is a valuable exploratory tool. It is also an extremely limited one, fixated as it is with the physical realm. To understand the true nature of the Moon, we have no other choice but to transcend the physical, the conventional, and the mundane. Think of this as an exciting journey into largely uncharted territory. You might be surprised what we discover along the way.

In 2009 an extremely important science-fiction film, *Moon,* was released. I say "important" because the film is tremendously relevant to the theme of this book, almost uncannily so. After studying the film, I have concluded it contains a profound and important message for humankind, albeit a subtle and symbolic one. Only by looking at the film from an occult perspective can its message be truly appreciated. This is not a job for the left brain, but for the right.

Moon was directed by Duncan Jones, son of the infamous British rock star David Bowie. An actor as well as a musician, Bowie appears in a number of supernatural and science-fiction films, perhaps the most famous being *The Man Who Fell to Earth* (1976). In this film, Bowie plays the lead role of Thomas Jerome Newton, an alien from a dying planet. *Moon* was the first film directed by Jones. His second was *Source Code,* released in 2011, a sci-fi thriller starring Jake Gyllenhaal.

But back to *Moon.* For those who haven't seen the film, let me here issue a "spoiler alert" as I describe the entire plot. The story takes place on a desolate mining base located on the far side of the Moon. Here, with the use of large, self-automated harvesting machines, valuable helium-3 is extracted from the lunar regolith and stored in canisters, then rocketed to Earth to be used as a clean source of fuel in the generation of fusion energy. Only one employee is needed to keep the base running. His name is Sam Bell, who is played by Sam Rockwell.

Only two weeks away from completing his three-year contract, Sam is looking forward to returning to Earth to be reunited with his wife and young daughter. Life on the base is lonely and boring. To speak with his family or employer, Sam must rely on pre-recorded messages, because direct communication with Earth is not possible. His only assistance is provided by an intelligent computer named GERTY.

Sam's physical and mental health begins to show signs of deterioration. One day he sees a hallucination of a young woman. Later, while driving his rover in close proximity to a harvester, he is distracted by a hallucination and crashes his vehicle, losing consciousness.

When Sam regains consciousness, he is back inside the base. GERTY explains that he suffered an accident but will soon be back to his normal self. Moments later he overhears part of a live and covert conversation between GERTY and his employer. Afterward GERTY tells Sam, for reasons that aren't made clear, he's been instructed to not allow Sam outside the base. Sam begins to realize that something suspicious is going on.

Tricking GERTY, Sam drives over to the scene of the accident. Here he makes a disturbing discovery: inside the crashed rover is a man with the same appearance as him; in fact, it is him. The man is unwell but soon regains consciousness. Now there are two Sams, and each one suspects the other of being a clone. At first the men detest each other, at one point engaging in a vicious physical fight. Later they begin to cooperate.

GERTY reveals that both men are clones and that the memories they share—of having a wife and child, for example—are implants taken from the real Sam Bell. When, earlier, GERTY told Sam that he'd suffered an accident, this was a lie to conceal the fact that he'd actually been awakened; that he was, in fact, a new clone. Meanwhile the original Sam Bell (the one from the crashed rover) continues to experience poor health; it's as if his body is falling apart.

The men discover that the base is equipped with radio-jamming antennas, which serve as an explanation for the lack of live communication with Earth. Also discovered are classified video logs taken by previous clones. The footage shows clones getting sick toward the end of their three-year contract—a clear indication that they "expire" around this time. The footage also shows what is believed to be a hibernation pod used to transport a person back to Earth and reveals the pod has another purpose entirely: it kills the sick clones while they're asleep.

The men discover a secret facility located beneath the floor of the base, stocked with thousands of unawakened clones. What's been happening on the base is obvious: Every three years, following the death of one clone, a new clone is awakened. Each clone believes he's the real Sam Bell and that he'll eventually be sent back to Earth. In reality, each clone expires toward the end of his three-year contract and is killed in the aforementioned pod, immediately after which another clone is awakened. And thus, the cycle continues.

Of course, the second Sam Bell was not supposed to have discovered the body of the original Sam Bell. Had they never encountered each other life on the base would have continued as normal. The clones would have remained asleep and ignorant, presumably for thousands of years.

So, what makes *Moon,* in my view, so important? And what is its message exactly? One interpretation is as follows: The clones of Sam Bell represent humanity. Once asleep, we begin to realize we are not who we thought we were. We become aware that we are someone else's property, that we are owned—and exploited—by an intelligence of lunar origin. This intelligence manipulates our minds, feeds us lies, and tampers with our DNA, so we will continue to provide it (or them) with energy.

I admit the message of the film could be interpreted in a number of different ways; no doubt it already has been by conspiracy theorists on Internet forums. At its most basic, the message is this: There is something sinister and artificial about the Moon; it cannot be trusted.

Jones, who co-wrote the script for *Moon* and is clearly fascinated by the Moon itself, commented in an article that appeared in the *Independent:* "For me, the Moon has this weird mythic nature to it. There is still a mystery to it. . . . We (humankind) have been there. It is something so close and so plausible and yet at the same time, we really don't know that much about it."[3] *Moon* was extremely well received by the public and critics alike, despite it being a low-budget production with less than impressive special effects. It won two film awards, including Best British Independent Film for 2009.

It's obvious Jones has more than a superficial understanding of the

Moon, one example being that the clones in his movie follow an eternal cycle of birth, death, and rebirth—and so too does the Moon. As we know, the Moon waxes (is born and grows), is full (reaches maturity), wanes (grows old), and then disappears (dies) for a period of three days. The cycle could be reduced to three visible phases: waxing, waning, and full. Three is the number of years of a clone's life span. Coincidence? I doubt it.

While we're on the topic of Moon-related movies, mention ought to be given to *The Truman Show* (1998), directed by Peter Weir. The film's protagonist is a thirty-something insurance salesman named Truman Burbank, played by Jim Carrey, who lives with his excruciatingly boring wife in a pleasant and perfect little town called Seahaven. Little does Truman know that his hometown is actually a giant TV set peopled entirely by actors and crew, and he is the subject of the show. Enjoyed by billions of ardent fans, it is broadcast live around the world twenty-four hours a day.

Since Truman was an infant, his every word and action has been captured by thousands of hidden cameras. His wife and best friend, even his mother and father, are all actors in the show. The show is scripted: Truman is placed in specific situations and made to exhibit certain emotions. Like a puppet, his strings are pulled this way and that, leaving him but the tiniest grain of free will. Almost every aspect of Truman's life is controlled; he is manipulated, for example, into marrying a woman whom he doesn't love.

The TV set, the only world Truman knows, is contained inside a dome so huge it's visible from space. In a sense, nothing inside the dome, except for Truman himself, is natural. The stars, the sky, the Moon, the weather, and the ocean are all artificial. Throughout the film Truman undergoes a process of awakening in which he begins to realize that his reality is false. What triggers this process is an event that occurs at the beginning of the film: a spotlight—one of the set's artificial stars—accidentally falls from the sky, smashing at his feet. (The event is covered up by local media.)

Significantly, Truman's "puppet master" is the director and producer of the show. A beret-wearing intellectual named Christof, he operates from a studio hidden inside the Moon called the "lunar room." Like the real Moon, of course, it is located high in the sky, making it a kind of watchtower. The lunar room is packed with all sorts of gadgetry. From here, Christof and his assistants are like gods, able to control Truman and the world in which he lives however they please. If they want to make it rain, they can do so. If they want to produce a lightning storm, they can do that too.

The symbolic significance of Christof's lunar room would not have escaped the reader's attention. In fact, the film is so packed with meaning—of a lunar nature and otherwise—one could probably write an entire book on the subject. Following its release thirteen years ago, *The Truman Show* is still the focus of a great deal of intellectual, spiritual, and philosophical analysis. Academics have called it an "existential masterpiece." Christians too have had a field day with the movie, identifying Christof (Christ?) as the almighty himself.

Both *Moon* and *The Truman Show* could be taken as examples of humanity's inability to trust the Moon. Other works of fiction, in the form of books and movies, communicate the same message. For example, in *The First Men in the Moon* (1901), written by H. G. Wells, two explorers, one a scientist and the other a businessman, discover the Moon to be inhabited by malevolent antlike creatures and are taken prisoner by them.

The creatures, called selenites (a reference to the Moon goddess Selene), have large heads, prominent black eyes, gray skin, and degenerated limbs—characteristics similar to what members of the UFO community call "greys." The selenites live beneath the surface of the Moon in a network of large, interconnected caves filled with air. Only rarely do they venture to the surface.

Sometimes a work of fiction contains information that could only have been obtained by highly intuitive means. It cannot be denied that Wells, known as the "father of science fiction" and "the man who invented tomorrow," was equipped with a prophetic mind. Perhaps, while writing his remarkable novel, he had a "psychic glimpse" of the Moon's hollow interior and the mysterious entities who dwell there.

Speaking of mysterious lunar entities, although it is considered a scientific fact that the Moon is entirely devoid of life, many continue to believe otherwise—and for sensible reasons too. There is some evidence that Apollo astronauts saw UFOs on and within close proximity to the Moon, perhaps indicating that the Moon is inhabited by extraterrestrial beings living in secret on its far side or under its surface.

In April 2011, I interviewed the author Nick Redfern, an expert on UFOs, conspiracies, and the paranormal. We discussed the possibility of an alien presence on the Moon, a topic on which he is well researched. Redfern accepts, like most people, that the Moon is a dead and sterile world with the faintest trace of an atmosphere, as we've been told by scientists and other authorities. "However," he explained, "that doesn't mean that advanced entities, perhaps deep underground, may not have constructed vast installations that have self-enclosed environments. . . . Even with the Moon being a vacuum, this wouldn't necessarily hinder advanced entities from inhabiting it with the right life-sustaining technologies."[4]

The idea of alien beings living in secret beneath the surface of the Moon is a strange but not illogical one. From an informal e-mail survey involving a group of friends and acquaintances of mine, I discovered that the vast majority of my small "sample" was willing to acknowledge the possibility of an alien presence on the Moon. The results of my (admittedly less than perfect) survey suggest that the average person regards the Moon with suspicion or, to put it another way, as a place of secrets and mysteries.

So we know there's something odd about the Moon—something suspicious and "out of place." But can we put our finger on it? The Moon is like a man dressed in a trench coat and sunglasses, loitering outside a jewelry store. He appears there day after day, but whenever you attempt to approach him he scuttles off down an alley.

It would not be unreasonable to suggest humanity is currently undergoing a kind of "lunar awakening." If true, it would mean a significant percent of the population is beginning to awaken to the realization that the Moon is not what it seems. It has been said that when humanity begins to awaken to a truth, the manifestations that occur as a result are at first implicit and creative. Books and films are produced, for example, and they tend to resonate strongly with the public, connecting with something deep inside of us. Could the production of the film *Moon* and, to a lesser extent, *The Truman Show* be taken as evidence of a lunar awakening? If such an awakening exists, evidence of it ought to naturally emerge as we continue our investigation of the Moon.

When we look at the Moon we see what we've been conditioned to see: a huge, dead rock. The late Carl Sagan was bold enough to call the Moon "boring." I suppose, in a way, the Moon is boring, but only because of our limited perception of it. It's a bit like looking at a painting from the side; one can barely see the image this way. With this book I hope to widen your perception of the Moon and show you what a fascinating place it truly is.

The Moon has two sides. The side turned away from Earth, the side we cannot see, is known as the far side. The side always visible from Earth, the one we can see, is known as the near side. So too the Moon possesses a physical (or visible) side and a nonphysical (or invisible) side. Before we explore the latter we must deal with the Moon we know. Only by approaching it this way, one lunar rock at a time, can we begin to perceive the Moon in its true form.

1

THE MOON WE KNOW

When the Sun is shining on the surface at a very shallow angle,
the craters cast long shadows and the Moon's surface seems very
inhospitable. Forbidding, almost. I did not sense any great
invitation on the part of the Moon for us to come into its domain.
I sensed more that it was almost a hostile place, a scary place.

MICHAEL COLLINS, FORMER U.S. ASTRONAUT

As I write this from my house in Melbourne, Australia, it's 6:30 in the evening, and the Moon outside in the sky is one day away from being full. Because the Moon is still low on the horizon, it looks extremely large. If I were to wait several hours and look at the Moon again, when it would be farther above the horizon, it would appear smaller. This mysterious optical effect, known as the "Moon illusion," remains unexplained by science, although various theories have been offered to account for it.*

---

*There is no consensus among scientists as to the cause of the Moon illusion. Explains *Encyclopedia Britannica:* "Considerable debate surrounds the source of the Moon illusion. Some explanations have attributed it to the paradoxical idea that the Moon at the horizon seems larger because the brain perceives it as being farther away than the Moon at the sky's zenith. Another explanation is that the lack of distance cues in the night sky causes the eyes to adjust to a near-focus position, which makes the high Moon appear smaller."[1]

11

So if you're thinking of doing some serious Moon gazing, I suggest you plan it for when the Moon has either just risen or is about to set. That way you'll catch it at its largest and most spectacular.

Currently, the right-hand edge of the Moon is in shadow. Tomorrow the shadow will be gone and the Moon will appear as a complete, unbroken disc—a full moon. Next it will wane; a shadow will creep across the Moon until it disappears entirely, becoming new. For a period of three days, the Moon will not be visible in the sky. Then, it will wax again, eventually becoming full. And on the cycle goes.

We know Earth orbits the sun with a revolutionary period of 365 days, or one solar year. We also know the Moon orbits Earth and together comprise a planet-satellite system. In case you're wondering why we only ever see one side of the Moon, the near side, this is because the Moon is in synchronous rotation with Earth. The rotation of the Moon—the amount of time it takes to spin once on its own axis—is equal to the amount of time it takes to complete one orbit of the Earth. This time period is called a sidereal month and lasts for 27.3 days.

Only 50 percent of the Moon's surface is visible from Earth at any one time (disregarding the lunar phase). Due, however, to effects known as liberations, we are able to see a total of 59 percent of its surface, leaving 41 percent permanently out of sight. So no matter how often you look at the Moon, you will never be able to see it in its entirety. If you wanted to catch a glimpse of the Moon's far side and the spectacular craters that cover it, you would need to orbit it in a spacecraft of some kind—something that most of us will probably never experience. It was Mark Twain who said, "Everyone is a Moon, and has a dark side which he never shows to anybody."*

---

*Mark Twain ought to have written "far side" as opposed to "dark side," though no doubt this would have sounded less poetic. The terms "far side of the Moon" and "dark side of the Moon" are often used interchangeably, when in fact they refer to two different things. The Moon's far side is the hemisphere that is permanently turned away from Earth. The dark side of the Moon is the hemisphere that is not presently illuminated by the Sun. If the Moon was half-full, for example, we would refer to the part of it in shadow as its "dark side." Of course, there is no actual dark side of the Moon—every one of its sides receives the light of the Sun.

The phases of the Moon depend on its position in relation to the Sun and Earth. A new Moon occurs when it is between the Sun and the Earth; a full Moon, when Earth is between the Sun and the Moon. In both cases, the Sun, Moon, and Earth lie in a straight line; this configuration is known as syzygy. When the Moon is at "first quarter" or "last quarter" (often called a "half-Moon"), the three bodies form an L-shape, with the Moon and the Sun at right angles to each other. A slightly different configuration results in a crescent-shaped Moon.

The term *terminator* refers not only to a cyborg assassin from the future but also to the boundary between the day and night sides of the Moon. Another name for the terminator is the "twilight zone." On Earth, the terminator is experienced as a sunset or sunrise. Contrary to common assumption, a full Moon is the worst time to view the Moon with a telescope. A lack of shadows, and therefore lack of contrast, means its features are difficult to discern. It's best to wait for when the Moon's terminator is present, such as at half-Moon. Within this transitional zone—this merging of darkness and light—long shadows are cast, and the Moon's features stand out.

Unlike the Sun, the brilliant life-giving star at the center of the solar system, none of the eight major planets produce any light of their own; they merely reflect the Sun's rays. This also applies to the Moon. The Moon's "glow" is an illusion, in the sense that it doesn't emit light; it borrows it. But no matter how impressive the Moon looks when it's full, being as it is the brightest object in the night sky, it is actually a very poor reflector of light.

The albedo of an object refers to its reflective power. Snow, which reflects light so strongly it can damage the eyes and skin, has an albedo of around 0.8, or 80 percent. The average albedo of the Moon, on the other hand, is 0.12, or 12 percent, similar to that of coal.

When positioned between the Sun and the Earth, a new Moon cannot be seen at all by anyone on Earth—there is no light to illuminate its near side. Conversely, when the Moon is "full"—with the Earth positioned between it and the Sun—its near side is exposed to the Sun, illuminating it. Only by illumination from the Sun, either partially or completely, is the Moon made manifest in the sky.

As you may have deduced already, a day on Earth and a day on the Moon are far from equal in duration. The former consists of twenty-four hours, which is the time it takes for the planet to rotate once on its own axis. We know the Moon takes 27.3 days (approximately 655 hours) to rotate once on its own axis. Therefore we know that one day on the Moon is equivalent to this period, which roughly equals one month on Earth.

Imagine living on the Moon and seeing the Sun rise every 655 hours! Imagine, too, experiencing around two weeks of daylight, followed by an equally long night. During the course of a day on the Moon, the part of its surface exposed to the Sun can reach up to 243°F. Then, at night, when shielded from the Sun, this same area can cool to around −272°F. On Earth, our atmosphere acts as a protective shield blocking some of the Sun's light as well as trapping heat, which helps to maintain a comfortable range of temperatures. But the Moon has little atmosphere to speak of, hence the extreme temperatures experienced there.

As mentioned before, the Moon orbits Earth every 27.3 days, yet the period between consecutive full Moons is more than two days longer than this. One important fact needs to be taken into consideration: at the same time the Moon is orbiting Earth, the Earth is also orbiting the Sun. Let's say the Moon was full and has now completed one revolution of the Earth (one sidereal month). The Moon in this position is not in line with the Sun and Earth. For it to be full again, the Moon has to travel slightly more than 360 degrees, adding more than two days to its journey. The period between consecutive full Moons (or new Moons, for that matter) is an average of 29.5 days. This time period is called a synodic month.

It's important to realize our planet isn't the only planet in the solar system to be accompanied by a natural satellite, or moon. Mars, for example, has two; Jupiter has seventeen; Uranus is orbited by a whopping twenty moons. (Neither Mercury nor Venus have moons.) There are several reasons why our Moon remains unique. First, Earth is the only planet in the solar system with exactly one Moon.

There was a time when distant, ice-covered Pluto was considered, like Earth, to have only one moon. Since 1978, we've known Pluto is orbited by a moon with a diameter measuring more than half of its own, named Charon. (Charon is about 52 percent the size of Pluto.) In 2005, two smaller moons were added to the list, named Nix and Hydra, giving Pluto a total of three moons. Then, in 2006, Pluto was officially classified as a mere dwarf planet, changing the total number of known planets in the solar system from nine to eight.

Considering Pluto's new status, our Moon is the largest natural satellite in the solar system relative to its primary planet, Earth. The mean radius of Earth is 3,959 miles, whereas the mean radius of the Moon is 1,079.40 miles. If you divide the latter by the former you get roughly 0.27, or 27 percent, meaning that the Moon is about one-quarter the size of Earth. (See plate 1.)

Because of the Moon's relative size, its presence in the sky is most conspicuous. It's even possible to identify some of its features without the aid of a telescope. Most noticeable of all are the dark splotches on its surface, which contrast greatly with the paler areas called highlands. The contrasting areas produce the image of a face; or, if you like, a rabbit. Some might refer to this image as the Man in the Moon.

The dark areas are called *maria* (Latin for "seas"), named as such because early astronomers mistook them for actual seas. They formed as a result of volcanic activity in the ancient past, which forced molten material from the mantle onto the surface, producing huge lakes of lava. Now solidified, these vast areas consist of basalt rock. They cover about 17 percent of the Moon's surface and are located mainly on the near side. The far side, which has very few maria, is rough and densely cratered. Due to their iron-rich composition, the maria are less reflective than the highlands; therefore, they appear dark to the naked eye.

Putting it mildly, the Sun is an extremely huge object, with a diameter of around 864,949 miles—more than one hundred times the size of Earth and approximately four hundred times the size of the Moon. Furthermore, the Earth and the Sun are separated by a mean distance of 93 million miles, which just happens to be more than four hundred times greater than the mean distance between the Earth and the Moon. Expressed another way, the Moon is $1/400$ of the distance between Earth and the Sun. These figures may seem meaningless at first glance, but they explain why the Sun and the Moon appear to be the same size from Earth. And this, indeed, is a remarkable coincidence.

"Whilst we casually take it for granted that the two main bodies seen in Earth's skies look the same size, it is actually something of a miracle," explain Knight and Butler.[2] So amazed by the phenomenon was Isaac Asimov (1920–1992), respected author and scientist, that he called it "the most unlikely coincidence imaginable."[3]

This "unlikely coincidence" is what makes total solar eclipses possible. Described by Patrick Moore, a well-known British astronomer, as "the grandest sight in all nature," they occur when the disc of the Moon completely covers the disc of the Sun.[4] More common, but less spectacular, are lunar eclipses, which occur when Earth is positioned between the Sun and the Moon, creating a shadow that blocks out the Sun's disc. Eclipses can be either partial or complete.

The ancient Greeks correctly recognized that the Moon orbits the Earth but incorrectly assumed that it does so in a circular fashion. In their view, the circle represented perfection. Because everything in the heavens is entirely perfect to the ancient Greeks, all celestial bodies must follow circular orbits. It was Johannes Kepler (1571–1630), the brilliant German astronomer and mathematician, who discovered the planets move around the Sun in ellipses. This motion also applies to the Moon. Kepler's first law of planetary motion, published in 1609, states this fact.

Given the Moon's elliptical orbit around the Earth, the distance between them varies. Its mean orbital distance is 238,600 miles. When at its closest distance to Earth we say the Moon is at *perigee;* at its farthest, *apogee.*

Some astronomers believe the Earth and Moon deserve to be classified as a double-planet system as opposed to a planet-satellite system. One reason for this is the Moon doesn't technically orbit Earth; more correctly, the two orbit each other around a common center of mass called the barycenter. In the case of a propeller blade, which has an equal amount of weight on each end, the barycenter is located directly in the middle. Such is not the case with the Earth and Moon, because the Earth has eighty-one times the mass of the Moon. As you can imagine, the balancing point, or barycenter, of the two bodies is far closer to the heavier one, the Earth. In fact, the barycenter is located *inside* the Earth (2,920 miles from its center).

You would assume that the Moon orbits the Earth at a constant velocity; however, according to Kepler's second law, the velocity of the Moon is faster the closer it is to Earth and slower when it's farther away. The same holds true for Earth as it journeys around the Sun. On average, the Moon hurtles around the Earth at a velocity of about 0.636 miles/sec (or 2,288.5 miles/hour). To give some sense of comparison, the cruising speed of a Boeing 747 is about 560 miles/hour—about four times slower than the mean orbital velocity of the Moon.

Depending on the Moon's distance to Earth, it can appear smaller or larger in the sky. The Moon's diameter at apogee is about $9/_{10}$ of what it is at perigee, which is equal to a decrease in size of about 19 percent. Certainly the distance between the Earth and Moon does vary significantly, not only because of the Moon's elliptical orbit but also because of the combined gravity of the Earth, Sun, and the planets. This distance can range between an average of 221,000 miles at perigee and 253,000 miles at apogee.

Within the last three decades of the twentieth century, for example, the Moon's apogee ranged approximately between 251,034 miles and 252,712 miles, whereas its perigee ranged between 221,519 miles and 230,156 miles. In 2011 the Moon reached its maximum perigee on March 19, at 221,567 miles, and its maximum apogee on April 2, at 252,684 miles.

On the evening of March 19, 2011, I was fortunate enough to get a clear look at the Moon, and what I saw was extremely impressive. On that date, as mentioned, the Moon was 221,567 miles away—making it appear quite large. This in itself wasn't special; however, what was special was that the Moon was also full. This is what's called a lunar *perigee-syzygy*, which occurs when a new or full Moon is at its closest distance to Earth. It is a unique alignment.

The March 19 perigee-syzygy was significant enough to be mentioned in all the major newspapers. Although a perigee-syzygy is not an uncommon occurrence as they happen about once every thirteen months, this one was special for two reasons. First, the most recent full Moon to have passed this closely to Earth was eighteen years ago, in March 1993, and second, the alignment between perigee and syzygy overlapped each other in about an hour's time. To be precise, the full Moon reached its peak one hour and thirteen minutes after attaining its maximum perigee. An alignment this close is a rare occurrence. The end result was a very spectacular full Moon, one of the biggest and brightest I've seen to date.

According to astronomers, we won't see another perigee-syzygy of roughly the same magnitude until November 14, 2016. I say "roughly" because the full Moon on that date will be just as close, in fact, slightly closer at 221,526 miles; however, the alignment between perigee and syzygy won't be as synchronized. The full Moon will reach its peak two hours and thirty minutes after attaining its maximum perigee. Provided the human race hasn't gone extinct by 2016, I'll certainly take a look at it.

So what would it be like to visit the Moon and to actually walk on its surface? Instead of thinking of the Moon and Earth as a duo, we must think of them as two separate worlds. These two worlds are so different in nature that they're beyond comparison. Of course, it's impossible not to compare the two in our current perception. We are earthlings, and, consequently, Earth is the only point of reference we have. We are limited to thinking of the Moon in strictly Earth terms.

Someone who has never traveled outside of their own country would have great difficulty accepting that other parts of the planet differ greatly in terms of geography, climate, culture, and so forth. I've never left Australia, for example, and so I think of the deserts in Africa as similar in appearance to the ones in Australia. Only by visiting the deserts in Africa would I come to realize that the two environments share as many differences as they do similarities. I might even decide that the deserts in Africa are so unique that they cannot be compared to any other environment.

Perhaps the first thing you'd notice on the Moon, in comparison to conditions on Earth, is the lower surface gravity. You would feel much lighter on your feet, allowing you to hop with ease. The mass and size of the Earth is such that the surface gravity is ideal. Most of us, with the exception of the sumo wrestler and the Indian yogi, don't have to deal with the problem of getting stuck to the floor or drifting off into space. But the Moon, remember, is one-fourth the size of Earth and eighty-one times less massive. The result is a surface gravity one-sixth that of the Earth.

The terms *weight* and *mass* don't mean the same thing; simply put, the former relates to gravity and the latter does not. The less gravity exerted on a person, the less their weight will be. If, for example, you weigh 132 lbs on Earth, your weight on the Moon would be about 22 lbs; however, your mass would remain the same. Neil Armstrong, the first human being to set foot on the Moon, described walking on its surface as "even perhaps easier than the simulations. . . . It's absolutely no trouble to walk around."[5]

Commenting on its surface, Armstrong used the words "fine and almost like a powder."[6] He was referring, of course, to the topmost layer of the lunar soil, or regolith. Like Earth, the Moon is also covered in a soil of sorts, a nonorganic substance devoid of moisture and air. Varying in depth from seven to thirty feet, it is primarily composed of very small particles in addition to some rocks. The regolith is continually bombarded by micrometeorites, which churn it up.

The topmost layer of the regolith, a fine dust, has been described by astronauts as an extremely sticky, abrasive, and annoying substance, capable of finding its way into every nook and cranny. Former astronaut Gene Cernan, who traversed the Moon in a battery-powered "lunar rover" buggy during the *Apollo 17* mission, found the surface dust a nuisance. "It gets into your spacesuit and all moving parts of your vehicle," he said. "The dust is so fine it even gets into the pores of your skin. It took me many weeks after my return to get rid of the last traces of it."[7] The dust, which carries an electrostatic charge, is capable of circulating above the lunar surface.

Speaking of the region above the lunar surface, there is but the faintest trace of an atmosphere on the Moon. It is of such little density that the term *vacuum* applies, meaning "a volume of space essentially devoid of matter." So high is this vacuum that it generally surpasses those created in laboratories on Earth. The tenuous nature of the lunar atmosphere can be explained in part by its weak gravitational pull. If, for example, the gravitational pull of the Earth was much less, our atmosphere would simply "leak away"; the planet would be incapable of holding on to it.

To understand this better, it helps to compare the Moon's escape velocity to that of Earth. In the case of a rocket, it has to reach a certain minimum velocity before it can escape from the gravitational field of the Earth, hence the term *escape velocity*. The escape velocity of Earth is 7 miles per second, and the Moon's is a mere 1.5 miles per second. At any speed less than 7 miles per second on Earth, the rocket would be unable to make it out; it would come crashing back to Earth.

The lower the escape velocity of a planet or moon, the easier it is to break free of its gravitational field. For this reason, the Moon would be a perfect location from which to launch rockets. Assuming the Moon had a more substantial atmosphere in the past, whatever gases were once present have since leaked off into outer space, because the gas particles must have attained (or exceeded) escape velocity.

On the gas giant Jupiter, the largest planet in our solar system and the one with the strongest gravitation field, the escape velocity is an incredible 37 miles per second. This figure explains Jupiter's extensive atmosphere, composed primarily of the light gases hydrogen and helium. Mars (3.1 miles per second), on the other hand, has quite a thin atmosphere, made up primarily of the slow and heavy gas carbon dioxide. To offer another example, Mercury (2.6 miles per second), the smallest planet in the solar system and the one closest to the Sun, hardly has an atmosphere.

The entire lunar atmosphere weighs no more than 33 tons, which is about the same weight as fifteen SUVs. If, somehow, you managed to compress the Moon's atmosphere so that it was equal to the density of the Earth's air at sea level, it would fill a cube with a side length of 67 yards, or 200 feet. What little atmosphere the Moon has is composed of the gases neon, hydrogen, helium, and argon. Some of these gases are released from the lunar surface through radioactive decay. For example, most of the argon on the Moon comes from lunar rocks containing potassium; when potassium undergoes radioactive decay, it releases argon. Another constituent of the lunar atmosphere, mentioned earlier, is the fine dust that circulates within it.

One obvious reason why astronauts were required to wear spacesuits on the Moon is because it lacks an atmosphere. Even if an atmosphere were present on the Moon, it wouldn't necessarily be a breathable one. Venus, for example, which is closer to the Sun than Earth is, has a dense atmosphere made up primarily of carbon dioxide and is surrounded by clouds rich in poisonous sulfuric acid. According to the British astronomer Patrick Moore, "Any astronaut unwise enough to land on the surface [of Venus] would promptly be squashed, corroded, poisoned and fried." He humorously adds that although the planet is named after the Goddess of Beauty, "conditions there are much more akin to the conventional idea of hell."[8]

The reason spacesuits are colored white is to reflect the Sun's rays. In space, temperatures in direct sunlight can be as high as 250°F; in the shade as low as −250°F. As mentioned previously, similar extreme temperatures exist on the Moon. After all, the Moon's atmosphere is a vacuum, like outer space is a vacuum, albeit to a greater extent. It could be said, then, that conditions in outer space are extremely similar to conditions on the Moon. One exception is the former is a weightless environment, whereas the latter is not.

To describe the Moon (or outer space) as a hostile environment would be putting it mildly. To prevent them from literally being cooked, Apollo astronauts wore spacesuits fitted with cooling systems. These consisted of a special garment, similar to a pair of long johns, with a network of tubing sewn into the fabric. By circulating cool water through the tubing, heat was transferred from the astronaut's body into the backpack of the spacesuit, then into outer space. The cooling systems used in spacesuits today, such as those worn by astronauts aboard the International Space Station, are extremely similar in design. As for a heating system, a spacesuit doesn't need one; heat is provided by the astronaut's own body.

A person cannot survive in a vacuum for long; thus spacesuits are pressurized. We've all seen what happens in science-fiction films and movies when a person gets ejected into outer space or their spacesuit fails: they instantly freeze and then dramatically explode into a thousand pieces. In reality, no such explosion would occur; nor would the body instantly freeze; nor would the blood boil; nor would consciousness be instantly lost. In fact, there would be no immediate injury.

According to a NASA website, in an entry written by a scientific expert, a person exposed to the vacuum of outer space would experience the following: "Various minor problems (sunburn, possibly 'the bends,' certainly some mild, reversible, painless swelling of skin and underlying tissue) start after ten seconds or so. At some point you lose consciousness from lack of oxygen. Injuries accumulate. After perhaps one or two minutes, you're dying. The limits are not really known." The website goes on to state a very interesting fact: "You do not instantly freeze because, although the space environment is typically very cold, heat does not transfer away from a body quickly."[9]

It's no mystery why, in the vacuum of outer space, "heat does not transfer away from a body quickly": a vacuum is an extremely effective insulator. Anyone who owns a thermos, or vacuum flask, would know this. Dorky but useful, a vacuum flask basically consists of two thin-walled bottles, one placed inside the other. The space between the bottles is literally vacuumed of air. The greater the vacuum, the fewer atoms it contains, and the less effective it is as a medium through which heat can transfer.

It is often assumed that space is a cold environment. But because space is not a medium, like air or water, it cannot be cold; nor can it be hot. Space itself is absent of matter, thus absent of a medium, and so temperature does not apply. For an object to be become hot in outer space it has to be exposed to the Sun's radiation. To become cold it has to be shielded from the Sun's radiation for a long period of time.

Everything in the universe is made up of energy and matter; the latter is composed of atoms and molecules (atoms bonded together). These atoms and molecules are constantly in motion. They vibrate and collide, producing thermal energy, or heat. The greater the rate of their motion, the warmer an object becomes; the slower, the cooler. Nothing is entirely devoid of thermal energy, no matter how cold. Absolute zero is the lowest temperature theoretically possible, equivalent to −273.15°C, also known as 0 degrees Kelvin and the point at which molecular motion ceases entirely. Scientists have managed to achieve a state very close to absolute zero.

The first law of thermodynamics states that energy can neither be created nor destroyed; it can only change form. This law makes possible the fact that rubbing two sticks together can generate enough heat to start a fire—something every outdoorsman has tried at least once.

In this example mechanical energy (friction) is converted to thermal energy (heat). Another example is the common household lightbulb, by which electrical energy is converted to light energy as well as some heat.

When you put energy into a system you speed up the rate of the molecules that comprise it, causing it to heat up. Take energy away, and it cools down. We've all experienced how uncomfortable it is to sit on a metal chair, such as a park bench, on a chilly night. The temperature of your body drops as the heat is absorbed by the seat. Energy is transferred between your warm body (which is high in thermal energy) and the cold metal surface beneath you (which is low in thermal energy). The energy flows from a place of high concentration to a place of low concentration. If you sit there long enough, a state of equilibrium will be reached—your body and the seat will be the same temperature.

The Sun emits electromagnetic (EM) radiation across a wide range of frequencies, including X-rays, ultraviolet waves, radio waves, and of course visible light. The spectrum of EM radiation consists of radio waves at one end, which have large wavelengths, and gamma rays at the other, which have small wavelengths. The longer the wavelength, the lower the frequency (measured in hertz); the shorter the wavelength, the higher the frequency. What we call visible light is but a tiny portion of the EM spectrum, with a range of approximately 430 trillion Hz to 750 trillion Hz.

When EM radiation is emitted by the Sun, it travels through space at the speed of light (nearly 186,282 miles per second), striking the Earth about eight minutes later. Because of the energy's interaction with matter, the Earth is warmed. This is a further example of energy being converted from one form to another, in this case solar energy to thermal energy. I wish to make it clear that the Sun does not warm the vacuum of space; heating can only occur when energy strikes matter—and, as we know, space is devoid of matter. Space, I repeat, is neither hot nor cold.

Another way of looking at this process is in terms of atoms, the building blocks of matter, and photons, which are single units, or quanta, of EM radiation. Each could be considered a single force. According to the philosopher and mystic G. I. Gurdjieff (1866–1949), who taught that humanity is "food for the Moon," every phenomenon, small or large, "is the result of the combination or the meeting of three different and opposing forces."[10] Gurdjieff called this the "law of three." Many people recognize, he said, the interaction of two forces—one active, the other passive—to produce a phenomenon, for example sperm and ova giving rise to fertilization. Yet, he added, there is actually a third force involved, which is neutralizing and not readily apparent. In fact, there must be a third force involved, because two forces alone cannot produce a phenomenon. To offer an example, take a simple circuit consisting of a battery connected to a light. The positive charge of the battery is the active force; its negative charge is the passive force. But what about the third force, neutral? This may be identified as the medium through which the electricity flows—that is, the wires connecting the battery to the light. If one of these forces were missing, no such "phenomenon" would be able to occur, and the lightbulb would not shine.

The law of three is not limited to physics. Allow me to relate an example from my own life. In March 2011, I attempted to get in touch with a man by the name of Donald K. Wilson, author of *Our Mysterious Spaceship Moon* (1975) and *Secrets of Our Spaceship Moon* (1979), both of which support the Vasin-Shcherbakov spaceship moon theory as well as provide evidence of UFO activity on the Moon. Both books, published by Sphere, are out-of-print and difficult to acquire. So far as I can tell, these are the only books written by Wilson.

I eventually came across a Donald K. Wilson on the social networking website Facebook, an elderly engineer with glasses and short gray hair who owned a photonics company in America. He had—as I expected the author of *Our Mysterious Spaceship Moon* to have—a scientific background. He looked about the right age too. From his website I was able to obtain his e-mail address. I decided to write to him.

The reply I received was polite, yet odd. He was, not surprisingly, the wrong Donald K. Wilson. After mentioning this, he explained that he'd lost contact with one A. Bain, whose sister-in-law, Susie, had recently died. He wanted to contact A. Bain, the husband of his deceased cousin, to pass on the sad news. A. Bain lived in Melbourne, he said, and his details could be found in the local phonebook, "because I was able to look him up when I was in Sydney some years ago." He continued, "I will appreciate if you can give [A. Bain] a call and ask him to e-mail me, or simply e-mail his address to me."

I found one "A. Bain" in the phonebook, and so I forwarded his details to Donald. I was surprised to discover that A. Bain's address was just a stone's throw away from my brother's, located in an area of Melbourne with which I am very familiar. And so, unwittingly, I ended up becoming an intermediary between two men I didn't know, one of them an American by the name of Donald K. Wilson, the other a fellow Melbournian. Thanks to my serving this role, news of the dead was able to pass from the former to the latter. In his last e-mail to me, Donald cheekily added: "Good luck with the hollow Moon!" As of writing this, I haven't managed to track down the right Donald K. Wilson. Perhaps, like A. Bain's sister-in-law, he is no longer among the living.

The experience consisted of three forces, of which I was the neutral, Donald was the active, and A. Bain was the passive. My role as the neutral force, or intermediary, enabled Donald, the active sender, to connect with A. Bain, the passive recipient. Even though none of us knew each other and did not interact in a physical sense, we each formed part of a trinity of sorts.

So how does the law of three relate to photons and atoms and the conversion of solar energy to thermal energy? Once again, two forces alone cannot produce a phenomenon. For example, photons can travel through the vacuum of space, unimpeded, without producing a phenomenon. We have an active force (photon) and a neutral force (space), but where is the third, the passive force? A third force is introduced when a photon encounters an atom—that is, when the sun's EM radiation strikes the atmosphere or surface of a planet (or moon), resulting in the production of heat.

But why, might you ask, have I classified empty space as a force? Scientists now accept that there is no such thing as an absolute vacuum. Even if you emptied space of all matter and energy, you would not be left with total nothingness. What would remain is the Zero Point Field (ZPF), a sea of subatomic activity, consisting of particles popping in and out of existence.

The reason I give special mention to Gurdjieff's law of three is because it makes it easier to understand a number of topics discussed in this book.

On Earth, during the day, we are used to the sky being light and the ground being darker in comparison—a contrast to which we are accustomed. On the Moon, you'd experience this in reverse: a dark sky and a light ground. Because it has no atmosphere, its sky is perpetually dark, even during the long lunar day. In many ways the Moon couldn't be more opposite to Earth.

The Moon landing hoax theory states that the U.S. Government faked the Moon landings in order to appear to win the Space Race against the Soviet Union. One argument used to support this theory is the absence of stars in photographs taken on the lunar surface. If the sky is always dark, shouldn't the stars be extremely visible? There is a simple reason why they aren't: glare created by the reflective lunar surface (rocks especially). If you wanted to perceive the stars while standing on the surface of the Moon during the lunar day, you would have to shield your eyes for a moment to allow them to adjust to the darkness. During the lunar night there would be no light to create glare, making the stars stand out like floodlights. All of the Apollo missions took place during the lunar day—hence, the absence of stars in photographs.

The Moon is by no means a vibrant, colorful place. "The Moon is a very light concrete color," commented *Apollo 12* astronaut Charles "Pete" Conrad Jr. while looking down on the surface of the Moon in 1969. "If I wanted to . . . look at something I thought was the same color as the Moon, I'd go out and look at my driveway."[11] But the color of the Moon isn't always the same as a certain astronaut's driveway; it can vary depending on the angle from which it is viewed. While facing the Sun it appears light brown; from the opposite direction, gray; and, close up, black.

Astronauts have found it's difficult to judge distance on the Moon. One reason for this is the lack of atmospheric haze. On the Moon objects and their shadows appear equally sharp, no matter how close or how far away. On Earth our atmosphere has a softening effect. Another fact to consider is the Moon is smaller than the Earth, and its surface is therefore more curved, resulting in a closer horizon. When on the Moon one can only see about half as far into the distance as one can when on Earth.

As a result of the Moon having virtually no atmosphere, and therefore no medium through which sound waves can propagate, it is a totally silent world. "In space," as the saying goes, "no one can hear you scream." So too no one can hear you scream on the Moon. Wind, seasons, and climactic zones are also absent on the Moon. It's fair to assume that if the Moon is one day colonized, there will be no need to hire weathermen and women. (Luckily for the citizens of the Moon, these individuals will be forced to remain on Earth.) The only real activity on the Moon (ignoring the possibility of an alien presence) is the occasional occurrence of moonquakes.

Moonquakes are mild in comparison to earthquakes, never exceeding more than a value of three on the Richter scale. If you happened to be standing on the epicenter of one, there's a fair chance you wouldn't even notice it. Unlike on Earth, where seismic disturbances are capable of flattening entire cities, there would be no risk whatsoever of seismic activity on the Moon causing damage to any kind of base or dwelling. As far as natural disasters are concerned, the Moon is a safe place to live. The only conceivable danger, perhaps, would be in the form of meteorite impacts.

When chunks of matter from space (fragments of asteroids and comets), called meteoroids, strike the dense atmosphere of the Earth, they are heated by friction and glow. Often these chunks are so small that they burn up and turn to dust within the upper atmosphere. We call such objects meteors and the streaks of luminosity they produce shooting stars. Meteoroids that, due to their larger size, manage to pass through the Earth's atmosphere without disintegrating entirely are called meteorites. Either the object breaks into fragments, scattering over a wide area, or strikes the ground with great force, leaving a crater. Major meteorite falls are rare. The largest meteorite recovered and put on display was discovered in Greenland by the polar explorer Robert Peary and weighs an incredible thirty-six tons.

On June 30, 1908, an explosion occurred in close proximity to the Podkamennaya Tunguska River, flattening some 770 square miles of pine forest and leaving a further forty square miles charred. One theory given to the mysterious Tunguska explosion is a huge meteorite impact. Considering the incredible magnitude of the explosion—which is estimated to have been a thousand times more powerful than the atomic bomb dropped on Hiroshima—it's fortunate to have taken place in central Siberia, Russia, one of the most remote and uninhabited regions on Earth. There were no human casualties. No doubt, however, animals living in the region, such as reindeer, would have been pulverized (or vaporized) into nothingness.

The meteorite thought to have caused the explosion has been estimated at 150 to 300 feet in diameter, technically making it an asteroid. It is believed to have exploded in the atmosphere high above the Earth's surface, producing a fireball and a blast wave, but no impact crater. Because no meteoric fragments have yet been found, it has even been suggested that the object was a fragment of comet—that is, a big ball of ice and dirt. The ice would have evaporated, of course, leaving scant trace of the object behind.

Meteorite strikes on the Moon occur on an infrequent basis and are rarely ever observed. They are said to be occasionally responsible for moonquakes. Many incorrectly assume that because of its tenuous atmosphere the Moon is constantly bombarded by meteorites, whereas the Earth, having a more substantial and consequently more protective atmosphere, is not. In actuality, both "mediums" are almost as equally ineffective when it comes to offering protection against meteorites; a bullet will pass through foam just as easily as it will pass through paper. In regard to shooting-star meteors, our atmosphere is dense enough to burn them up, whereas the Moon's atmosphere is not. Another difference is that the lunar surface does get struck, and fairly often, by micrometeorites, which end up constituting part of the regolith.

Seismographs left behind on the surface of the Moon by Apollo crews have shown that moonquakes number as many as 1,300 per year and fall into two main categories: deep and shallow. The former and most frequent originate at a depth of five hundred to six hundred miles below the surface; the latter at a depth of 31 to 124 miles. The most common time for them to occur is at perigee, when the Moon is closest to Earth—a clear indication that some of them are triggered by the gravitational pull of the Earth. Another cause of moonquakes is the substantial variations in temperature that take place as the Moon's surface is subjected to the heating and cooling of day and night, which results in the expansion and contraction of rocks.

While studying the structure of the Moon, scientists have subjected it to several man-made quakes. The quakes were created by dropping the discarded ascent stages of lunar modules and other objects onto its surface and measuring the resultant shockwave. The experiments showed the Moon "rang like a bell," and could therefore be hollow. But before we explore this topic in greater depth, we have a long stretch of lunar terrain to cover first.

The purpose of this chapter is to sum up everything we know about the Moon—the most obvious information—focusing especially on its movements and phases, its place in the solar system, its atmosphere, and some of its most superficial features. We're now going to delve a little deeper and discover what lies beneath the lunar surface.

2

A CLOSER LOOK AT THE MOON

*It's anything but boring. We don't even begin to
understand it. To think that twelve guys went there and
we've figured it out, that's crazy.*

JOHN YOUNG, FORMER U.S. ASTRONAUT

In the previous chapter we established the Moon is a world far different from Earth, but in many ways just as special, with its own unique characteristics. It's now time we took a closer look at the Moon's impressive features: its craters, maria, mountain ranges, domes, and rilles. So that we don't get lost while exploring it—and stumble by accident into a deep and potentially electrified crater—it's necessary to first lay down some basic geographical facts.

Would you believe the Moon seen in Australia, in the Southern Hemisphere, looks different from the one seen in the United States, in the Northern Hemisphere? It's the same Moon, only viewed from a different perspective: what was once North is now South, and what was once East is now West. Put simply, one is an inverted version of the other.

In the Northern Hemisphere the waxing Moon grows larger from right to left, and the waning Moon diminishes in the same direction. The Northern Hemisphere Moon is a clockwise Moon, for one would see the terminator move in this direction. From this perspective, most of the Moon's maria are positioned in its Northern Hemisphere, with Mare Crisium in the northeast, Mare Imbrium and Mare Serenitatus in the north, and the huge crater Tycho in the south.

If you live in the Southern Hemisphere, everything I've just described is opposite. The movement of the terminator is also opposite; a Southern Hemisphere Moon is a counterclockwise Moon. Because most of my readers live in the Northern Hemisphere, I will from now on describe the Moon as seen from this perspective. So, for us Australians, Brazilians, and South Africans, a bit of inversion will be required.

Most maps of the Moon show it from a Northern Hemisphere perspective. What I just described, with Tycho in the south, is exactly what you'd find on a typical lunar map. Also on such a map, the top of the Moon points north and the bottom points south, which, might I add, wasn't always the case. Prior to the age of spaceflight, when studying of the Moon was limited to telescopic observation, the Moon was always drawn with north pointing down; the reason being that most astronomical telescopes give an inverted image. (In terrestrial telescopes, like binoculars, the image is inverted once more to give the correct orientation.)

As we already know, the only features of the Moon that can be seen clearly from Earth with the naked eye are its maria. Without them, the Moon would look rather plain and have much less character. Certainly there would be no "man in the Moon." It's rather odd, don't you think, that there happens to be an abundance of maria on the side that we can see and very few maria on the side that we can't? This question may seem pointless to many, but it's one that a shaman would be inclined to take seriously.

In ancient times, when shamanic consciousness was the norm instead of the exception, people saw the natural world as meaningful and alive. They came up with colorful stories (what we now call myths) to account for its myriad wonders, including how the Sun and the Moon came into being. Numerous lunar myths offer an explanation as to how its face became "stained" or "dirty." How, in other words, it obtained its maria. We won't deal with these just yet.

The names of the maria are all in Latin. My personal favorite is Lacus Mortis, meaning Lake of Death. There is also Mare Nectaris (Sea of Nectar), which couldn't sound more romantic and poetic. It's no wonder these vast, dark splotches were once mistaken for actual seas, because that is exactly what they look like when viewed from an earthly perspective. There are nine main maria on the near side of the Moon, in addition to a number of smaller ones. And, like the seas of Earth, most of them are connected. The far side, of course, has few maria to speak of, and none of any significant size.

Many of the Moon's major maria are reasonably circular when viewed from directly above, despite appearing elliptical to us. One example is Mare Crisium, which stands alone in the northeastern quadrant of the Moon. In one direction it measures 260 miles; in another, 335 miles. But, contrary to how it appears from Earth, the larger diameter runs from east to west. Located in the same quadrant is Mare Serenitatis, another reasonably circular, well-bordered mare. Its size is similar to that of the United Kingdom.

Attached to Mare Serenitatis is Mare Tranquillitatis, the site of the first manned Moon landing. More specifically, the *Apollo 11* crew touched down on the southwestern part of Mare Tranquillitatis, on a rocky area called Cat's Paw, about 3.7 miles from where they intended to land. Five potential landing sites were identified for the mission—all of them spread out along the Moon's equator—but this one just happened to be chosen in the end. *Tranquillitatis* means "tranquillity"—a good choice if you believe names hold meaning.

Maria cover about 17 percent of the Moon's entire surface. The rest of its surface, the lighter portion, consists of highlands, called as such because of their higher elevation. There are some important differences between the two terrains. In general, the lunar highlands are rough and heavily cratered, whereas the maria are smooth and sparsely cratered. The older something is, the more wear and tear it has sustained. With that in mind, you can easily believe the highlands predate the maria.

Early in its history, about 3.9 billion years ago, the Moon underwent a period of violent and cataclysmic bombardment, which dramatically altered its surface. Meteorites of various sizes, some of them inconceivably huge, cratered the Moon beyond all recognition. The event, aptly called the Great Bombardment, "obliterated most of the direct evidence of the first half billion years of lunar history," explains Kenneth Lang, professor of astronomy at Tufts University.[1] The Moon's large impact basins, such as the Nectaris and the Imbrium, were formed during the final stages of the Great Bombardment.

After that the lunar interior gradually heated up due to the radioactive decay of uranium, thorium, and other unstable elements. The increase in temperature resulted in a period of volcanic activity lasting seven hundred million years, during which time the impact basins were flooded with molten basaltic rock from the Moon's interior. This produced the dark, circular maria that can be seen today. According to Lang, the maria weren't formed "in a single quick pulse of volcanism, but by repeated outpourings that gradually filled the . . . basins." He further explains that the lava flowed "with about the consistency of motor oil," spreading "for hundreds of kilometers before hardening into a thin veneer, only a few hundred meters or less in thickness."[2] What were once solely impact basins are now impact basins filled with solidified lava. The Nectaris basin, for example, which has a diameter of 530 miles, contains the Mare Nectaris.

When a large meteorite strikes the Moon, the first thing to occur is the disintegration and vaporization of material, leaving behind a huge, bowl-shaped depression in the ground. At the same time, shock waves create fractures in the rock beneath the basin. Additional fractures are formed by the Moon's rebounding mantle. Because the impact removed material from the crust, making it thinner, there's less mass resting on the mantle, and so it naturally bulges upward. In the case of an impact basin flooded by lava to produce a mare, the aforementioned fractures in the rock acted like channels, allowing magma to flow from the Moon's interior to the surface. This is now an impossible occurrence, because the Moon is no longer volcanically active.

You'd expect such an impact and resultant blast to push up surrounding material, and this is exactly what happens. Thus, impact basins are surrounded by mountains. Take the Imbrium basin, for example, which has a diameter of 930 miles and contains the Mare Imbrium. Separating the Mare Imbrium from the nearby, smaller Mare Vaporum is a spectacular range of mountains called the Apennines. They stretch approximately 620 miles. Mount Huygens, the highest peak in this range, has an elevation of about 20,000 feet. Other peaks vary from about 15,000 to 12,000 feet. The range ends in the north near Mount Hadley, not far from where the *Apollo 15* astronauts touched down on the lunar surface.

Bordering Mare Imbrium are several other mountain ranges, including the Alps, the Jura, and the Carpathian—all of which, I should add, are extremely dissimilar to those found on Earth. The Moon's mountain ranges are not mountain ranges in the true sense of the term, like our Himalayas or Rockies, for example, but are really just crater walls. Moore cautions, "The Moon's ranges are of the greatest significance in theoretical studies, but we must beware of drawing too close an analogy with the familiar mountain chains of Earth."[3]

Many writers make the mistake of describing the mountains of the Moon as "Alpine-like" and with "jagged peaks." In one book I came across the author compared the Apennines to the American Rockies. But nothing could be further from the truth. During visits to the Moon, astronauts have frequently commented on the barren and featureless nature of the lunar landscape. The *Apollo 15* astronaut David Scott, who drove across the foothills of the Apennines and caught a good view of one of its peaks, Hadley Delta, called it a "featureless mountain."[4] A photo of Hadley Delta shows exactly that: a smooth mound entirely devoid of "jagged peaks."

The mountains of Earth look the way they do because of the weathering to which they've been exposed—that is, the action of wind, rain, frost, and so forth. On the Moon, where there is no atmosphere, and therefore none of the above, the only erosion possible is in the form of meteorite bombardment. There is also an absence of tectonic activity on the Moon, meaning tectonic plates are not colliding and buckling upward to produce mountain ranges.

When we talk about lunar mountains we're really talking about crater walls. Even so, there are some isolated mountains on the Moon. The most impressive of these to behold are those on the maria. A good example is Mare Imbrium's triple-peaked mountain, Pico, which has a height of 8,000 feet. Pico is located 100 miles south of the crater Plato, a hotspot for transient lunar phenomena (TLP), which are short-lived lights, colors, or changes in appearance on the surface of the Moon. The Mare Imbrium is home to another isolated mountain of considerable height called Piton. It is slightly shorter than Pico, at 7,000 feet. Few of the Moon's isolated mountains are higher than Pico or Piton.

Similar to these features, though certainly not as tall, are the mare domes. These are circular or oval shaped landforms with a diameter between 1.2 and 15 miles and range in height from 295 to 1,640 feet. The tops of them are convex, gently angled at 2 to 3 degrees. About eighty domes have been mapped on the surface of the Moon, and most of them are concentrated on the Oceanus Procellarum. Located there is Marius Hills, the largest volcanic dome field on the Moon. The domes are thought to be volcanic swellings, created as a result of unreleased pressure, like bubbles in flatbread that haven't burst. Some believe the domes to be artificial structures, placed there by the supposed inhabitants of the Moon.

While we're on the topic of volcanic landforms on the Moon, an important question needs to be addressed: Why does the Moon's far side have so few maria and the near side so many? The crust of the Moon is disproportionately thick, much like the white of a hard-boiled egg. It's thinner on the near side and thicker on the far side. The near-side crust is an average of 31 miles thick, while the far side crust is about 9.3 miles thicker. This is a notable difference, enough to give the Moon a center of mass that is offset from its geometric center by about 1.2 miles. When the Moon was volcanically active, the thinner crust on the near side allowed the magma beneath to penetrate to the surface more easily than the thicker far side. The Moon's uneven crust is one of its many mysteries.

When you think of the Moon you might invariably think of craters, and that's because its surface is riddled with them. The Moon's numerous impact basins, some of which are filled by mare, are nothing more than craters of a very large size. The impact basins are, in fact, the largest form of crater that can be found on the Moon, ranging in diameter from 87 miles to more than 620 miles. Given the depth and size of these depressions, it's no wonder they underwent substantial flooding billions of years ago by the dark lava that emerged from the Moon's interior.

The Moon's craters aren't quite visible with the naked eye; however, with a pair of binoculars it's possible to perceive some of its larger ones. One example is the very impressive Copernicus, which sits in the middle of the Oceanus Procellarum. Dubbed "the Monarch of the Moon," it has a diameter of 58 miles and an estimated age of eight hundred million years, making it one of the youngest and best-preserved craters on the near side. A telescope is mandatory if you wish to catch a glimpse of the Moon's smaller craters, many of which, you'll notice, overlap.

It was the invention of the telescope that brought the Moon's craters to the attention of humanity. Before the telescope, the Moon was thought to be a perfectly smooth sphere, closer to "heaven" than Earth and devoid of Earth's "imperfections." Galileo, one of the first scientists to study the Moon with the aid of a telescope (with weak magnification), made some important observations regarding the Moon's features and produced some impressive lunar sketches. He correctly identified that its surface "is not perfectly smooth, free from inequalities and exactly spherical . . . but that, on the contrary, it is full of inequalities, uneven, full of hollows and protuberances, just like the surface of the Earth itself. . . ." He also wrote about the Moon's "spots"—that is, craters—and described them as "thickly scattered" across the Moon's entire surface, "especially the brighter portions of it" (the highlands).[5]

As soon as the Moon's craters were noticed, they became the focus of a great deal of speculation as scientists and astronomers tried to understand how they were formed. The most widely accepted theory, even up until the mid-1900s, is that the craters were volcanic in origin like the majority of those found on Earth. Because it seemed to be the most simple and obvious explanation, few people questioned it.

The British astronomer Sir John Herschel (1792–1871), son of astronomer Sir William Herschel (1738–1822), was a strong advocate of the volcanic theory. He likened lunar craters to volcanoes known in Europe, saying, "they offer, in short, in its highest perfection, the true volcanic character, as it may be seen in the crater of Vesuvius."[6] (Vesuvius is an active volcano in southern Italy.)

The word *crater* is derived from the Greek word for "bowl," and the most common definition of a crater is "a bowl-shaped depression forming the mouth of a volcano." Originally, the term referred exclusively to a depression of volcanic origin; it wasn't associated with meteorite impacts. In fact, meteorite impacts, not to mention meteorites themselves, were considered something of a myth until the early 1800s. Because there were no reliable reports of "rocks falling from the sky," scientists were naturally very skeptical of the phenomenon. The word *meteorite* only dates back to 1824. The first recorded fall occurred in the United States on December 14, 1807. Called the Weston meteorite, the object fell above the town of Weston, Connecticut, and fragments of it were collected and analyzed by two Yale professors. The president at the time, Thomas Jefferson, allegedly wrote, "I would more easily believe that two Yankee professors would lie than that stones would fall from heaven."[7]

A German astronomer and physician named Franz von Paula Gruithuisen (1774–1852) is recognized as being the first to suggest that the Moon's craters were caused by meteorite impacts. He further believed the Moon was inhabited and that the Schröter crater was the site of a city. As an unorthodox thinker, he also attributed Venus's "ashen light," a subtle glow seen to emanate from its night side, to fire festivals held by the planet's inhabitants. Gruithuisen proposed that Venus is covered in thick, fast-growing jungle, which must be burned periodically—hence the fire festivals—to make way for farmland. Sadly for Gruithuisen, many of his ideas (including his less outlandish ones) were met with considerable skepticism.

Following Gruithuisen's death, the meteorite theory underwent a brief revival during the 1870s, thanks to the English astronomer Richard A. Proctor (1837–1888). Proctor had misgivings about the volcanic origin theory and instead suggested the craters were formed "by the splash of meteoric rain," which fell at a time when the Moon was "in a plastic condition."[8] Eventually Proctor discarded the theory, and it went out of vogue for some time before being revived again in 1892 by the American geologist Grove Karl Gilbert (1843–1918).

Gilbert spent most of his career working for the U.S. Geological Survey (USGS), during which time he undertook a study of Coon Mountain, today known as Meteor Crater. One of the most spectacular of its kind, Meteor Crater is a huge, rimmed, bowl-shaped pit located in the desert of northern Arizona. When driving between the towns of Flagstaff and Winslow via Highway 66, it's possible to catch a glimpse of the depression. It is 4,000 feet in diameter, 600 feet deep, and its rim rises about 200 feet above the surrounding plain.

Gilbert began to study the crater in 1891, the same year it was discovered, and wrongly concluded it had been formed by a volcanic steam explosion. He erroneously reasoned that, if it were caused by the impact of a meteorite, part of the meteorite should still be there, buried beneath the crater floor and naturally taking up space. And, because meteorites are known to have a large component of iron, he further reasoned that the iron, having the ability to attract a magnet, would create a disturbance in the local magnetic field. This would cause the needle of his compass to deflect; however, it didn't. Clearly there wasn't a large mass of iron buried beneath the crater floor; therefore, no meteoric fragment. Gilbert argued that the only possible explanation for the depression was volcanic activity. As for the evidence that suggested otherwise (namely consisting of meteoric fragments discovered within close proximity to the crater), Gilbert chose to ignore it; he treated the fragments as a coincidence.

That the crater was caused by the impact of a meteorite was first suggested in 1903 by Daniel M. Barringer, a mining engineer and former lawyer. Barringer was known to be hot-tempered and impatient, yet also charismatic. Apparently his wife affectionately referred to him as "half-gentleman, half savage." The initial response to his theory by members of the scientific community was one of general skepticism. Some of Gilbert's colleagues saw it as an attempt to undermine the authority of a much-respected scientist, and so targeted it with unfair criticism. In a letter to a friend, the inventor and engineer Elihu Thomson, Barringer stated, "They know in their heart of hearts that I have got them beaten and yet they are not men enough to admit it."[9]

Over time, support for Barringer's theory grew. The geologist George P. Merill discovered that the rock beds below the crater were undisturbed, proving that whatever force created it had to have come from above, not below. Also, in an experiment that involved firing bullets into mud at various angles, Barringer demonstrated how a high-speed projectile traveling at an oblique (slanted) angle can in fact produce a round hole. The results of the demonstration contradicted the commonly held assumption that in order for the almost perfectly circular crater to have been caused by a meteorite, the object had to have struck the ground from directly above.

Barringer believed the meteorite to be buried somewhere beneath the south rim of the crater. As for its weight, his initial suggestion was a figure of more than one million tons; later he suggested ten million. If indeed the object existed, it would be rich in precious iron and nickel—making it worth a fortune. Motivated by greed and an obsessive desire to see his theory proved, Barringer organized a mining operation. He used his own fortune as well as hundreds of thousands of dollars provided by investors to try to recover the supposed object.

Extensive drilling proved futile; it began to look as though the object, if indeed it existed at all, was smaller than what Barringer predicted. When asked to give his opinion as to the size of the meteorite, an astronomer named F. R. Moulton stated a figure of only 300,000 tons, significantly smaller than Barringer's estimate. Moulton further opined that the object would have vaporized on impact, and thus little trace of it would remain. This disappointing news halted the mining operation.

Of course, the initial reaction of Barringer and his colleagues was to doubt Moulton's conclusions, resulting in months of debate. But before long the evidence presented by the astronomer proved too convincing to ignore. On November 23, 1929, Moulton backed up his conclusions with a lengthy report featuring a mathematical analysis of the impact. Barringer realized he wasted a massive amount of time and money (including most of his own fortune) trying to uncover an object that didn't exist. The realization must have been more than Barringer could handle, because he died of a heart attack about a week later.

Barringer's theory—that the crater was caused by a meteorite impact—wasn't proved conclusively until decades after his death. Today we know the meteorite responsible for the crater weighed 300,000 tons (exactly as suggested by Moulton), and it was traveling at a speed of 7.5 miles per second (27,000 miles per hour) and, upon striking the ground, exploded with the force of about 2.5 million tons of TNT. This is about 150 times more powerful than the atomic bomb dropped on Hiroshima. The event itself occurred approximately 50,000 years ago, during the last ice age, when the area wasn't dry and arid as it is today but covered in lush, green forest. As with any event of this kind, animals living in the area, such as mammoths, mastodons and giant sloths, wouldn't have known what hit them.

Barringer was right in so many ways and very much ahead of his time; however, his greatest mistake was to assume the meteorite (or part of it at least) had survived the impact. Scientists now know that the object, a chunk of nickel iron with a diameter of 160 feet, not only produced an explosion upon impact but also generated such intense heat that it almost entirely melted. All that really remained of the object was a very fine mist of molten metal, which quickly dispersed across the landscape. Some of this material became mixed up, of course, with the millions of tons of limestone and sandstone that were "blasted out" by the impact. It was these fragments of meteoric iron that Gilbert entirely dismissed, but which Barringer took as evidence to support his theory.

Although popularly known as Meteor Crater, its scientific name is Barringer Crater. The crater is privately owned by the Barringer Crater Company, which is run by the Barringer family, and those who wish to visit the site are required to pay an admittance fee. On the official Meteor Crater website the crater is described as "the most well known, best-preserved meteorite crater on Earth."[10] Interestingly, the crater was used as a training ground for Apollo astronauts during the 1960s and '70s, as were a number of other locations in the deserts of northern Arizona.

It's true that Meteor Crater is one of the "best-preserved" of its kind on Earth, but it is certainly not the largest. There are about 175 known impact craters on our planet, and some of them measure well over one hundred miles in diameter. The oldest and largest yet found is South Africa's Vredefort Crater, with a diameter of 155 miles and an age of about two billion years.

It is indeed ironic that Gilbert, despite having erroneously concluded Meteor Crater was volcanic in origin, was fully convinced that the craters of the Moon were caused by meteorite impacts, stating as such in a well-known 1892 paper. In fact, Gilbert is acknowledged as being one of the first scientists to identify the correct process by which the Moon's craters were formed. Another lunar crater pioneer was the American amateur astronomer Ralph B. Baldwin, author of the seminal *The Face of the Moon* (1949). Baldwin found compelling similarities between the Moon's craters and holes left by WWII bomb explosions, concluding that "the great majority of the lunar craters were born in gigantic explosions . . . [and] that these explosions were caused by the impact and sudden halting of great meteorites."[11]

In his remarkable book *The Big Splat,* science writer Dana Mackenzie highlights the fact that "a meteorite impact is first and foremost an *explosion,*" comparable to, but bigger than, the kind produced by a typical nuclear bomb.

> Like a nuclear bomb, the explosion would first produce a blinding spherical shockwave, then a mushroom cloud. The shockwave is what does most of the damage. It travels down into the ground as well as up into the air, subjecting the rock to immense pressure. It isn't really accurate to think of the meteorite "digging" a crater; instead, it compresses, fractures, pulverizes, and even melts the rock, and launches it in all directions.[12]

Because of Baldwin's book, which demonstrated that meteorite impacts are comparable to artificially produced explosions, geologists began to scour supposed impact sites for materials that could only have been formed if extremely high pressures were involved. At the time, experiments in laboratories showed that if, for example, graphite is subjected to immense pressure (about thirty thousand atmospheres) it will change into diamond, and that quartz, subjected to less pressure (twenty thousand atmospheres), will change into a mineral called coesite. When, in 1953, coesite was first artificially produced, its existence in nature had not yet been confirmed.

The first location on Earth where coesite was discovered was none other than Meteor Crater; specifically, in the crater's sandstone floor. The discovery was announced in 1960 by geologists Eugene Shoemaker and Edward Chao. The existence of coesite—a mineral that does not occur naturally in the Earth's crust but can only be produced under extremely high temperature and pressure—was the final piece of evidence needed to prove the crater's meteoritic origin. How else could coesite be present on the floor of the crater if the crater hadn't been fashioned by an explosion? Clearly something released a great deal of pressure (through a shock wave, for example), in addition to a tremendous amount of heat and changed the quartz into coesite. That something had to have been a meteorite impact. No other explanation was possible.

The discovery of coesite at Meteor Crater meant geologists had a reliable way to identify impact craters. By the end of the 1960s the total number of craters on Earth to be identified as impact in origin was 47. Today it's 157.

Eugene Shoemaker (1928–1997) was the first geologist to positively identify impact craters on Earth. Also an astronomer, one of his most famous discoveries, together with his wife, Carolyn Shoemaker, and David Levy, was the comet Shoemaker-Levy 9. Ever since his twenties, when he realized scientists would visit the Moon within his lifetime, Shoemaker dreamed of being chosen for such a mission. His interest in the Moon was a deep and passionate one. Having studied and proven the existence of terrestrial impact craters, he wanted nothing more than to extend his research to lunar impact craters in the most authentic way possible—by actually walking on its surface. Possessed of a brilliant intellect, Shoemaker graduated from the California Institute of Technology (Caltech) at the age of nineteen, earned a master's degree a year later, was employed by the USGS, and eventually ended up working for NASA.

Shoemaker was one of the lead scientists involved in project Ranger, a pioneering attempt by NASA during the early 1960s to send a series of unmanned probes to the Moon. He also assisted with the training of astronauts. He did everything he could to become the perfect candidate for a very special job: a scientist-astronaut. But something unexpected prevented Shoemaker from ever fulfilling his dream. He was diagnosed with Addison's disease, a rare disorder that affects the adrenal glands.

Though he was successfully treated for the illness, it rendered him unable to qualify for a pilot's license. So instead of being the first geologist to visit the Moon, the honor belongs to Harrison Schmitt, who went there aboard *Apollo 17.*

On the afternoon of July 18, 1997, while driving along an unpaved road in the Tanami Desert, Northern Territory, Australia, Shoemaker, accompanied by Carolyn, collided head-on with another vehicle. Shoemaker was killed instantly in the collision, but Carolyn was taken to a hospital and survived. Shoemaker and Carolyn took frequent trips to Australia to examine impact craters, and at the time of their accident they were on their way to Wolfe Creek. He was only sixty-nine when he died.

Shoemaker did in fact make it to the Moon, but not in the way he intended. Some of his ashes, sealed in a special capsule, were carried aboard the NASA space probe *Lunar Prospector,* launched on January 6, 1998. *Prospector*'s main mission was to search for evidence of ice at the poles of the Moon, a subject about which I'll have more to say shortly. On completion of its mission the probe was designed to crash into a crater in the south polar region of the Moon, which is exactly what it did on July 31, 1999. The collision dispersed Shoemaker's ashes across the lunar surface. He is the only person to have been buried in such a fashion.

The Moon's craters vary markedly in size, shape, and complexity. To quote Moore, "there is endless variety."[13] Some, as I said, are so inconceivably huge that to try to compare them with the largest known impact craters on Earth would be pointless. At least 30,000 of the Moon's craters have a diameter greater than 3,280 feet. (Meteor Crater, remember, is 4,000 feet in diameter.) Some are tiny pits. Others are so small as to be microscopic. They can consist of a simple bowl shape, feature a central peak, have scalloped walls and a flat floor, have sharp, well-defined edges or blunt, eroded edges—the list goes on. The characteristics of craters are numerous, and no two craters are identical.

When you visit the beach on a crowded day the sand can be so heavily trampled that it's impossible to tell where one footprint begins and another ends. The once smooth surface has been rendered uneven and chaotic. The same effect is present on the surface of the Moon, with craters of various sizes overlapping. Some of the small craters that surround larger ones are actually secondary craters. Often arranged in clusters or aligned in rows, they are formed by the impact of ejecta. As the name implies, *ejecta* is "ejected material"—material, largely consisting of crushed and broken rocks, thrown out by a meteorite impact. This material forms a kind of symmetrical blanket around the crater called an ejecta blanket, with the thickest part of the blanket at the crater's rim.

Given its tenuous atmosphere and low surface gravity, ejecta blankets on the Moon extend farther than they do on Earth. In fact, large meteorite impacts on the Moon have scattered material for hundreds of thousands of feet, in some instances leaving behind rays. Rays are streaks of fine ejecta that extend like spokes on a wheel from the center of craters. On the Moon they stand out because of their high reflectivity (high, that is, in comparison to the rest of its surface).

Whether a crater's rays are still visible gives a good indication of its age. Exposure to micrometeorites and particles from the solar wind can cause ray systems to fade over time and eventually disappear altogether; therefore, craters with their rays intact are among the Moon's youngest. A good case in point is the prominent crater Tycho, which is located in the southern highlands, south of Mare Nubium. It has a diameter of 53 miles, a depth of 2.5 miles, and its bright system of rays extend across the lunar surface for a remarkable 1,600 miles. One scientist called it "the metropolitan crater of the Moon." Tycho and its rays are extremely well-preserved on account of it being formed only 108 million years ago. The aforementioned crater Copernicus, though much older than Tycho, also exhibits a bright system of rays.

Two important discoveries concerning the Moon's polar craters have been made in recent years. First, some of them appear to contain significant amounts of water ice; and, second, they may be extremely electrified environments. Allow me to deal with the latter first. A stream of charged particles, consisting mainly of electrons and hydrogen ions (protons), continuously emanates from the outermost part of the Sun's atmosphere, or corona. These charged particles, which travel at speeds of hundreds of miles per second, constitute a kind of wind, hence the term *solar wind*. The solar wind streams out from the Sun in all directions, filling interplanetary space and blowing past the planets in our solar system and beyond. During a total solar eclipse, when the disc of the Moon completely covers the disc of the Sun, the solar wind is rendered visible for a moment as a halo of white light against the darkened sky.

Earth is protected from the solar wind by its atmosphere and magnetic field; however, the situation is rather different when the solar wind encounters the Moon. The Moon, after all, has virtually no atmosphere; nor does it possess, like the Earth, a global magnetic field. The Moon does have magnetized regions, in fact many hundreds of them, but they are extremely weak. And so, like a tree situated on a riverbank during a heavy flood, the Moon is overwhelmed by the full solar wind. Some of these solar particles are absorbed by the lunar regolith, meaning, in a sense, the Moon collects the sun.

In many ways the solar wind behaves like the wind on Earth. It flows across the lunar surface in such a direction as to be almost horizontal to the poles. It also flows along the terminator. Computer simulations have shown that as the solar wind encounters natural obstructions at the Moon's poles, such as the rims of craters or a deep valley, it flows downward, entering the depression. A separation occurs between the negatively charged electrons and the positively charged hydrogen ions within the solar wind. Because the electrons are far lighter than the ions, they travel more quickly and enter the depression first, resulting in the buildup of an electron cloud and, consequently, a negative electric charge of hundreds of volts. In the case of a crater, the charge is believed to accumulate mainly on its floor and inside rim.

Whether the Moon's polar craters are electrically active environments has yet to be accurately determined. If it turns out they are, then special precautions will need to be taken by future lunar explorers, human or robotic. There could be some danger, for example, of a static discharge causing damage to sensitive equipment.

That the Moon's polar craters are of such immense interest to NASA and other space agencies (as well as to private companies, as I'll explain shortly) relates to the fact that they seem to contain large deposits of ice—an extremely valuable resource for future lunar colonists. But how is it possible, you ask, for water to exist on the surface of the Moon? Wouldn't the intense sunlight to which the Moon is exposed, coupled with the fact that it has no atmosphere, boil away any liquid water (or ice) on its surface? While it's true that the Moon is, generally speaking, a bone-dry world with an extremely hot daytime temperature, its polar regions pose something of an exception to this rule.

The *ecliptic* is a term used to refer to the plane of the Earth's orbit around the Sun. The angle between the Earth's equator and the ecliptic plane is called the *obliquity*. It is this tilt, which varies no more than 23.5 degrees, that gives the Earth its seasons. But the Moon has no seasons, of course, because its rotational axis is oriented nearly perpendicular to the ecliptic plane. As the Moon rotates, it stays almost completely upright, like a high-speed spinning top. The Earth, on the other hand, is like a spinning top that has lost much of its momentum; instead of spinning upright, it tilts from side to side.

If the Earth didn't tilt on its axis, the north and south poles would receive less sunlight, and temperatures in those regions would be far colder than they already are. It would also be hotter than it already is near the equator. This gives us an idea of what it's like on the Moon. Its poles are so starved of sunlight that the bottoms of polar craters are in constant shadow; therefore, they are extremely cold. They are among the coldest regions in the solar system, ranging in temperature from −369.67°F to −333.67°F, which is not far from absolute zero. "Any ice deposits in these frozen reservoirs would be preserved indefinitely in the eternal dark and cold," writes Lang.[14]

Evidence of ice in lunar polar craters first emerged in 1994 as a result of the *Clementine* mission. Part of a joint venture between NASA and the U.S. Strategic Defense Initiative Organization, the purpose of the mission was both scientific and militaristic. The probe used in the mission was equipped with a vast array of remote-sensing instruments. Launched from the Vandenberg Air Force Base on January 25, 1994, it spent more than two months mapping the Moon at resolutions never before obtained, focusing particularly on the polar regions. Also during this time, some military equipment attached to the probe was tested under space conditions.

After leaving lunar orbit, the probe was supposed to rendezvous with the near-Earth asteroid Geographos, but a computer malfunction prevented this from happening. Overall, the mission was a success. On December 3, 1966, the Pentagon announced the probe had detected ice in craters at the Moon's south pole, causing great excitement within the scientific community. As for the quantity discovered, it was said to be enough, when melted, to fill a lake two square miles and thirty-five feet deep.

Clementine was followed by *Prospector,* the same lunar probe that carried Shoemaker's ashes to the Moon. *Prospector* launched from Cape Canaveral, Florida, and was solely a NASA project. Like *Clementine, Prospector* was equipped with a special device, a neutron spectrometer, used to detect the presence of hydrogen. And, like *Clementine,* it found it—this time at both poles. According to NASA, the amount of frozen water detected by the probe was somewhere between 2,600 million and 80,000 million gallons. Because hydrogen on its own is hardly water, you might be wondering where oxygen comes into the equation: hydrogen had combined with oxygen to produce water. NASA's William Feldman stated, "There are a bunch of craters filled up with water ice. This is a significant resource that will allow a modest amount of colonisation for many years. Water can now be mined directly on the Moon, instead of having to be shipped from Earth."[15]

To conclude that frozen water had been found on the Moon was extremely premature of Feldman and others. The results obtained by *Prospector* and *Clementine* left plenty of room for doubt. Among those in the skeptics' camp was former scientist-astronaut Harrison Schmitt, Ph.D. As the only geologist to have been to the Moon, his opinion carried weight. Perhaps, Schmitt and other skeptics argued, the spectrometers aboard *Clementine* and *Prospector* were picking up not water but solar wind, which does contain hydrogen ions. And how is it even possible, they wondered, for water to have formed on the Moon in the first place? All of the rock and soil collected from the Moon by both Apollo astronauts and Russian unmanned probes was found not to contain the slightest trace of water.

When, in 1999, *Prospector* was deliberately crashed into a crater at the Moon's south pole, it was hoped that the impact would throw up a plume of debris containing traces of water. Telescopes and Earth-based spectrometers aimed at the impact site failed to detect any such plume. The results of the experiment couldn't have been more disheartening. Even so, NASA wasn't ready to give up on the promise of precious lunar water. About ten years later, on October 9, 2009, NASA launched its lunar crater and observation sensing satellite (LCROSS). Carried on the same Atlas rocket was another spacecraft, the *Lunar Reconnaissance Orbiter* (LRO), which is still in orbit around the Moon today, making low-altitude passes of its polar regions for mapping purposes.

LCROSS's mission was similar to that of *Prospector*'s, in the sense that it too was deliberately crashed into the Moon. Prior to this it deployed a "bomb"—the Centaur upper stage of the Atlas rocket. The 2.2-ton object separated from LCROSS on October 8. Its target was Cabeus, a large, permanently shadowed crater about 60 miles from the Moon's south pole. Traveling at a speed of 5,600 miles per second, it made an impact about ten hours later, creating a debris plume that rose above the lunar surface. LCROSS, following closely behind, swept through the debris plume to analyze its composition. It also crashed on the Moon. The results of the experiment were nothing short of impressive. Remarked LCROSS project scientist Anthony Colaprete, "Indeed, yes, we found water. And we didn't just find a little bit, we found a significant amount." Colaprete added that the water "would be water you could drink, water like any other water. If you could clean it, it would be drinkable water."[16]

India's space agency, the Indian Space Research Organization (ISRO), has supplemented NASA's findings concerning the existence of frozen lunar water. On October 22, 2008, ISRO launched *Chandrayaan-1,* India's first unmanned lunar probe. (*Chandrayaan* is Hindu for "Moon craft.") Before its mission came to a premature end August 28, 2009, due to the cessation of radio contact, the spacecraft mapped the Moon in infrared, visible, and X-ray light. Using imaging instruments contributed by NASA, it also scanned the Moon's polar regions for evidence of ice.

One of the instruments, called a miniature synthetic aperture radar (Mini-SAR), uses the polarization properties of reflected radio waves to characterize surface properties, and hence detect the presence of ice. The Mini-SAR found strong evidence of ice in forty small craters at the Moon's north pole, making an estimated sum of 600 million metric tons. "After analyzing the data, our science team determined a strong indication of water ice, a finding which will give future missions a new target to further explore and exploit," commented NASA's Jason Crusan.[17]

Evidence that the Moon's polar craters contain significant deposits of ice (useable ice that is realistically extractable) is growing stronger every day. Private companies are already making plans to mine it. But until we explore these dark and freezing regions of the Moon, either with robotic probes or during manned missions, and attempt to extract some of the supposed ice for study, we won't know for sure what we're dealing with.

Being the place of illusions that it is, could it be that the Moon harbors not real water but "fool's water"? And even if it's proved to be the genuine article, we must ask ourselves how willing the Moon would be to surrender its precious resources. Perhaps we're forgetting the Moon is not as gentle and yielding as the Earth. If we attempt to "explore and exploit it" like a bunch of reckless Texas oilmen, chances are it might bite back. The Moon is universally portrayed as a deity not to be taken lightly; one which, when offended, can be extremely unforgiving.

One firm that's gazing at the Moon with dollar signs in its eyes is Texas-based Shackleton Energy Company (SEC), a subsidiary of Stone Aerospace. The company plans to mine the Moon for water, methane, ammonia, and other resources, and it is determined to accomplish this very soon. In early 2011 the company stated it aims to send robotic scouting missions to the Moon's poles within the next four years, after which it intends to establish a permanent lunar outpost where mining operations will take place. Of particular attraction to the company is the promise of large deposits of lunar ice. Once found, the resource will be extracted, split into its component hydrogen and oxygen, and then transformed into rocket fuel. The fuel will be sold in "fueling stations" that orbit the Earth, making it possible for spacecraft to refuel while in space.

Bill Stone, founder of Shackleton Energy, commented:

> A major issue in making access to space cheaper is that every space mission must carry its own fuel for in-space operations, since in-space refueling does not currently exist. Even if it did, that fuel would have to be lifted and stored on orbit in fuel depots at even higher prices. To avoid this high-cost barrier to real progress, a means to provide cheaper propellants in space has to be developed. We have the answer: water-derived propellants from the moon.[18]

SEC is confident their liquid propellant fueling stations will be orbiting the Earth within the decade. If true, it will lessen the cost of spaceflight to a radical extent, exactly as Stone said. As we already know, launching a rocket from Earth into outer space is an extremely costly venture because a great deal of fuel is required to blast the rocket free of the Earth's powerful gravitational pull. If, instead, spacecraft could be launched from the surface of the Moon, or if spacecraft already liberated from the Earth or Moon had a way to periodically refuel (without having to land), we could avoid the Earth's "fuel sucking" gravitational pull.

As it is, spacecraft are chained to the Earth and unable to venture very far. Fueling stations in Earth orbit would provide the necessary support to further humanity's reach into outer space—kind of like the first rung in a ladder to the stars. According to research done by SEC, it's fifteen times cheaper to launch a mass from the Moon and bring it into Earth orbit than it is to launch a mass from the Earth. Meaning, of course, that if we wish to venture farther into outer space, mining the Moon's ice reserves may be our only option.

How the Moon's frozen water came to be, if indeed it exists, is a question that has many scientists scratching their heads. The most popular theory put forward is the water was deposited there by the impact of water-bearing comets and other bodies. Another theory is, instead of originating from somewhere else, the Moon produces its own water when the solar wind interacts with lunar rocks and soil. The theory states that as hydrogen ions within the solar wind strike the lunar surface, they break apart oxygen bonds in soil materials. The free oxygen atoms and hydrogen ions then combine to produce minute amounts of water or hydroxyl (one oxygen and one hydrogen chemically bonded). As of yet, neither theory has been proved.

Continuing with our exploration of the Moon's features, there is one type that commonly gets overlooked, despite remaining just as mysterious as ever: rilles. *Rille* means "groove" or "furrow" in German. Rilles are valleys or trenches on the surface of the Moon that vaguely resemble dried-up riverbeds on Earth. They are classified into two main categories: sinuous (winding) and straight. They were first discovered during the seventeenth century, when telescopic observation was still in its infancy, by the brilliant Dutch scientist Christiaan Huygens (1629–1695).

The first astronomer to chart the Moon's rilles in detail was Johann Schröter (1745–1816), a German who has both a lunar crater and a rille named after him. Called Schröter's Valley, it was discovered by Schröter himself in 1787, and is the largest sinuous rille on the Moon. Located on the Oceanus Procellarum, it winds its way across the landscape for more than 100 miles. It has a maximum depth of more than 1,000 feet. Shaped like a snake, its "head," the widest part of it, measures about 6.2 miles. At the very tip of its "head" is a crater with a diameter of about 3.7 miles. Its "body" is finely tapered so that its "tail" gets thinner and thinner until it eventually disappears.

Another sinuous rille worthy of noting is Hadley Rille, which is located at the southeastern edge of Mare Imbrium. It winds its way along the foot of the Apennines for more than 60 miles. Like most rilles, its walls are a reasonably steep V-shape. On average, it is 0.9 mile wide and 1,300 feet deep. One reason that makes it significant is that it's the only rille on the Moon to have been visited by astronauts. During their three-day exploration of the Hadley-Apennine region, *Apollo 15*'s David Scott and James Irwin drove their rover to the very edge of it. What they encountered was by no means a gaping precipice. In fact, Scott managed, on foot, to safely cross the rim of the rille and descend into it a short distance. To descend any farther was out of the question as the risk was too great, especially for a man in a cumbersome spacesuit. No astronaut has yet ventured into a lunar rille.

Sinuous rilles are thought to be volcanic in origin and similar to flow channels created by lava flows on Earth. It is thought too that some of them might be collapsed lava tubes. Lava tubes are formed when the top of a lava flow begins to solidify, forming a crust, while the lava beneath drains away.

As the name implies, straight rilles are relatively straight. Some of them consist of crater chains—that is, craters arranged in a row, often without dividing walls. Others are believed to be grabens, which are sections of the crust that have collapsed between two parallel faults.

How some of the Moon's rilles were formed has yet to be adequately explained. They are among the most anomalous features on its surface, and Schröter's Valley is no exception. "In some way that remains to be accounted for, hundreds of cubic kilometers of fluid and excavated mare material vanished," to quote *Encyclopedia Britannica*.[19]

If, as would seem most likely, a large proportion of the Moon's rilles are collapsed lava tubes, could there also be some that haven't collapsed? The answer is yes—in fact, they've already been detected by both Indian and Japanese lunar probes. In a *New Scientist* article dated October 2009, it was reported that a team of Japanese scientists searched through images taken by the Japanese spacecraft *Kaguya* and found what appears to be a "skylight" on the surface of the Moon; in other words, the entrance to an underground tunnel.

The 213-foot hole, which extends at least 260 feet down, sits in the middle of a rille located in the Marius Hills, one of the Moon's most volcanically interesting areas. (An area, as I said, known for its volcanic domes.) It is thought that the hole might lead into a lava tube as wide as 1,213 feet. "The discovery strengthens evidence for subsurface, lava-carved channels that could shield future human colonists from space radiation and other hazards," stated the article.[20]

In February 2011, ISRO scientists announced the discovery of a huge underground chamber on the Moon. It was found while examining images captured by *Chandrayaan-1*. Located on the Oceanus Procellarum, the buried, uncollapsed, and near-horizontal lava tube measures 0.8 mile long and is already being considered as a possible habitat for future lunar colonists. "Such a lava tube could be a potential site for future human habitability on the Moon for future human missions and scientific explorations, providing a safe environment from hazardous radiations, micro-meteoritic impacts, extreme temperatures and dust storms," reported the team of Indian scientists.[21]

Considering that one uncollapsed lava tube has already been found on the Moon (along with what appears to be a partially collapsed one), it's fair to assume others will eventually be discovered as well. For all we know there could be a vast network of well-preserved lava tubes beneath the surface of the Moon, enough space to create a thriving underground city.

An uncollapsed lava tube, such as the one found by Indian scientists, would indeed make an ideal dwelling for future lunar colonists.

Whereas the temperature of the Moon's surface fluctuates wildly, from chillingly cold at night to sweltering hot during the day, the temperature inside a lunar cave is a constant minus 20°C (−4°F). Another advantage of using a lunar cave as a dwelling is that its walls offer effective shielding against cosmic rays and solar flares, which have the potential to damage DNA, causing cancer and other maladies. Although not a single Apollo astronaut who visited the Moon fell ill from radiation sickness, none of them stayed longer than three days. What might happen as a result of prolonged exposure to these kinds of radiation is consequently difficult to tell. Furthermore, a lunar cave would be a safe place to reside in the event of a meteorite impact. Presumably too such an environment would be completely free of lunar dust, which is not only abrasive and clingy but also potentially damaging to the respiratory system and other parts of the body.

With regard to exploring the Moon, humanity has barely scratched the lunar regolith. Just about anything could lay hidden beneath its surface, from a vast network of caves to something beyond our wildest imaginings; or something, for that matter, with a great deal of mass.

In 1968, while studying the movements of the U.S. spacecraft *Orbiter 5,* two American astronomers, P. Muller and W. L. Sjogren, made a most unusual discovery. They kept a close eye on the spacecraft as it completed eighty consecutive revolutions around the Moon. Each orbit took a total of three hours and eleven minutes. They noticed that instead of traveling at a constant velocity, as one would expect, there were brief moments when the spacecraft sped up by the slightest amount. Further observation showed these periods of increased velocity were connected to various areas of the Moon—the maria in particular.

So what was causing the spacecraft to change its speed? The astronomers knew there could only be one explanation: gravitational fluctuations. We already know as the Moon orbits the Earth in an ellipse its velocity is always changing; it's faster when closer and slower when farther away. The same law applies to a spacecraft orbiting the Moon. In the case of *Orbiter 5*, it kept speeding up as a result of being pulled closer to the Moon. Clearly the gravitational field of the Moon was not smooth like a basketball but irregular like a potato. It had "bumps," or regions where the gravity was stronger than usual. And these bumps had to correspond to dense masses beneath the lunar surface. Gravity, after all, increases with mass (and distance). These regions of excess gravitational attraction came to be called *mascons,* short for "mass concentrations." (See plate 2.)

Mascons produce what scientists call a "positive gravitational anomaly," meaning a greater pull than average. By tracking the velocities of spacecrafts that have orbited the Moon (in the same manner as Muller and Sjogren), scientists have been able to produce detailed gravity maps of its surface. The maps show the most positive gravitational anomalies exhibited by the Moon all correspond to its largest, lava-filled impact basins, or maria; namely, the Mare Imbrium, Mare Serenitatis, Mare Crisium, Mare Humorum, and Mare Nectaris.

Now, considering these regions consist of high concentrations of dense solidified magma, you could assume the existence of mascons can be easily accounted for; however, the matter is rather more complicated than that. Contrary to initial belief, mascons aren't restricted to the Moon's maria-rich near side. Data gathered by *Prospector* has shown that mascons also exist beneath a number of impact basins on the far side—basins, that is, which were never filled by lava.

So if large deposits of solidified lava aren't responsible for these mass concentrations, how do we explain them? According to one theory, they are due to the fact that, following the huge meteorite impacts that produced the basins, the Moon's mantle bulged upward beneath the crust to produce raised basin floors. And, because mantle material is of a higher density than crustal material, these regions exert a greater gravitational pull.

However pleasing this explanation is, it still leaves a lot left unanswered. Even NASA itself is willing to admit that "lunar mascons are a mystery. . . . It's unclear how much of the excess mass is due to denser lava material filling the crater or how much is due to upwelling of denser iron-rich mantle material to the crust."[22] In other words, we know practically nothing about mascons, except that they are associated with impact basins and lava-filled impact basins (maria) especially.

It's easy to overlook the fact that only one side of the Moon is visible from Earth. We all know the Moon has an unseen far side, but it's something we tend not to think about. This is not surprising, for we generally pay little attention to the things we cannot see. But even though the far side of the Moon is ignored, and it's not quite as mysterious as it used to be having been revealed in photographs and film footage, it nonetheless continues to arouse the human imagination. We might find ourselves asking the question: "If I had the chance to explore it, what mysterious wonders might I find there?"

No one had any idea what the far side looked like until October 24, 1959, when images of it were sent back to Earth by the Russian spacecraft *Luna 3*. This represented a first in the history of unmanned spaceflight, as did *Luna 3*'s circumnavigation of the Moon. The probe took twenty-nine black and white photographs of the Moon's far side, about seventeen of which were successfully transmitted back to Earth. Today it's easy to find fault with the quality of the photographs. But at the time they were taken—now more than half a century ago—they meant something of a true technological breakthrough.

The far side of the Moon was first seen by human eyes about a decade later during the *Apollo 8* mission. The three-man crew, consisting of Frank Borman, James Lovell, and William Anders, left Earth December 21, 1968, and reached the Moon three days later. Shortly after maneuvering their craft into lunar orbit, it began to make its journey around the far side, severing radio contact with Earth. (As I mentioned earlier, the far side of the Moon is shielded from radio transmissions from Earth.)

Given the close range of their orbit, coupled with the Moon's lack of atmosphere, the astronauts were able to gaze upon its surface with considerable sharpness and clarity. They were struck by its bleak and lonely appearance. "It looks like a big beach down there," commented Anders. Added Lovell, "The Moon is essentially gray. No color. Looks like plaster of Paris, or sort of a grayish beach sand."[23] The crew was treated to a spectacular view of Tsiolkovsky Crater, as discovered by the Russians from photographs brought back by *Luna 3*. At 112 miles in diameter and flooded by dark mare material, it's one of the most prominent impact craters on the Moon's far side.

Apollo 8 completed a total of ten lunar orbits, each roughly two hours in duration. As you'd expect, the astronauts spent a large portion of that twenty-or-so-hour period with their still cameras and movie cameras aimed toward the lunar surface. The surface itself was about 68 miles away. Anders, the crew's principal photographer, was given an extensive list of targets to photograph, and for hours on end he worked through his list. Of the hundreds of photographs taken, some of them were used to identify potential landing sites for upcoming Apollo missions. Others, no doubt, were purely scientific and to be studied by geologists and such. But what if not all of the photographs taken were mundane in nature? According to some, the *Apollo 8* astronauts found irrefutable proof of an extraterrestrial base on the far side of the Moon. Photographs of the base were taken, they say, yet have not been released to the public.

On December 25, the last day of the mission, *Apollo 8* completed its final orbit of the Moon, emerging from behind it for the last time. Just as soon as radio contact had been reestablished with Houston, Lovell transmitted a bizarre message to fellow astronaut and CAPCOM Thomas Mattingly, the meaning of which remains a mystery to this day. "Please be advised there is a Santa Clause," stated Lovell. Mattingly's response only adds to the mystery: "That's affirmative. You are the best ones to know."[24]

There are two main interpretations as to the meaning of Lovell's "Santa Clause" message. One, he meant it as a carefree Christmas joke; and two, it was coded. The former is fairly self-evident, so allow me to address the latter. Only the very naive would deny that the Apollo program was first and foremost a military—and hence partly clandestine—undertaking. It should come as no surprise, then, that the Apollo astronauts made use of code words when communicating information of a confidential nature over the airwaves. So assuming "Santa Clause" was a code word, what could it have possibly meant? One theory states that Lovell used it to confirm the existence of either UFO activity or an extraterrestrial base on the far side of the Moon.

The idea of an extraterrestrial base on the far side of the Moon is one UFO believers and conspiracy theorists get very excited about. Why, they ask, weren't astronauts sent to the Moon's far side during one of the Apollo missions? Why only the near side? Is it because the Americans were ordered to keep away by a hostile extraterrestrial presence, making the far side out of bounds, so to speak? Indeed it is unusual that of all the Apollo Moon landings, beginning with 11 and ending with 17, not a single one of them took place on the far side. (*Apollo 13* ought to be excluded, of course, since the mission was a failure.) You would think that NASA would have been eager to explore the Moon's far side with a human crew and would have aimed to do so for at least one of the Apollo missions. Yet, no such mission was ever planned, let alone considered.

Had NASA followed the advice of Harrison Schmitt, our understanding of the Moon's far side would probably be less lacking than it is today. In 1970, prior to Schmitt being selected as a crew member of *Apollo 17*, he tried his hardest to persuade NASA administrators to designate Tsiolkovsky Crater as the landing site for the mission. The floor of the crater would be a perfect location to land, he said, because it would not only allow astronauts to obtain a sample of far side mare material but also of samples from deep within both the far side crust and the crater's central peak. Schmitt's idea was bold and brilliant, though it did have a few complications, not to mention some added costs. In order to solve the radio communication problem that arises by being on the Moon's far side, it would have been necessary to place two relay satellites in lunar orbit.

Schmitt lobbied hard, almost aggressively, to have his idea accepted by NASA administrators until eventually he was told to stop badgering them. According to the official version of the story, his idea was rejected on the grounds it would have been too costly and because of the added risks it entailed, such as the risk of one or both relay satellites malfunctioning. While these reasons are certainly legitimate, one can't help but wonder if there wasn't some hidden motive for why NASA chose not to send astronauts to the Moon's far side when the opportunity was ripe.

The far side of the Moon, on account of its lack of maria, may appear dull in comparison to the near side, but it's by no means less interesting geologically. While orbiting the Moon in 1998, *Prospector* drew the attention of scientists to a highly reflective region on the far side, located between the Compton and Belkovich craters. The region was found to exhibit a high concentration of thorium and other silicate rocks, indicating the presence of rare, silicic volcanic deposits—different from basaltic ones. The existence of the features remained unproven until more than a decade later. NASA's *LRO* spacecraft obtained high-resolution images of the region. These proved that, indeed, the region is home to a number of intriguing—now extinct—volcanic features. Dome-like and with steeply sloping sides, they vary in diameter from 0.6 to 3.1 miles.

The domes form an isolated field measuring 21.7 miles across, at the center of which is a presumed caldera. "This small volcanic complex occurs far away from the part of the Moon where most of the volcanic activity was concentrated, and where other silicic volcanism occurred," commented Brad Jolliff of Washington University. "That's a puzzle."[25] To add to the puzzle, the volcanic deposits are believed to have formed only 800 million years ago, making them a good deal younger than the Moon's basaltic deposits, the majority of which formed between three billion to four billion years ago. The very youngest of the basaltic desposits is about one billion years old. It would seem the Moon was volcanically active for longer than previously thought.

Scientists are reasonably certain the era of volcanism on the Moon that produced the mare basalts (which is thought to have ended 3.2 billion years ago and to have lasted for 700 million years) was in no way related to the formation of the silicic volcanic deposits recently discovered on the far side. After all, the latter are far too young to be associated with heating activity generated by the decay of radioactive deposits in the Moon's core, these deposits having presumably decayed long before the volcanoes were formed. How, then, did they come to be? According to Jolliff, "the Moon may still have a molten outer core" that is capable of generating "pulses of heat."[26]

3

ORIGIN UNKNOWN

The English naturalist Charles Robert Darwin (1809–1882), founder of the theory of evolution by natural selection, needs little introduction; however, few of us realize Darwin wasn't the only "high achiever" in his family. He belonged to an extremely aristocratic bloodline called the Darwin-Wedgwood family, and from it emerged a number of famous and highly influential men. Overshadowed by Darwin, their names are largely forgotten. By examining the Darwin-Wedgwood family and the lives of its most important members, a strong lunar theme becomes apparent. I would even go so far as to say that the history of the family is marked by a pervasive lunar thread—a thread that runs through every branch of the family tree.

At the top of the family tree sits Erasmus Darwin (1731–1802) and Josiah Wedgwood (1730–1795). Wedgwood was a pottery designer and manufacturer who is recognized as the "father of English pottery." Descended from a long line of potters, he began learning the craft from a young age and served for some time as an apprentice under his brother. He survived an attack of smallpox, but the disease damaged his health to the extent that his right leg had to be amputated. No longer able to work the foot pedal of a potter's wheel, he turned his attention to the science and design of pottery making. He later established his own pottery factory called Etruria, where he devised and implemented special techniques and materials to improve the quality and appearance of his wares. He invented an apparatus called a pyrometer that made it possible to measure high temperatures inside furnaces and kilns, earning him Royal Society membership. As an interesting side note, his wife, Sarah Wedgwood, was also his third cousin.

Erasmus Darwin, also a member of the Royal Society, was a good friend of Wedgwood. A physician, botanist, poet, and inventor, he encouraged Wedgwood to invest in steam-powered engines for his factory. With the engines installed in 1782, the manufacturing of pottery became a rather more industrialized affair. Erasmus's interests and achievements were diverse. He wrote a number of important books, the best known of which is *Zoonomia* (or *The Laws of Organic Life*). Of great interest to modern medical historians, it was an ambitious attempt to classify facts about animals, to define laws concerning organic life, and to catalogue diseases and their treatments. He also wrote *A Plan for the Conduct of Female Education,* in which he argued that girls ought to be educated in schools rather than privately at home.

Erasmus Darwin. Painted by Joseph Wright.

Erasmus was one of the founding members of the Lunar Society of Birmingham (originally called the Lunar Circle), a group of powerful and wealthy men, consisting of industrialists, natural scientists, engineers, and social reformers, who met every month between 1765 and 1813 in and around Birmingham for dinner and conversation. As to why they called themselves the Lunar Society, it's apparently because their meetings occurred on the Monday nearest the full Moon; the Moon's light would allow them to walk home safely in the dark. (Gas street lighting wasn't introduced in Birmingham until 1818.)

As a joke, members referred to themselves as "lunatics." Lunatics included Josiah Wedgwood; the engineer and manufacturer Matthew Boulton; the Scottish instrument maker and inventor James Watt, who helped develop the steam engine; the polymath Joseph Priestly, who is credited with the discovery of oxygen; William Murdock, the inventor of gas lighting; the author and abolitionist Thomas Day; the clockmaker and scientist John Whitehurst; and the physician and scientist William Withering. Working together, these men made a huge contribution to what became known as the Industrial Revolution and have even been described as the "think tank" behind it.

Erasmus married twice, had at least one extramarital affair, and fathered a total of fourteen children. (Much mention has been made of his scandalous, womanizing nature.) His first wife, Mary Howard, died at the age of thirty-one from alcohol-induced liver failure, but not before giving birth to five children, two of whom never survived past infancy. The eldest, Charles Darwin (not to be confused with Charles Robert Darwin), showed great promise as a medical student at the University of Edinburgh and would have gone on to become a successful physician, only he died unexpectedly at the age of twenty. His cause of death was believed to have been a meningococcal disease contracted after sustaining a cut while performing an autopsy. Another tragic event in the family was the suicide of Erasmus's second son, Erasmus II. A wealthy solicitor, he drowned himself in the river Derwent at the age of forty.

Robert Waring Darwin (1766–1848), who would end up being

the father of the famous Charles Darwin, was born roughly six years after Erasmus II. Like his brother Charles, he studied medicine at the University of Edinburgh. At about the age of twenty-two he was elected a member of the Royal Society. He was a successful physician with a very large practice and, according to some accounts, a very ruthless businessman. He made money as a landlord and was also a major stockholder in the Trent and Mersey Canal.

At least one portrait exists of Robert Darwin, and it's not exactly flattering. It shows a smug, bald, pale-faced giant with a plump upper body and long, spindly legs. He apparently weighed around 336 pounds and stood about six foot two inches. His weight was such a concern that, when it came to entering the house of a patient, his coachman had to first inspect the flooring to make sure it offered sufficient support. Also, unable to board his carriage by normal means, he made use of a set of custom-built stone steps. According to James Foard, he had a "prodigious appetite, often carrying provisions of food for a fort-night's journey stuffed under the seat of his buggy while driving about the countryside as he visited patients . . ."[1] In a more positive vein, he was "keenly alive to the emotional problems of his patients, especially ladies, for whom he became, according to Charles, 'a sort of Father Confessor.'"[2]

At the age of thirty Robert Darwin married Josiah Wedgwood's eldest and favorite daughter, Susannah, as had been the long-standing arrangement between the two families. Josiah passed away about a year prior to the marriage, leaving Susannah an inheritance of £25,000 (the equivalent of tens of millions today). This, combined with Robert Darwin's own heavy purse, made for a very lucrative partnership. Before she died in her early fifties from a gastrointestinal illness (possibly cancer), Susannah bore six children to Robert: four girls and two boys. Charles Robert Darwin was the second youngest. "My mother died," wrote Charles Darwin, "when I was a little over eight years old, and it is odd that I can remember hardly anything about her except her death-bed, her black velvet gown and her curiously constructed work-table."[3]

Following the death of his mother, the young Charles Darwin was raised by his three elder sisters. For someone who would eventually take the scientific world by storm, he was most unremarkable in his youth. Pudgy, given to daydreaming, slow to develop, and more than likely spoiled, he showed more interest in physical than intellectual activities and attained poor grades in school. Although also very fond of catching rats, his favorite pastime as a young man was game shooting, and he targeted birds, especially. In his own words: "I do not believe that anyone could have shown more zeal for the most holy cause than I did for shooting birds . . . if there is bliss on Earth, that is it."[4]

In 1825, at the age of sixteen, Charles Darwin was sent by his father to Edinburgh University to study medicine—a subject in which he had no interest at all. The sight of blood made him acutely squeamish, and he failed to apply himself to his studies. Not sure what to do with his aimless, rifle toting, rascal of a son, Robert Darwin had him switched to Christ's College, Cambridge, in 1828. Despite being distracted by his favorite pastimes, which at the time consisted of shooting, drinking, horse riding, and beetle collecting, he nonetheless managed to complete his Bachelor of Arts degree in theology. The qualification would have enabled him to embark on a respectable career as an Anglican parson. But of course, as the famous story goes, he instead embarked on an epic, five-year voyage around the world aboard the *HMS Beagle*. At the end of the journey he formulated his revolutionary ideas about evolution, which decades later he introduced with the publication of *On the Origin of Species* (1859).

The real story, however, is not quite so romantic. Anyone who's taken the time to scrutinize Darwin's life and work would realize he wasn't a genuine scientific pioneer in that the theory of evolution by natural selection wasn't his alone; at least half of the credit belongs to the brilliant British naturalist Alfred Russel Wallace (1823–1913), who was working on the theory at the same time as Darwin. The two scientists corresponded with each other, and it's clear Darwin pilfered many of Wallace's ideas and passed them off as his own.

What also ought to be acknowledged is how much Darwin was influenced by the work of his grandfather, Erasmus. Darwin closely studied his grandfather's writings on evolutionary theory and gained much insight from them. The following reads like an excerpt from *On the Origin of Species,* but in fact it was written by Erasmus Darwin. "The final course of this contest among males seems to be, that the strongest and most active animal should propagate the species which should thus be improved."[5]

Despite being fully aware of the dangers of inbreeding, Darwin chose to marry his first cousin, Emma Wedgwood. Being cousins, they shared Josiah Wedgwood as a grandfather and were related in other ways too. They had a total of ten children: six boys and four girls, three of whom died before they reached maturity. Two of them were only infants when they passed away, while the third, Annie—Darwin's second and favorite child—died at the age of ten after contracting typhoid fever. The loss of this bright and happy child had a devastating effect on Darwin, causing him to lose the "last shreds of his belief in Christianity."[6]

Darwin's children suffered from poor health overall, and there's little doubt that inbreeding was a major contributory factor. Anxious about the health of his children, he wrote in a letter to a friend that "they are not very robust."[7] It is difficult to ignore inbreeding was rife within the Darwin-Wedgwood family with numerous instances of marriage between cousins, especially given the ironical nature of the fact. We all know that inbreeding weakens the gene pool and that offspring produced by means of inbreeding tend to be less healthy and less fertile than "normal" offspring. Although inbreeding was not uncommon within the Victorian upper classes, the Darwin-Wedgwood family was excessively inbred, and they were, generally speaking, a sick and infertile lot.

Although Darwin's children were certainly not the "fittest," of the seven who did survive four of them went on to have distinguished careers. Sir George Darwin (1845–1912), an astronomer and mathematician, came up with the fission theory of the formation of the Moon, which we'll examine in just a moment; Sir Francis Darwin

(1848–1925), a botanist of some note, conducted botanical research in collaboration with his father; Sir Horace Darwin (1851–1928), a civil engineer, founded the Cambridge Scientific Instrument Company in 1881 and was mayor of Cambridge from 1896 to 1897; and, lastly, Leonard Darwin (1850–1943), first a major in the British Army then a member of the Liberal Unionist Party, served as president of the Royal Geographical Society and was later president of the Eugenics Education Society.

Of Darwin's four successful sons, Leonard Darwin was the least outstanding and the only one whose name is associated with controversy. Unlike his brothers, he was never knighted and so didn't carry the title "Sir," nor was he elected a member of the Royal Society. In view of the achievements of his brothers, he considered himself to be among the least intelligent of Darwin's children. His wife, Charlotte Mildred Massingberd, was the daughter of Edmund Langton and Charlotte Wedgwood, making her also his first cousin. They never produced any children—a symptom, perhaps, of inbreeding. As mentioned, Leonard served as president of the Eugenics Education Society (later called the Eugenics Society and today known as the Galton Institute) from 1911 to 1928, remaining a member of the institute right up until his death.

The term *eugenics* was coined in 1883 by Charles Darwin's cousin, the explorer and natural scientist Sir Francis Galton (1822–1911). Inspired by Darwin's theory of natural selection, he saw eugenics as a system that would provide "the more suitable races or strains of blood a better chance of prevailing speedily over the less suitable."[8] Eugenics was once a respectable and serious scientific field, comparable to something like economics. However, it acquired a tainted reputation during the 1930s and '40s, when the Nazis made use of eugenics in an attempt to exterminate the Jews and other "inferior" races. Galton became president of the Eugenics Education Society in 1908, one year after it was formed. And since blood is thicker than water, the honor was then passed to Leonard Darwin, a devout eugenicist if ever there was one.

In a statement that appeared in the *New York Times* in 1912, in which he outlined the "objects and methods of eugenic societies," Leonard Darwin wrote:

> There will no doubt always remain a class quite outside the pale of all moral influence, and of these there will be a small proportion who, if they become parents, are certain to pass on some grievous mental or bodily defect to a considerable proportion of their progeny. Here and here only must the law step in. As to whether surgical sterilization should ever be enforced on such persons we have still an open mind . . . Unquestionably these unfortunates must be treated with all practical consideration . . . yet sufficient control must be maintained over them in institutions or elsewhere to prevent them from breeding.[9]

Coming from a man whose parents were cousins, married his own cousin, and therefore carried unhealthy genes, it's difficult to take the words of the great Major seriously!

Earlier I explained that the history of the Darwin-Wedgwood family is marked by a pervasive lunar thread. The thread of course begins with Erasmus Darwin and Josiah Wedgwood, both key members of the Lunar Society, an organization that played an instrumental role in bringing the Industrial Revolution into being. They were friends and business partners who shared an enthusiasm for science, technology, progress, and the attainment of wealth and power.

Being secretive in nature, much remains unknown about the Lunar Society, such as the exact year it was established and for how long it operated. Most historians state 1765 as the year it was established and 1813 as the year it was disbanded. We cannot rule out the possibility that the society continued to exist well after 1813, only in a more underground form. If the society produced some kind of membership list, constitution, or other formal documentation, it must have been hidden or destroyed, because nothing of this nature has ever been found.

Everything we know about the Lunar Society is what has been gleaned from the surviving letters and notes of its members.

One needn't be a conspiracy theorist to realize the Lunar Society was more than just a group of intelligent, well-bred British gentlemen who gathered every month to smoke their pipes and talk science and industrial progress after sitting down to a delectable meal of roasted stuffed pheasant. They were a small but powerful secret society; their impact on history is a significant one. That the society got its name because meetings were held in accordance with the full Moon so members could travel home safely by moonlight is not, I suspect, the entire truth. It's possible the name had less to do with the prevention of stumbling over a rock in the dark and more to do with the Egyptian lunar deity, Thoth.

Not only was Thoth a god of the Moon, but of reckoning, learning, and writing. In art one finds Thoth represented as a man with the head of an ibis or as a baboon (see plate 3). A scribe, interpreter, and advisor of the gods, Thoth was credited with the creation of writing and languages as well as with the invention of mathematics and science, including botany, astronomy, geometry, astrology, medicine, and theology. He was a god of the intellect and of rational thought—a left-brain god, if you like. Thoth was known, among other names, as "the thrice great." Having recognized the similarities between Thoth and their own god Hermes, the Greeks combined the two and worshipped them as one—a figure they called Hermes Trismegistus (Hermes the Thrice-Great). Some sources state Hermes Trismegistus was an actual man: an Egyptian priest and magician who authored the sacred Hermetic literature.

When you consider that Thoth was a god of science, mathematics, and rationalism, does it not make perfect sense why Erasmus Darwin and his buddies—this group of science and technology enthusiasts— opted to call themselves the Lunar Society? There could be further occult significance in the fact that the society consisted of between twelve to fourteen members. Twelve, thirteen, and fourteen are all lunar numbers. In any given solar year there are either thirteen full moons and twelve new moons, or thirteen new moons and twelve full moons.

The lunar calendar, moreover, is made up of twelve, sometimes thirteen, synodic months.

If you wanted to simplify matters by dividing the synodic month in half, which gives you roughly fourteen days, you could say the Moon waxes for fourteen days and wanes for fourteen days. The ancient Egyptians divided the Moon's phases in this way. In the Ptolemaic Temple of Horus at Edfu, Egypt (built between 325–30 BCE), there is a painting of Thoth that shows the lunar deity standing beside a lotus pillar that supports heaven, on top of which is a crescent. Within the crescent is the Utchat of Thoth, which represents the full Moon. There are fourteen steps leading to the top of the pillar, each representing a day of the waxing Moon. On each step stands a god or goddess specific to that day. One of the deities present is Isis, the wife and sister of Osiris. Osiris, according to the well-known legend, was murdered by his brother Set and dismembered into fourteen pieces—this being a reference, of course, to the number of days of the waning Moon.

Did the Lunar Society consist of twelve to fourteen members for a specific, esoteric reason—one related, perhaps, to ancient Egyptian mythology and symbolism? Although it's impossible to provide a conclusive answer to this question, there's no doubt that some, maybe all, of those who belonged to the Lunar Society were well versed in matters of an esoteric nature. Several members of the Lunar Society were known to have been freemasons, including Erasmus Darwin, who, as I said, was a founding member of the group. "Before coming to Derby in 1788, Dr. [Erasmus] Darwin had been made a Mason in the famous Time Immemorial Lodge of Cannongate Kilwinning, No. 2, of Scotland," states one reliable source.[10] The Lunar Society member Josiah Wedgwood was also a freemason. Not only was there established in Staffordshire, England, in 1887, a Masonic lodge bearing his name, called the Josiah Wedgwood Lodge, but many of the designs on his works of pottery are explicitly Masonic in nature. Believing the Lunar Society and the freemasons shared a close affiliation is not an improbable supposition.

Continuing with the lunar theme I identified earlier in relation to the Darwin-Wedgwood family, there is the interesting fact that almost all the men in the family had professions in either science or medicine. Both branches of knowledge are associated with the lunar deity Thoth. Thoth is also associated with writing, and we find there were numerous writers in the family. By far the most notable in this respect were Erasmus Darwin and Charles Darwin, both of whom wrote seminal scientific books. If you trace the branches of the tree even further to include the distinguished Galton family so that we now have the Darwin-Wedgwood-Galton family, the lunar theme of which I speak becomes even more apparent.

Earlier I pointed out that Francis Galton and Charles Darwin were cousins. They were, to be precise, half cousins, both grandsons of Erasmus Darwin. Francis Galton's mother, Francis Ann Violetta Darwin, was the daughter of Erasmus Darwin and Elizabeth Pole. (Elizabeth was Erasmus Darwin's second wife, not to be confused with his first wife, Mary Howard, who was the mother of Robert Darwin.) Francis Galton's father, Samuel Tertius Galton, a businessman and scientist, was the eldest son of Samuel John Galton, a Quaker, arms manufacturer, Lunar Society member, and Fellow of the Royal Society.

Samuel John Galton, who inherited the arms business from his Quaker father, "became a contractor on a large scale for the supply of muskets to the army during the great war [WWI]."[11] The business prospered under his management, and he managed to amass a great deal of wealth. As an author and scientist like his Lunar Society colleague, Erasmus Darwin, he wrote a book called *The Natural Life of Birds,* plus various technical papers, and dabbled in scientific experimentation. In 1804 the Galton family took up banking (a profitable business if ever there was one) and, under the direction of Samuel Tertius Galton, ceased manufacturing arms. As a businessman, Samuel Tertius Galton lacked his father's zeal and ambition, retiring early to live comfortably off the proceeds of his vast estate. Francis Galton was the youngest of his seven children.

Much could be written about Francis Galton—one of the great high priests of mechanistic science—but I haven't the space to include it here. Allow me, therefore, to give a short summary of his work. As well as having pioneered eugenics, Francis Galton made significant contributions to a number of scientific fields, including statistics, meteorology, anthropology, and forensic science. Obsessed with standards of measurement and the improvement thereof, he was the first to apply statistics to the study of human intelligence. A very prolific author, he wrote hundreds of papers and many books throughout his lifetime, including an unpublished novel about a utopian world in which the population follows a eugenics-based religion. (Sadly, significant portions of the novel were burned by his niece, who found its sexual content offensive.) He was a Fellow of both the Royal Society and the Royal Geographical Society and was knighted in 1909.

The lunar characteristics of the Galton family are fairly self-evident. The fact that they were involved in arms manufacturing, science, banking, writing, and statistics is reminiscent of the lunar deity Thoth. Don't forget too that Samuel John Galton was a member—in fact a founding member—of the Lunar Society. As to why I've included arms manufacturing in my list, it ought to be known that a number of lunar deities (though not so much Thoth) were fierce and warlike in nature. I'll have more to say about this in chapter 5.

If the Darwin-Wedgwood-Galton family was an embodiment of all things lunar, it makes perfect sense why Charles Darwin's second son, George Darwin, obsessively tried to explain how the Moon came into being, formulating the once popular, but now defunct, fission theory. Born in 1845 in Downe, Kent, England, George Howard Darwin developed an interest in science at a young age. While other children spent their time playing sports, he spent his time tinkering with lenses. The lenses were a gift from his father, who did his best to support the interests of his children.

Being the son of a scientist, the environment in which he was raised was a considerably intellectual one. He was fourteen years old when his father's *On the Origin of Species* was published, and he enjoyed listening to the long and stimulating discussions that took place when friends and scientific colleagues of his father came to visit. "I do not know that I have ever heard anyone whose sentences so often contained some infraction of grammatical rule," he once wrote in recollection of his father.[12]

George Darwin studied at Cambridge University, where he demonstrated considerable talent in mathematics. After graduating in 1868 his health began to suffer—poor health, of course, being common within the Darwin family. He studied for the bar and could have gone on to become a lawyer but instead chose to further his education in science and mathematics at Cambridge. While trying to find his calling, so to speak, he dabbled in statistics, sociology, and economics. Inspiration didn't arrive until 1875, when he applied his mind to a compelling geological puzzle: the cause of ice ages.

George Darwin

A popular theory at the time held that ice ages were caused by shifts in Earth's axis (so that the poles had actually changed locations on Earth) and these shifts were possibly brought about by the rising of new continents out of the sea. By tackling the problem using mathematics, George Darwin demonstrated that for the poles to shift by even the minutest amount, it would take a massive continent, a quarter the size of the Northern Hemisphere, to rise by an incredible ten thousand feet (which is equal to the height of the Tibetan Plateau). Having proved the theory extremely untenable, he began to wonder whether a different kind of shift was responsible for ice ages; namely, an increase in the obliquity of Earth's rotation (which, as mentioned, varies no more than 23.5 degrees). This led him to theorize that the Earth's tilt had increased as a result of tides—not ocean tides, but tides in the very body of the Earth.

Just as the Moon pulls upon the oceans, creating ocean tides, so it also pulls upon the solid body of the Earth, creating land tides. Though we tend to think of the Earth as perfectly rigid, like steel, it is in fact surprisingly flexible. Like an elastic band that's stretched and then released, the Earth quickly returns to its original shape after being distorted by the Moon's (and to a lesser extent the Sun's) gravitational pull. Whereas tides are between 0.4 inch to 1 foot in height midocean and can build up to tens of feet when they reach the shore, land tides amount to a minuscule 4.5 inches (on average).

In a paper published in 1878, George Darwin demonstrated land tides caused by the Moon had affected the Earth in such a way as to cause the tilt of its axis to increase. Following the paper, George Darwin went on to become a highly respected scientist and one of the world's leading authorities on tides. He was elected a fellow of the Royal Society in 1879, and in 1883 he became a professor of astronomy at Cambridge University, a position he held until his death in 1912 at the age of sixty-seven. He married an American woman named Martha du Poy, with whom he had four children. (The two weren't related.)

Their eldest son, Sir Charles Galton Darwin (1887–1962), was a physicist and eugenicist who participated in the Manhattan Project (a U.S. government research project that occurred during WWII and led to the development of the first atomic bomb). He served not only as president of the Eugenics Society but also wrote a eugenics-themed book: a dry, philosophical essayistic work called *The Next Million Years* (1952). A treasure trove for conspiracy theorists, an excerpt from it reads: "It always comes back to the same point, that to carry out any policy systematically in such a way as permanently to influence the human race, there would have to be a master breed of humanity, not itself exposed to the conditions it is inducing in the rest."[13]

The German philosopher Emmanuel Kant (1724–1804) was the first to put forward the idea of tidal friction, which states that the Moon, because of the tides it creates, is gradually slowing Earth's rotation. Yet it was George Darwin who developed the idea mathematically and calculated that tidal friction is causing the Moon to gradually recede from Earth. But before we take this idea further, let me explain more about the tides.

If you've spent time observing the ocean you will have noticed that the period between high tides is about half a day; or, to be precise, twelve hours and twenty-five minutes. But why not exactly a half day? Why the additional twenty-five minutes? This is because the Moon orbits the Earth in the same direction the Earth rotates on its axis; the twenty-five minutes is the "catch up" time needed for the Earth to rotate enough so the same point on its surface is once again facing the Moon. If, on the other hand, by some incredible act of God, the Moon were suspended in its orbit around the Earth while the Earth continued to rotate on its axis, the tides would occur precisely every twelve hours.

As the Moon orbits the Earth, it pulls Earth's ocean water toward it, creating a tidal bulge. We experience this bulge, this rising of the ocean, as a high tide. The Moon creates not one tidal bulge, but two: one facing it, but ever so slightly ahead of it, and the other behind it, on the opposite side of the Earth. What this means is the Earth's ocean water takes on the shape of an egg, with the two "ends" of the egg corresponding to high tides and the two narrow parts of the egg corresponding to low tides. Obviously the gravitational distortion to which I refer relates not only to the Earth's ocean water but also to the solid body of the Earth. In this instance it is ocean tides, not land tides, on which we ought to be focusing our attention.

To clarify, there are always two high tides and two low tides on Earth. This means within a period of twenty-four hours and fifty minutes, no matter where you are in the world (just as long as you're positioned near the ocean!) you'll see two high tides and two low tides. Though it all sounds simple, it would be a mistake to think the tides are a straightforward matter, or that their behavior is easy to predict. Aside from the obvious fact that the oceans are of various shapes and depths, giving rise to complicated local effects, there is the additional fact that tides, although a product of the Moon's (and Sun's) gravitational pull, are out of synch with the presence of the Moon. A high tide will occur not when the Moon is directly overhead, but rather before the Moon has reached this position. Depending on location and other factors, this delay can vary from minutes to hours. In a moment I'll explain why the tides are out of synch with the Moon.

Further complicating matters is the fluctuating distance between the Moon and the Earth. The tides at perigee, when the Moon is closest to the Earth, are considerably higher than those at apogee, when the Moon is farthest from the Earth. Would you believe the difference in height can be as great as 20 percent? Another matter to consider is the Sun. Although the Moon is the major driving force behind the tides, the Sun plays an important role in tidal formation too. In comparison to lunar tides, solar tides are less than half as powerful; meaning, gravitationally speaking, the Moon is the dominant of the two when it comes to generating tides.

The Moon and the Sun are engaged in a constant game of tidal tug-of-war by which they alternately reinforce and interfere with each other. When the Moon is either new or full, and in alignment with the Sun and Earth, its gravitational pull combines with the Sun's, and the result is an exceptionally high tide—what we call a spring tide. When, on the other hand, the Moon is at first quarter or last quarter, and is therefore at a right angle to the Sun, its gravitational pull interferes with the Sun's, and the result is an exceptionally low tide—called a neap tide.

So why aren't the tides synchronized with the presence of the Moon? Or, expressed another way, why are Earth's two tidal bulges, rather than being in line with the Moon, positioned ahead of the Moon? The reason for this is due to the fact that the speed of Earth's rotation far exceeds the speed at which the Moon orbits the Earth. As the Earth rotates on its axis it tries to carry its two tidal bulges along with it. Earth succeeds to some extent while the Moon, unable to keep up, drags behind the nearer tidal bulge.

Because of this "out of step" alignment, if you like, between the Moon and Earth's two tidal bulges, an interesting gravitational phenomenon occurs. At the same time the Moon pulls the more distant bulge forward, in the same direction of Earth's rotation, it also pulls the nearer bulge backward, against the direction of Earth's rotation. And so here we have two competing forces: the former of which accentuates Earth's rotation, and the latter of which opposes Earth's rotation. Because gravity increases with proximity, it is the nearer bulge, or "opposing force," that dominates in this competition. The result is the speed of Earth's rotation is very gradually diminishing.

But of course we must not forget gravity is a two-way process. The Moon is not only pulling on the Earth's two tidal bulges; they are also pulling on the Moon. Once again realizing the nearer bulge carries the most "force," we find the Moon is being pulled forward in its orbit, causing it to slowly spiral away from Earth. What we are left with, then, is an interesting situation in which the tides produced by the Moon are causing the rotation of the Earth to slow down, while at the same time causing the Moon to retreat from Earth.

To better understand this idea, it helps to picture the Earth as a rotating wheel and the Moon a kind of brake that works by applying resistance. The Earth wants to carry the tides around with it, but the Moon prevents this from happening by pulling back, so to speak, on the nearer tidal bulge, as though by means of a strong, invisible cord. The result is the tides remain fixed in place as far as the orbit of the Moon is concerned. The Earth, meanwhile, continues to rotate. So if the Earth's ocean water gets held back by the Moon while the Earth underneath rotates, wouldn't the two rub against each other and consequently generate a great deal of friction? The answer is yes, they do. And it is this friction—caused by waves grounding on the bottom of oceans and on seashores—that slows the Earth's rotation.

If you impede the motion of a bicycle wheel by applying pressure to it with your hand, it produces friction and your hand and the wheel become hot. It is too with the Earth's ocean water, which is heated ever so slightly by means of tidal activity. With the example of the bicycle wheel we are referring to a process by which one form of energy is converted to another: kinetic energy to thermal energy. A similar process takes place with the Earth's tides, except not all of the energy is converted into heat; the rest of it goes toward making the Moon recede from Earth.

Because Earth's rotation is diminishing in speed and the Moon is gradually receding, the month is getting longer, and so too is the day. George Darwin, being the keen mathematician that he was, determined the day is lengthening more rapidly than the month. His calculations showed that when they eventually catch up to each other, reaching a point of equilibrium, both will equal fifty-five of our present days. Imagine that—a day not twenty-four hours long but an incredible six hundred hours long! At that time the Moon's distance from Earth will have increased by a factor of 1.6, and the situation will be such that the Earth will always show the same face to the Moon—as distinct from the current situation in which the Moon constantly presents the same face to the Earth.

So as to gain an accurate measurement of the distance between the Earth and the Moon, laser reflectors were set up on the lunar surface by the crews of *Apollo 11, 14,* and *15.* The concept behind this is simple. By shooting a beam of light from Earth at the reflector and seeing how long it takes to return, it's easy to calculate the distance the laser beam covered. (When the result is divided in half, it gives you the distance between the Earth and the Moon.) Experiments of this kind have determined the Moon is receding from Earth by 1.5 inch per year. This is hardly something to lose sleep over. It will take about fifteen billion years for the Earth and the Moon to achieve a state of equilibrium, whereby the Moon will have reached its final resting point. And in any case, the Sun will have become a red giant long before that happens, causing the destruction of the Earth.

When we examine the effects of tidal friction in greater detail, we find that the day is lengthening at a rate of two milliseconds (0.002 seconds) per century, which is equivalent to one second every fifty thousand years. This means that tomorrow will be sixty-billionths of a second longer than today. It also means yesterday was shorter than today. Tidal friction is a phenomenon of relevance to the past as well as the future. Studies that examined the differences in growth increments between fossilized corals and modern corals have confirmed the day was shorter in the past. Four hundred million years ago, for example, the day was twenty-two hours.

Picture the Earth and the Moon in their current configuration: separated, as they are, by a very large distance. Try to go back in time so that the distance between them diminishes. Imagine the two bodies in a young and molten state, so close to each other that they're practically touching, both undergoing distortion because of the immense tidal forces involved. The Moon, at the beginning, was reasonably spherical and now has more of a cigar shape. If you went back even further in time you would see the Earth and the Moon merge together to form a single body.

Although the first part of the above scenario has been confirmed by science, in that we know the Moon was once far closer to the Earth and has been receding ever since, the rest of it belongs to the realm of pure speculative thinking. It was by engaging in such kind of thinking that in 1879 George Darwin came up with the fission theory of the Moon's birth. In a paper published that year, he wrote, "These results point strongly to the conclusion that if the Moon and the Earth were ever molten viscous masses, then they once formed parts of a common mass."[14]

We know that Charles Darwin's scientific work was guided to some extent by that of his grandfather Erasmus Darwin. Interestingly, it would appear that the same holds true of George Darwin's scientific work, especially in regard to his fission theory. Erasmus Darwin, as I said, was a poet, and a very accomplished one. He used the medium of poetry to communicate some complex and original scientific ideas. In a poem called *The Botanic Garden* (1792) he paints a powerful picture of the Moon's birth by means of the ejection of matter from Earth.

> Gnomes! how you shriek'd! when through the troubled air
> Roar'd the fierce din of elemental war;
> When rose the continents, and sunk the main,
> And Earth's huge sphere exploding burst in twain.
> Gnomes! how you gazed! when from her wounded side
> Where now the South-Sea heaves its waste of tide,
> Rose on swift wheels the Moon's refulgent car,
> Circling the solar orb, a sister star,
> Dimpled with vales, with shining hills emboss'd,
> And roll'd round Earth her airless realms of frost.[15]

We can assume George Darwin read the above poem and that it helped inspire him to formulate his fission theory. The theory, now defunct on account of its many flaws, remained popular among scientists until about 1930. It's one of the more elegant and logical theories put forward to explain the Moon's birth, in that it involves a steady process of growth and evolution, while to accept it requires no great stretch of the imagination. (Although there are Erasmus Darwin's shrieking gnomes to consider.) Furthermore, the idea that the Earth gave birth to the Moon, which would mean that the Moon is a kind of daughter planet, is appealing in the sense that it carries mythological significance.

You would think that if George Darwin's theory was correct you could expect to find evidence on Earth—a birth scar, if you will—of the separation that occurred in the past when the Moon was "flung off." Thinking along these lines, an English geologist named Osmond Fisher (1817–1914) suggested in 1882 that the Pacific Ocean basin was just such a scar. Although we now know this theory has no basis in fact, it proved very popular when first proposed. The Pacific, after all, has an average depth of 2.7 miles and is by no means a large enough depression from which the Moon could have been torn free. Yet it was so popular it ended up being combined with George Darwin's fission theory, despite his having never adopted it himself. The combined version of the theory "continued to be taught in schools long past the time when scientists had lost faith in it," explains Dana Mackenzie.[16]

George Darwin applied the mathematics of tidal friction as a means to look back in time as far as he possible could, eventually reaching a point where he found the molten Earth and the molten Moon to be separated by a distance of about five thousand miles. The Moon was orbiting the Earth once every five hours and thirty-six minutes, and the rotational speed of the Earth was the same. (Note that the mean orbital distance between the two is presently 238,600 miles, almost fifty times greater than the figure given above.)

He next attempted to deduce the amount of time it would have taken for the Moon to have drifted from its original position to its current one and came up with a conservative estimate of fifty-four million years. When he came up with this figure in the late nineteenth century the age of the Earth (now thought to be about 4.6 billion years old) was generally believed to lie somewhere between twenty million and forty million years. This conflicted with George Darwin's theory, according to which the Moon (and by implication the Earth) had to be much older. From a modern perspective, the fission theory is impossible to fault on the basis that the Moon lacked the necessary time to reach its present orbit; 4.6 billion years is more than enough time.

So if, as the fission theory says, the Earth and the Moon once formed a "common mass," how did they come to separate so that the Moon began to orbit the Earth at a distance of around five thousand miles? In other words, how do we account for the massive and unsightly gap between these two very different situations? George Darwin, who can't be accused of lacking determination, spent thirty years working on this question, jotting down on paper what would have amounted to thousands of mathematical calculations. According to the model he developed, the molten, proto-Earth was spinning extremely quickly and, as it gained momentum, underwent a number of changes in shape: first to that of a beanbag, then to a cigar, then to a pear. At each stage it was getting more and more unstable. Finally, having reached a state of "unstable equilibrium," the two ends of the pear separated to form the Earth and the Moon.

George Darwin realized that in order for such a split (or "fission") to have occurred, the proto-Earth would have had to rotate at a very high speed. And this is where he encountered a problem. Because, as the mathematics of tidal friction had shown, the newly born Moon would be orbiting Earth at a rate of once every five and a half hours and the Earth would be rotating at the same rate. If fission had been successfully achieved, the rotational speed was nowhere near as fast as it should have been. How, then, did the proto-Earth undergo fission if it lacked the necessary rotational speed?

George Darwin came up with a fascinating answer to the problem by suggesting the gravitational pull of the Sun had helped bring about the birth of the Moon; however, this was no ultimate solution. For how could the tidal force of the Sun (which, remember, is only half as strong as that of the Moon) possess the strength required to tear two planets apart? Clearly it did not. Unless, wondered Darwin, the Sun's tidal force had undergone amplification by means of resonance. Explained simply, resonance is a phenomenon that occurs when something is stimulated to vibrate at the same frequency it naturally vibrates (called its natural frequency), resulting in an amplification of those vibrations. A good example of resonance in everyday life is when you push a child on the swing with just the right timing, so that each small, effortless nudge causes the swing to climb higher and higher. Resonance is what enables some singers to shatter wineglasses using nothing but their voices.

What George Darwin was suggesting is the solar tides, working in harmony with the natural frequency of the whole Earth, had caused the proto-Earth to resonant and consequently break in two. It was a beautiful idea, but only an idea. Robert Darwin knew this, calling it "wild speculation." He wrote, "But the truth or falsity of this speculation does not militate against the acceptance of the general theory of tidal friction, which . . . throws much light on the history of the Earth and the Moon, and correlates the lengths of our present day and month."[17]

In addition to its failure to explain how the proto-Earth could have split in two to bring the Moon into being, the fission theory falls short in other ways. If the Moon formed from the same material as the Earth, the two ought to be practically identical in composition; however, they're not. Compared to Earth, the Moon is rich in refractory substances and highly depleted in volatiles. A refractory substance is one with a high boiling point and includes such metallic and nonmetallic elements as titanium, aluminum, silicon, and calcium, all of which can be found in great abundance on the Moon. A volatile is the opposite of a refractory in that its boiling point is very low. Examples include hydrogen, nitrogen, and carbon, all of which are very scarce on the Moon.

While it's true that the basaltic material forming the maria does contain a fair amount of iron, the Moon is generally iron poor. The Earth, in contrast, is very iron rich. Although this hardly helps the fission theory, it doesn't greatly harm it either. It's been suggested that the fission in question occurred after most of the Earth's iron had settled to its center. Another important point to address when it comes to assessing the plausibility of the fission theory is the fact that the plane of the Moon's orbit lies not within the equatorial plane of the Earth. The two are separated by a significant degree. According to the fission theory the Moon ought to occupy such a position since it was launched from the proto-Earth's equator. It's worth mentioning, incidentally, the Moon's orbit is very peculiar.

When you bear in mind that George Darwin's fission theory was the first real attempt to explain the origin of the Moon, and that no other scientist before his time (except Erasmus Darwin, perhaps) thought of its formation as a separate problem to the formation of the rest of the planets, you realize what an admirable attempt it was. Had his theory stood the test of time, he would probably be remembered today as a scientist comparable in greatness to his father. But, the truth is, his name has been gradually forgotten since his death about a century ago, and his only true scientific accomplishments were those within the area of tidal friction.

Prior to the Apollo missions of the late 1960s and early '70s, which brought about an entirely new outlook on the Moon, there were three main contending theories as to the Moon's formation. They are commonly known today as the "three classic theories of lunar origin." One of them we've already dealt with—the fission theory. The remaining two are known as the "capture theory" and the "coaccretion theory."

The capture theory views the Moon as a maverick—an alien world, if you like—that formed in another part of the solar system, and which, having strayed too close to Earth, was seized by the Earth's gravitational field. Given the sensational nature of the theory and its ability to pique the imagination, it's perhaps no wonder that the astronomer who founded it, Thomas Jefferson Jackson See, was controversial, rebellious, and egotistical beyond compare. Appropriately named after two confederacy heroes, Jefferson David and Stonewall Jackson, See was born just outside of Montgomery City, Missouri, in 1866. An extremely rural

Thomas Jefferson Jackson See (T. J. J. See)

area, the farm on which he grew up consisted of some six hundred acres covered in wild prairie grass.

Thanks to a book published in 1913 and ridiculously titled *The Biography and Unparalleled Discoveries of T.J.J. See* (written by William Webb, though no doubt with a generous amount of input from See), plenty is known about the family background and early years of "the greatest astronomer in the world, and one of the greatest of all time." The book gives an almost godlike impression of the young See. We are told "as a baby [he] was large and vigorous, weighing nearly ten pounds," could count by the age of two, and, "as a child . . . learned his letters so early no teacher was required." Also as a child he showed signs of having a "prodigious memory" and was "every inch a natural philosopher."[18]

Although descriptions such as these are most certainly exaggerations (or outright fabrications), there's little doubt See was exceptionally bright and academically inclined. It was while studying at the University of Missouri, where he had access to the school's 7.5-inch telescope, that he fell in love with astronomy and decided to pursue a career in it. Of the thirteen students in his class, he graduated in first place. He completed his formal education at the University of Berlin, where, with the use of the Royal Observatory's 9-inch refractor, he devoted much of his time to discovering and studying double stars, which are pairs of stars that orbit within close proximity to each other.

Once back in America, See was offered a position in the Astronomy Department of the then recently established University of Chicago. It was an exciting time, because a new, state-of-the-art observatory (what became the Yerkes Observatory in Wisconsin) was to be built by the university, and See partook in the planning of the project. The future couldn't have looked brighter for the gifted young astronomer; however, the situation turned sour. See began to clash with his superior, George Ellery Hale, the head of the astronomy department. See thought he deserved to be given equal rank with Hale, stating in a long letter written to the president of the university, "I am in every respect entitled to the position of [full] Professor of Astronomy."[19] See was promoted, but only to that of assistant professor. Dissatisfied with the situation, he resigned.

See landed a job at the Lowell Observatory, but once again his ego got the better of him. After only two years he was fired. His colleagues accused him of being careless and overconfident with his observations. His next job, at the U.S. Naval Observatory in Washington, D.C., didn't work out either, this time due in part to poor health. Shortly thereafter he was more or less "exiled" to a naval base observatory on Mare Island, California, where his job was to keep the standard time for the West Coast. It was here, on a lonely island base with no direct access to the mainland, that he lived and worked until virtually the end of his life. And since the observatory was equipped with only a very small telescope—a 5-inch refractor—it was impossible to undertake any meaningful research. So See turned his attention to theoretical matters instead; namely, the origin of the solar system.

See's exciting capture theory of the origin of the Moon was announced June 26, 1909. The story, rather than being reported in scientific journals, was taken up by newspapers like the *New York Times*. Stated the newspaper: "He [See] rejects entirely the long-accepted theories of Laplace and Sir George Darwin ascribing earthly origins to the Moon, and asserts that his discovery is supported by rigorous mathematical proof . . ."[20]

At the time See made the announcement, those within the mainstream scientific community, who at one point had been willing to lend him their ears, had just about ceased paying him attention. Respectable publications like the *Astronomical Journal* had long since banned him from publication. As the years went on, his position as an outsider became even more pronounced. The media remained somewhat receptive to his theories, partly on account of their sensational nature and partly because See had impressive credentials and knew how to tell a good yarn. Explains Mackenzie: "He was the crazy old guy that the *Vallejo Evening Chronicle* always sent its cub reporters to interview, knowing they would return with their heads spinning from an incomprehensible two hour rant."[21]

From the time he arrived on Mare Island until his death in 1962 at the age of ninety-six, See wrote prolifically on all manner of scientific subjects, from the cause of earthquakes to the nature of the ether. And although, from a conventional scientific point of view, many of his ideas and theories are simply too wild to consider, it cannot be doubted he did score a number of impressive "hits." For example, he was one of the first scientists to adopt the idea (at the time considered ridiculous) that the Moon's craters were formed by meteorite impacts, not by volcanic activity. (A brilliant chapter on the subject can be found in his 750-page tome, *Evolution of the Stellar Systems, vol II, The Capture Theory.*)

See knew the only way the Earth could have captured the Moon is if, somehow, the Moon was first made to slow down. In a somewhat feeble attempt to address this problem, he suggested some kind of "resisting medium" (a substance analogous to the ether) had formerly pervaded space. Just as a meteorite begins to lose energy as soon as it encounters the resisting medium we call the Earth's atmosphere, so would have the Moon before taking up orbit around the Earth. A completely logical notion, but it did leave the capture theory at a dead end, which is where it remains to this day. The author and scientist Isaac Asimov explains, "It's too big to have been captured by the Earth. The chances of such a capture having been effected and the Moon then taking up nearly circular orbit around the Earth are too small to make such an eventuality credible."[22]

Sadly, it is Harold Urey (1893–1981), not See, who is commonly credited with founding the capture theory. Urey, a Nobel Prize–winning American chemist, played a decisive role in the Manhattan Project during WWII, after which he developed a strong interest in the Moon. He became one of the foremost supporters of the capture theory in the years leading up to the Apollo missions. Because the Moon, in his opinion, was far older than the Earth, he believed it ought to contain a record of the early history of the solar system—making it "the Rosetta stone of the planets," to quote the late Robert Jastrow, a leading NASA scientist.

We now come to the final classic theory of lunar origin: coaccretion. This theory holds that the Moon is neither the Earth's "daughter" nor some ancient captive from another part of the solar system, but the Earth's "sister." It suggests the two were formed from the solar nebula at the same time and in the same region; meaning they've always been gravitationally linked. Now, as the widely accepted nebula hypothesis states, the solar system began as a huge cloud (or nebula) of gas and dust that started to rotate and collapse in such a way that its center became the Sun (hence the term "solar nebula"). From this solar nebula a number of rings of material (Laplacian Rings) arose, and the material within each ring accreted to form a planet.

The meaning of the term *coaccretion*—the prefix *co-* meaning "with or together"—should be fairly clear by now. The coaccretion theory is also known as the double-planet theory, because if, as stated by this theory, the Earth and Moon formed concurrently, they would naturally make up a double-planet system. In other words, the coaccretion theory and the idea of the Moon as Earth's twin planet go hand in hand. Thus, to accept the coaccretion theory requires viewing the Moon as a planet as opposed to a satellite.

The fission theory and the coaccretion theory both paint a picture of the Earth and the Moon being formed from essentially the same material, making them relatives in the most direct sense. Contradicting this view is the fact that the two are so different in composition. The Moon is iron deficient, for example, which is another way of saying that it possesses a small core. Scientists realized this when they worked out its density to be 1.93 ounces per cubic inch. This is significantly less than the Earth's density at 3.19 ounces per cubic inch. If the coaccretion theory were true, you would expect the two bodies to have a similar, even identical, value in density.

It was hoped that at the end of the Apollo program—with its yield of about 842 lb of lunar material, consisting of rocks, core samples, soil (regolith), and so on, collected from six different exploration sites— enough would be known about the Moon for scientists to fit together the pieces of the puzzle and work out, once and for all, how it came into being. Ironically, more questions were raised than answered. As Don Wilson eloquently put it, "Instead of giving Earth's scientists a clear picture of the origin and nature of our neighbor, the Apollo missions merely added mystery upon mystery. . . ."[23] Some of the fascinating yet deeply puzzling lunar facts uncovered by the Apollo program are as follows:

- All of the rocks collected from the Moon were found to be extremely ancient. "Nearly every Moon rock is older than nearly every Earth rock," explains Mackenzie.[24] One sample, derived from the highlands, was dated at an incredible 4.5 billion years, indicating the Moon is around the same age as the Earth. To offer a terrestrial comparison, the oldest known rocks on Earth, discovered in northern Quebec in 2001, have been dated at 4.28 billion years. (These were a remarkable discovery. In fact, even to find a rock on Earth as old as 3.5 billion years is considered extraordinarily rare.) Furthermore, when a comparison was made between lunar rocks and the soil collected from near those rocks, the latter was found to be around 1 billion years older. For example, 3.5-billion-year-old rocks collected from the Sea of Tranquillity were resting on soil dated at 4.5 billion years.

- Scientists found similar quantities of oxygen isotopes (light and heavy oxygen atoms) in Moon rocks and Earth rocks, indicating the Moon and the Earth formed at much the same distance from the Sun. An object's oxygen isotope ratio, or signature, provides a clear indication as to where in the solar system it originated. The oxygen isotope signature of a Moon rock is different from that of a meteorite from Mars, for example, or of a meteorite from the outer asteroid belt. This finding dealt a serious blow to the capture theory, because it showed the Moon was no alien from another part of the solar system; but rather, it shared a common ancestry with the Earth.

- Moon rocks were found to be highly depleted in siderophile ("metal-loving") elements, such as cobalt or nickel. Siderophiles tend to be common in rocks containing iron. Although Earth rocks are about as siderophile poor as Moon rocks, there happens to be a very good reason for this: these elements were drawn to the Earth's large iron core during the process of its formation. The mystery, then, is how it's possible for the Moon, whose core is a great deal smaller in comparison to Earth's, to be so lacking in siderophile elements. Where did they go?

- Paradoxically, lunar rocks were found to be both very different in composition to Earth rocks, yet at the same time quite similar. Why do the Earth and the Moon share so many key elements such as oxygen, silicon, calcium, magnesium, aluminum, and iron, to name but a few, and yet also possess such radically different concentrations of these elements? Scientists wondered. The composition of the Moon has been compared to that of the Earth's mantle, the layer between its crust and outer core.

- As for lunar rocks being refractory rich and volatile poor, this led scientists to conclude that the Moon formed at an extremely high temperature and was perhaps once covered by a magma ocean with a depth possibly as great as 620 miles. Scientists speculate that, within this magma ocean, lightweight minerals like plagioclase were made to rise, while dense ones like pyroxene and olivine were made to sink.

- Speaking of refractory substances and high temperatures, samples of lunar maria were found to contain high concentrations of titanium, chromium, and zirconium. Not only do these substances have melting points above 1,850°C (3,362°F), and are thus extremely heat resistant, they're also very strong and resistant to corrosion. Because of its many admirable qualities, titanium is alloyed with other metals to produce strong, lightweight parts for aircraft, spacecraft, missiles, and ships. "Some of the Apollo samples proved to be ten times as rich in titanium as the most titanium-rich rocks ever found on our planet Earth," observed Wilson.[25] The discovery of an excessively high concentration of titanium in lunar rock samples (found in the form of ilmenite, which is also rich in iron) greatly puzzled scientists, causing Urey to confess, "I just don't know how to account for the titanium."[26]

Astonishing discoveries such as these forced scientists to realize all three theories of lunar origin—fission, capture, and coaccretion—were deeply and shockingly flawed. First of all, the fission theory was essentially disproven about fifty years before the Apollo program, when it was realized the proto-Earth could not have been rotating fast enough to throw off the Moon. And where was the evidence to show that the Earth had even begun its life in a molten state? There was none. So what about the capture theory? Evidence that the Moon had once possessed a magma ocean coupled with the identical match of oxygen isotope signatures between the Earth and Moon had essentially blown it out of the water. As for the coaccretion theory, the discovery of vast differences in composition between the Earth and the Moon had rendered it extremely untenable.

In the wake of the Apollo program, a new theory was needed to explain the origin of the Moon. Thus, in the mid-1970s, along came the giant impact hypothesis (also known as the big whack or big splat theory), which is today widely accepted despite its many weaknesses. It was first introduced in 1946 by Canadian geologist Reginald Aldworth Daly (1871–1957) in a short paper published in the *Proceedings of the American Philosophical Society*. Perhaps, suggested Daly, "the main part of the Moon's substance represents a planetoid which, after striking the Earth with a glancing, damaging blow, was captured." So great was the impact, he suggested, that fragments of the Earth were "torn off by the visitor," generating heat "of at least 1,000 degrees centigrade."[27]

Daly's paper, far from having a "giant impact" on the scientific community, had almost no impact at all, and in time it was virtually forgotten. Many decades later, in 1975—and without being aware of Daly's paper—American astronomers William K. Hartmann and Donald R. Davis published a paper in the science journal *Icarus* suggesting that early in the history of the solar system a large body (possibly more than one) collided with the Earth, producing a cloud of debris that coalesced to form the Moon. The impactor, with an estimated diameter of 620 miles, possessed a small molten core. This was swallowed by the Earth and explains why the Earth is so iron rich and the Moon so iron poor.

Another pair of scientists, Alfred Cameron and William Ward, improved upon the theory by making the hypothetical impactor an object of not 620 miles in diameter but a planet in its own right, roughly the size of Mars. (Mars, with a diameter of approximately 4,200 miles, is slightly greater than half the size of Earth.) They realized that only an impactor of such great size could, first, have caused an explosion big enough to launch a sufficient amount of material into orbit to form the Moon; and, second, have given rise to the current angular momentum of the Earth-Moon system. The latter concerns the rate of rotation of Earth and the rate of revolution of the Moon.

During a scientific conference held in Kailua-Kona, Hawaii, in 1984 devoted to the origin of the Moon, the giant impact theory came out on top. The consensus was it passed more tests of plausibility than any other theory. Since that time it's remained the most widely accepted lunar origin theory yet has undergone much revision, so it's difficult to tell whether a final, "glitch-free" version of the theory will ever emerge. The theory could be compared to an old boat owned by a poor fisherman. It manages to float okay but only if its leaks are regularly patched. The fisherman would like to replace it with something better but hasn't the necessary funds to do so. So he continues to use the boat day after day, all the while knowing, in the back of his mind, that it's just a matter of time before it sinks.

Ever since the 1980s scientists, such as astrophysicist Robin Canup of the Southwestern Research Institute in Boulder, Colorado, have attempted to simulate the events of the big whack theory using computer models but have so far achieved only minimal success. This problematic and highly experimental research has spawned several versions of the giant impact theory. One version even suggests not one but two giant impacts were involved. This states that after the first impactor struck the Earth and left it spinning rapidly, the second impactor, which came from the opposite direction, slowed the Earth's spin just enough so it now precisely matches the current rate. Computer models have since been tweaked to make the theory workable without having to include a second impactor.

The latest version of the giant impact theory involves a young, partly molten Earth possessed of a fully formed core and mantle and a Mars-size body with a mass possibly greater than Mars itself, also with a fully formed core and mantle. Some scientists refer to the impactor as Theia; Theia being the mother of the Greek lunar deity Selena. The event itself is said to have occurred about 4.45 billion years ago, during the final stages of the formation of the solar system, when large, violent collisions were the rule rather than the exception.

When the object struck Earth—not directly but in a sideswiping manner—it knocked it from an upright to a leaning position, which is why our planet has an axial tilt of 23.5 degrees. What is more, it caused an explosion with the equivalent power of a trillion hydrogen bombs. A portion of the debris ejected by the explosion, primarily consisting of mantle material from the impactor and the Earth, remained in Earth orbit. There, it took on the form of a full or partial ring, which, over the period of about a century, coalesced to form the Moon. It took tens of millions of years before the Moon's magma ocean finally solidified.

Not surprisingly, the giant impact theory has sparked mixed reactions among scientists. Some find it too contrived while others admire how it manages to incorporate "the best points of all three classical theories while patching up their most obvious flaws," to quote Dana Mackenzie.[28] Like any theory—especially one that's still being developed—it possesses both strengths and weaknesses. One of the biggest strengths of the theory is that it seems to shed light on why the composition of the Moon is similar to the Earth's mantle: the debris that accreted to form the Moon consisted primarily of both Earth and impactor mantle material. Also, if the core of the massive projectile was devoured by the Earth upon impact, it would explain why the Moon has such a small core and the Earth such a large one.

As for the theory's weaknesses, there are many; it would be impossible to address them all here. For the theory to have any basis in fact it would mean that the Earth was once covered in a magma ocean and that so too was the Moon. We are told the giant impact in question inflicted such immense damage to the Earth that its shape was left greatly distorted, that it "spilled its guts," and for billions of years afterward its surface remained covered in a magma ocean. Yet not a single shred of evidence exists to support the idea that the Earth possessed a magma ocean early in its history. That said, we know little about the early geological history of our planet because plate tectonics have erased such evidence.

Scientists are extremely certain that the Moon, on account of it being volatile poor and refractory rich, possessed a magma ocean early in its history; or, put more simply, it formed at an extremely high temperature, its birth the result of a violent and devastating fiery collision. In the words of Kenneth Lang, "The searing heat of such a collision would explain why the Moon holds no appreciable amounts of water and few volatile elements. All were boiled away."[29] It is because the Moon shows strong evidence of being formed at an extremely high temperature that scientists are so determined to hold on to the giant impact theory. After all, no other lunar origin theory, except for the now defunct fission theory, is consistent with the evidence of a magma ocean early in the Moon's history.

According to evidence that recently surfaced, the interior of the Moon appears to be just as wet as the Earth's upper mantle, thereby casting doubt on the Moon's "fiery birth" and, by extension, the giant impact theory. The traces of water were detected inside melt inclusions enclosed within tiny bubbles of volcanic glass, which were brought back from the Moon by *Apollo 17* astronauts. The colored glass beads, rich in titanium and iron, were discovered by Harrison Schmitt in the vicinity of a volcanic crater named Shorty, located in the Moon's Taurus-Littrow Valley. The glass beads stood out because they formed a patch of "orange soil." Closer inspection revealed that not all of the beads were orange; some were crimson, while others, which were found deeper beneath the surface, were black. (Similar beads of glass—in this case green in color—were discovered during the *Apollo 15* mission.)

The lunar glass beads were the product of a particular type of volcanic eruption, called a fire fountain. A fire fountain occurs when lava containing dissolved gases sprays out of a volcano with great intensity, much like liquid out of a soda can that's been shaken and then uncapped. The glass beads would have started out as molten rock from deep within the Moon's mantle. Once released from the volcano, the lava would have separated into droplets. Each droplet sailed effortlessly for hundreds, perhaps thousands, of miles through the lunar vacuum before cooling into tiny spheres and raining down on the ground.

Because the glass beads from the Moon were found to contain melt inclusions, they were an extremely valuable discovery. To clarify, melt inclusions are tiny globules of magma encased within solid crystals.

Astronaut Harrison Schmitt standing adjacent to a large boulder in the Moon's Taurus-Littrow Valley during the Apollo 17 *mission on December 13, 1972. Photo courtesy of NASA.*

Being encased in crystal means the pieces of magma (and any water they contain) remain preserved. Melt inclusions enable scientists to measure the pre-eruption concentration of water in the magma. A team of researchers, which included Alberto Saal of Brown University in Providence, Rhode Island, and Erik Hauri of the Carnegie Institute of Washington, painstakingly examined thousands of glass beads before finding a handful that contained melt inclusions. They studied the samples using a specialized ion microprobe and found water contents ranging from 615 to 1,410 parts per million. By comparison, the level of water in the Earth's upper mantle ranges from 500 to 1,000 parts per million.

"You would really not expect, based on what we know about this [giant impact] model, to have any water present in the Moon at all," commented Hauri, in an article that appeared on the SPACE.com website in May 2011. "The fact that these samples have terrestrial levels of water is really a stunner." It "requires us to think hard about understanding the giant impact process at a level that's anything more than superficial."[30] Added Saal, in an article that appeared in *New Scientist:* "If the whole Moon has the amount of water equivalent to what we analyzed, then the giant impact theory is in a little trouble. How the water got there is a question people need to work on."[31]

Not only does the study have massive implications in terms of how the Moon came into being, scientists say it also raises the possibility that some (maybe all) of the Moon's frozen reserves of water are internal rather than external in origin. Perhaps, early in the Moon's history, volcanic eruptions created a water-rich cloud that rained down on the surface of the Moon, the moisture freezing only in the darkest and coldest regions, such as at the poles. To test this theory, scientists would need to collect some frozen lunar water and compare it with the water found in the melt inclusions. If the samples were found to have a similar isotopic composition, it would lend support to the volcanic explanation.

The fact that Moon rocks and Earth rocks share identical oxygen isotope signatures is another reason to be doubtful of the giant impact theory. After all, the theory presupposes that the impactor originated from somewhere else in the solar system and hence possessed a unique oxygen isotope signature. That the data shows otherwise puts the theory on extremely shaky ground. Are we to believe that the impactor shared a similar orbit to the Earth, occupied that orbit for many millions of years, and then somehow ended up striking the Earth? As the British geologist Ted Nield explains in an article discussing the giant impact theory, "[I]t would be extremely unlikely that a wandering impactor would just happen to possess exactly the same oxygen isotope profile as the object it hit."[32]

The giant impact theory predicts that the Moon was formed from about 70 percent impactor material and about 10 percent Earth material. This means that the Moon and Earth ought to exhibit very different oxygen isotope signatures. Yet, as we've seen, this is not the case. Jay Melosh, Ph.D., a distinguished American geophysicist, addressed this isotopic anomaly at the 72nd Annual Meteoritical Society Meeting in 2009. He explained that isotopic measurements have increased in precision over the past decade, enabling scientists to determine that the Earth and Moon are "isotopically indistinguishable from one another at the level of five parts per million . . ." He went on to state, "Unless the isotopic compositions of the proto-Earth and projectile [impactor] were nearly identical by some fortuitous coincidence, there should be detectable differences between the isotopic composition of the present Earth and Moon . . ." He concluded, "Increasingly precise measurements of the isotopic ratios of elements composing the Earth and Moon have brought us to a new crisis in the still-unresolved problem of the Moon's origin."[33] There is simply no way to explain, in view of the giant impact theory, why the oxygen isotope signature of the Moon should be identical to that of the Earth.

Many scientists have pointed out problems with the giant impact theory. In a paper that appeared in the *International Geology Review* in 1998, scientists Alex Ruzicka, Gregory A. Snyder, and Lawrence A. Taylor examined the giant impact theory from a geochemical perspective. They found "no strong geochemical support for either the Giant Impact or Impact-triggered Fission hypotheses." They add, "This [hypothesis] has arisen not so much because of the merits of [its] theory as because of the apparent dynamical or geochemical shortcomings of other theories."[34]

In their book *Who Built the Moon?* Knight and Butler offer an accurate appraisal of the giant impact theory, calling it "the least impossible explanation for a celestial body that has no right to be there."[35]

This leaves us, then, in a very strange position, in which we are forced to accept the fact that science cannot explain the origin of the Moon. Since at least the mid-1970s scientists have been grasping at straws in an effort to understand the Moon and how it came into being—to the point where, today, the situation is almost embarrassing. So far we've examined all four theories of lunar origin, beginning with the fission theory and ending with the giant impact theory, and we've found that every one of them has a number of major flaws. What is needed, it would seem, is a fresh and radically different approach to the problem— an approach that's open-minded enough to appreciate how truly bizarre and mysterious the Moon is.

The only way to solve a puzzle, any puzzle, is by asking the right questions. Otherwise no progress can be made. The reason, in my opinion, why scientists have failed to solve the riddle of the Moon's origin is because they've been asking all the wrong questions. Instead of asking how the Moon was created, or how the Earth managed to acquire the Moon, we should be asking how the Moon came to exist in the state it's in now, or how the Moon managed to acquire the Earth. We should think of the Moon not as secondary in importance to the Earth but as equal in importance to the Earth. When we think along these lines an interesting question arises: Could it be that the Moon is not a moon at all, but the Earth's twin planet? Could the two of them have existed as companions since the very beginning, exactly as suggested by the coaccretion theory, but that somehow, at some point in the distant past, the Moon was "hijacked" and "tampered with"?

4

THE MYSTERY DEEPENS

*The more you study the Moon, the more you will become
aware that it is an orb of mystery—a great luminous
Cyclops that swings around Earth as though it were
keeping a celestial eye on human affairs.*

FRANK EDWARDS, *STRANGE WORLD* (1964)

Launched on April 11, 1970, at 13:13 central standard time (CST),
Apollo 13 was the third manned mission to the Moon. The objective
of the mission was to visit the Fra Mauro highlands and determine
once and for all if the Moon was geologically active. Of the three crew-
members, James Lovell, Fred Haise, and John Swigert, it was Swigert
who was the odd one out. He was not an original member of the crew
but a replacement for Thomas Mattingly, who, just three days prior to
launch, was grounded by the flight surgeon for fear he may have contracted German measles. (As it turns out, he didn't.)

Early on the evening of April 13, when the spacecraft was about

198,840 miles from Earth, Swigert, following an order from mission control, activated a switch to turn on the hydrogen and oxygen stirring fans. Less than two minutes later, a loud explosion was heard, followed by a complete loss of pressure in one of the service module's two cryogenic oxygen tanks. The other oxygen tank began to leak as well, though much more slowly. This was a serious problem, because the oxygen was needed to feed the craft's three fuel cells. The fuel cells were needed to generate electrical power, drinking water, and oxygen for breathing. Later, following an investigation into the accident conducted by the *Apollo 13* review board, an electrical fault was identified as the cause of the explosion. "The accident was not the result of a chance malfunction . . . but resulted from an unusual combination of mistakes, coupled with a somewhat deficient and unforgiving system," wrote the members of the board.[1]

With very little power remaining in the spacecraft's command module, called *Odyssey,* the astronauts had no other choice but to use the lunar module, called *Aquarius,* as a "lifeboat." As for landing on the Moon, this was out of the question; all that mattered was the survival of the crew. It was decided that rather than change course in mid-flight the better option was to slingshot around the Moon, after which the astronauts would transfer back into *Odyssey* and use its emergency battery power for executing re-entry into the Earth's atmosphere. And so, with the Moon less than a day's journey away, the astronauts set themselves up in *Aquarius.*

A heavily disabled *Apollo 13* swung around the Moon on the evening of April 14, its trajectory bringing it farther from Earth than any manned spacecraft before or since. During the homeward journey the astronauts took measures to keep their oxygen and electricity consumption down to an absolute minimum, which involved shutting down all of *Aquarius*'s non-critical systems. Conditions inside the cabin were cold (apparently as low as 3°C [37°F]), dark, and moist due to a buildup of condensation.

To prevent a buildup of carbon dioxide within the cabin, and thus prevent the astronauts from being poisoned by their own breath, *Aquarius* was fitted with lithium dioxide cartridges, which simply absorbed the gas from the air. They began to reach their limit, threatening the lives of the astronauts. Although *Odyssey* contained fresh cartridges, they were of the wrong shape. Fortunately, with the assistance of engineers at mission control, the astronauts managed to construct a makeshift adaptor that enabled them to attach the fresh command module cartridges to the lunar module hoses.

To fast-forward to the end of the *Apollo 13* saga, the command module containing the astronauts—alive but somewhat worse for wear—splashed down safely in the South Pacific Ocean in the early afternoon of April 17.

It's worth taking a moment to analyze the occult significance of the *Apollo 13* mission. Of the seven manned missions to the Moon that took place as part of the Apollo program, all of which were obviously very risky, it does seem somewhat unusual that the only one to have undergone a serious disaster—disaster so great the crew could easily have died—was *Apollo 13*. We all know the number thirteen is synonymous with bad luck and a great many number of superstitions surround it. The thirteenth Major Arcana in the tarot deck is the Death card, and it typically shows the image of a skeleton riding a horse surrounded by the dead. The card signifies permanent change, whereby the death of a situation gives rise to new life, often not without some degree of suffering involved. The *Apollo 13* crew members encountered death, and experienced great suffering, but managed to pull through the situation and emerge from it with more knowledge than they had originally. The mission, in Lovell's words, was a "successful failure."

Aquarius, the name of the crew's lunar module, which served the role of a lifeboat, is laden with occult meaning. In a book called *All We Did Was Fly to the Moon,* written by Dick Lattimer, we find this interesting quote from Lovell: "Contrary to popular belief, [the lunar module] was not named after the song in the play, 'Hair,' but after the Egyptian God, Aquarius. She was symbolized as a water carrier who brought fertility, and therefore, life and knowledge to the Nile Valley, and we hoped our Lunar Module, Aquarius, would bring life back from the Moon."[2]

The theme of bringing life or knowledge back from the Moon is also present in the *Apollo 13* mission badge, which was designed primarily by Lovell. It shows three horses dragging the Sun along with them, representing the sun god Apollo, with the Earth and the Moon in the background. Contrary to tradition, Lovell and his crewmen chose not to include their names on the patch. Instead, where their names would normally be, is the following Latin motto: *Ex Luna Scientia,* which means "From the Moon, Knowledge." (See plate 4.)

Missing from many popular accounts of the *Apollo 13* mission, including the movie starring Tom Hanks, is an important event that deserves mention here.

Just after the spacecraft had passed around the Moon and started along its path back to Earth, its earlier discarded S-IVB (the third stage of the *Saturn V* launch vehicle) crashed onto the Moon. The 16.5-ton object struck the Moon with the equivalent force of 12 tons of TNT, triggering an artificial moonquake. The experiment, which had been planned all along, was an attempt by scientists to shed some much-needed light on the nature of the lunar interior.

Mission control radioed to the *Apollo 13* crew: "*Aquarius,* we see the results now from 12's seismometer. Looks like your booster just hit the Moon, and it's rocking a little bit." Replied Swigert: "Well, at least something worked on this flight . . . I'm sure glad we didn't have an LM impact, too!"[3] The seismometers referred to, which had been set up by the *Apollo 12* crew, were located approximately 90 miles from the impact site. Amazingly they picked up reverberations that lasted for three hours and twenty minutes and which traveled to a depth of twenty-two to twenty-five miles. The Moon, according to NASA, "reacted like a gong."[4]

A similar phenomenon was observed the previous year, on November 20, 1969, during another artificial moonquake experiment. This time it was triggered by *Apollo 12*'s lunar module ascent stage. When the object struck the Moon, about forty miles from where the seismometers were located, the resultant vibrations persisted for more than an hour. Seismologists found it odd that the waves started small, built to a peak after about seven to eight minutes, then died out over such a long period. "The records are utterly different from any obtainable observations on the Earth," commented one scientist.[5] The geophysicist Maurice Ewing, who helped direct the puzzling seismic experiment, told reporters at a news conference: "As for the meaning of it, I'd rather not make an interpretation right now. But it is as though someone had struck a bell, say, in the belfry of a church. . . ."[6] Another geophysicist, Frank Press, Ph.D., of MIT, was equally puzzled by the strange results: "None of us have seen anything like this on Earth. In all our experience, it is quite an extraordinary event."[7]

Since the Moon was found to react like a gong or bell when struck, scientists began to wonder if it might be partially hollow. In fact, scientists suspected that the Moon might be hollow as early as the 1950s, when the space age was only just beginning. In his book *Our Moon* (1958), the famous British astronomer Hugh Percy Wilkins (1896–1960) offers an explanation for the Moon's low density as compared to Earth's. He suggests parts of its crust are likely to be hollow "within some 20 or 30 miles of the surface." He explains, "It thus appears that hidden from us are extensive cavities, underground tunnels and crevasses, no doubt often connected with the surface by fissures, cracks or blowholes."[8]

Wilkins envisioned the Moon's interior as follows:

The cavernous interior of the world within the Moon must be a strange place. Immersed in impenetrable darkness and absolute silence, the walls doubtless studded with numerous crystals, these gloomy caverns, branching and winding, here and there connected with the surface by a half choked pit or an open crack, may contain surprises for the first space-travellers to land on the Moon.[9]

There is good reason to believe that Wilkins was entirely correct—that the Moon does contain "underground tunnels and crevasses," and some of them are connected with the surface. Earlier we discussed how the Moon is thought to harbor uncollapsed and partially collapsed lava tubes, as indicated by recent photographic evidence.

In 1962, while NASA was engaged in project Ranger, a report appeared in *Astronautics Magazine,* written by NASA scientist Gordon MacDonald, Ph.D., stating that the Moon appears to be "more like a hollow than a homogenous sphere."[10] The finding was based on astronomical data alone. But rather than conclude the Moon is hollow, MacDonald suggested that either the data or observations are incorrect. Later, as satellites began to reveal that the Moon is gravitationally uneven, Harold Urey suggested some of its negative gravitational anomalies (as distinct from positive gravitational anomalies, or mascons) represent huge areas beneath its surface where "there is either matter much less dense than the rest of the Moon, or simply a cavity."[11] He called these areas "negative mascons." Sean C. Solomon, Ph.D., a former professor of geophysics at MIT, now working at the Carnegie Institute in Washington, must have been thinking along the same lines as Urey when he wrote, "The Lunar Orbiter experiments vastly improved our knowledge of the Moon's gravitational field . . . indicating the frightening possibility that the Moon might be hollow."[12]

To describe the possibility of a hollow Moon as "frightening" does seem rather strange. Actually, it relates to a statement made by the famous American astronomer Carl Sagan in a book he wrote in 1966 in collaboration with the Russian astrophysicist and astronomer Iosif Shklovsky, called *Intelligent Life in the Universe.* Schklovsky, having closely studied the unusual orbital motion of Mars' moon, Phobos, concluded the object was of such small density as to consist of little more than a very hard shell, its interior practically hollow. This led him to speculate that Phobos was an artificial satellite launched by a past Martian civilization. Sagan supported the idea, stating, "A natural satellite cannot be a hollow object."[13]

Mars' other moon, Deimos, is almost just as strange as its companion. Being so dark and small, they remained undetected until 1877. Or perhaps not entirely undetected; in Jonathan Swift's novel *Gulliver's Travels,* published in 1726, Mars is described as having two closely orbiting moons: one at six Martian radii, the other at ten. Amazingly, these figures are not far removed from fact, with Phobos at 2.7 and Deimos at 6.9. At the time *Gulliver's Travels* was written, telescopes had nowhere near the magnification ability required to detect objects as tiny and distant as the moons of Mars. I would say that Swift's prediction was more than just a lucky guess.

Phobos means "fear" in Greek—a name that couldn't be more appropriate. It orbits Mars so quickly and at such close range—only 5,827 miles (on average) above the surface—that if you were standing on the red planet you would see it rise and set three times in a single day. (A Martian day is only slightly longer than an Earth day.) Like our Moon, it has an unseen far side, in that it always keeps the same face to its host; its orbital period is equal to its rotational period. Also like our Moon, its orbit is reasonably circular. However, whereas our Moon has an orbit that lies far from Earth's equatorial plane, that of Phobos lies only one degree from Mars' equatorial plane.

Phobos, whose shape could be compared to that of a potato, measures 16.5 miles at its widest point (see plate 5). Much of its surface is covered in remarkably linear grooves, about 330 feet wide and 65 feet deep, the existence of which is difficult to explain. They were once thought to be related to the most dominant feature on the surface of Phobos—a six-mile-wide impact crater named Stickney. Nowadays scientists tend to think of them in terms of crater chains, formed by means of material ejected into space by impacts on the surface of Mars. Another possibility is that they're steam vents, or perhaps holes or crevices that were opened by tidal forces.

One of the strangest features on the surface of Phobos is its "monolith." This massive, bright piece of rock juts out from the surface in an extremely perpendicular fashion, casting a prominent shadow.

Official sources state that the object is probably a piece of impact ejecta. However, it does look oddly geometric as its shape is reminiscent of a steeply sided pyramid. During a 2009 cable-television interview, former astronaut Buzz Aldrin referred to the object as "a very unusual structure." He went on to say, "When people find out they will say, 'Who put that there? Who put that there?' Well, the universe put it there. If you choose, God put it there."[14]

It's been observed that Phobos is spiraling toward Mars at a rate of about 6 feet per century, so that, tens of millions of years from now, it will either be ripped apart by the planet's gravity or smash into its surface. So what, you ask, is pulling the moon toward unavoidable destruction? It was by attempting to answer this question that Schklovsky was led to conclude that Phobos is "a hollow, empty body, resembling an empty tin can."[15] He could think of only one thing that could be causing the object to lose energy and thus drop gradually closer to Mars: atmospheric friction, or air drag. In order to produce so much air drag, and considering the rarefied state of the Martian atmosphere at such an altitude, the object would need to be extremely light, and thus hollow. Stated Schklovsky: "Can a natural celestial body be hollow? Never! Therefore, Phobos must have an artificial origin and be an artificial Martian satellite. The peculiar properties of Deimos, though less pronounced than those of Phobos, also point toward an artificial origin."[16]

Schklovsky's theory has, in part, been refuted. It is now known that tidal friction, not air drag, is causing the moon's gradual descent. Phobos produces two tidal bulges in the solid body of Mars, the closest of which drags it down, draining its energy, if you will. The situation is similar to what happens between the Earth and the Moon, except opposite. The mean density of Phobos is 1.10 ounces per cubic inch, which, although nowhere near as low as Schklovsky's estimate, is strikingly low. In fact, it's been calculated that Phobos is not solid throughout but porous like a sponge, with a porosity value between 25 and 35 percent. In other words, about one-quarter to one-third of the object is hollow!

Scientists know practically nothing about the composition of

Phobos, except that it appears to be similar to a group of asteroids called carbonaceous chondrites, which are essentially big piles of rubble held together by gravity. If such is indeed the case, it would mean that the void space in Phobos is nothing remarkable—a few cracks here and there between rocks of various sizes. Being similar in composition to a carbonaceous chondrite would imply that Phobos is a captured asteroid—and this, indeed, is the prevailing theory; however, it's far from proven. New evidence suggests that Phobos may have formed in a similar fashion to our Moon, or in other words, that it accreted (more correctly, "re-accreted") from material that was ejected into orbit by a large impact on Mars. In the final analysis, everything about Phobos is a puzzle. As the chief scientist for the NASA Mars Exploration Directorate, David Beaty, was brave enough to admit, "Phobos is a funny object. It's kind of a mystery why Phobos is there and where it came from."[17]

Though I wish we could take a closer look at the many mysteries of Phobos, it's time we returned to the main theme of this chapter: the possibly hollow, possibly artificial nature of our own Moon. So far we've come across evidence that the Moon reacts like a "bell" when large, heavy objects are dropped on its surface, indicating it has to be at least partially hollow. This may sound completely farfetched, but the evidence for it is conclusive. Even in *The Cambridge Guide to the Solar System,* written by Kenneth Lang—a book that represents the absolute mainstream of scientific opinion—we are told that there is something very peculiar about the interior of the Moon. Lang explains, "While tremors on the Earth start suddenly and persist for only a few minutes, the moonquake waves build up gradually and continue for more than an hour, suggesting that the body of the Moon is an almost perfect medium for the propagation of seismic waves."[18]

I may not have been educated at the illustrious Cambridge University, but this I know with absolute certainty: the only way the Moon could be "an almost perfect medium for the propagation of seismic waves" is if parts of it were hollow. It's as simple as that. The next question we need to ask is this: Exactly how hollow is the Moon? On

May 13, 1972, a rare and extraordinary event occurred on the Moon that helped scientists answer that question. A meteorite weighing 2,425 pounds crashed just north of Mare Nubium, and with the equivalent power of two hundred tons of TNT, leaving a crater as "large as a football field." The impact occurred about two weeks after the completion of the *Apollo 16* mission. It was picked up by a total of four seismometers that showed the seismic waves penetrated the Moon to a considerable depth, traveling beyond the crust into its inner layers. No waves were reflected back. "Transverse seismic waves were not reflected back because it is felt that the soft structure of the middle of the Moon absorbed them," commented Urey.[19]

This incident, along with other incidents of seismic activity, indicated to scientists that the outer half of the Moon is cold and rigid and the inner half is warm and partially molten. Time and time again, moonquake waves were observed losing energy at depths greater than 620 miles. By combining all the available data, scientists have managed to construct a picture of the lunar interior. Beneath the Moon's low-density crust, which ranges in thickness from forty to sixty miles, is its mantle, a rocky layer about 620 miles thick. From here on things become rather less certain. The Moon is believed to possess a small, iron-rich metallic core with a radius of about 250 miles at most, which is equal to 20 percent of its radius. The Earth's core, in comparison, is equal to 50 percent of its radius. There is a further hypothetical layer within the Moon: a partially molten zone surrounding its core.

Earlier I explained moonquakes fall into two main categories: deep and shallow. Allow me to deal with each separately. Deep moonquakes occur within the Moon's lower mantle. They originate at a depth of between five hundred and six hundred miles at the boundary between the Moon's solid mantle and its partially molten zone and are thought to be triggered by tides. They are vastly more common than shallow moonquakes. Now, just to clarify, a total of six seismometers were placed on the Moon between 1969 and 1972 at the landing sites of *Apollo 11, 12, 14, 16,* and *17.* For some unknown reason, data gathered by the *Apollo*

11 and *17* seismometers tend not to be included in scientific studies. Thus it is generally recognized that four seismometers were placed on the Moon. These were in operation between 1969 and 1977. During this eight-year period they recorded more than 12,500 seismic events, a small number of which were related, of course, to the deliberate crashing of the discarded ascent stages of lunar modules and S-IVBs. As for the remainder of the events, the ones that occurred naturally within the Moon, about twenty-eight of them were of the shallow variety, having emanated from the Moon's upper mantle, while the rest were deep.

There's a link between deep moonquakes and the tidal forces exerted on the Moon by the gravitational pull of the Earth, which is well recognized by scientists. "Moonquakes reoccur again and again in the same locations at the same time of the month, which seems to link them to the monthly tidal cycles," explains NASA geophysicist Bruce Bills. And yet, "so far nobody has been able to construct a physical model or clear pattern to explain the relationship."[20] These specific regions of recurrent seismic activity, which incredibly and consistently produce identical seismograph records, have come to be known as "nests," and they have scientists completely baffled. Considering that the tides exert the same pressures from month to month on the entire Moon, one would expect moonquakes to be more general in nature and to occur throughout the Moon, not within specific pockets only. So why are moonquakes so predictable and mechanical? Researchers believe the tides cannot be the only force responsible for generating moonquakes. "Clearly there's something else here involved besides the tides," commented Bills.[21]

What that "something else" might be remains difficult to determine. Whereas the mantle of the Moon is cold and static, the mantle of the Earth is hot, plastic, and active; it churns slowly in convective motions. The Earth's lithosphere, the ridged, rocky outer layer made up of the crust and the outermost layer of the solid mantle, is broken into about a dozen separate blocks, or plates. Known as tectonic plates, they drift around on the mantle like icebergs on the ocean. It is the interaction between these plates that give rise to earthquakes, volcanoes, and other geological phenomena. Because the Moon lacks active plate tectonics on account of its mantle being cold and static, what forces could possibly be generating deep moonquakes apart from those exerted on it by the gravitational pull of the Earth? The question has scientists stumped.

We now come to shallow moonquakes, which are almost just as baffling as deep ones. While researching moonquakes I came across two contradictory figures as to their estimated depth. Some sources give a figure of 12 to 19 miles, which would indicate that they occur within the crust. Other sources, including Patrick Moore's *The Date Book of Astronomy* (2000), give a depth of between 31 and 124 miles. This would place them within the upper mantle. The distinguished geologist Ronald Greeley, in his *Introduction to Planetary Geomorphology* (2013), rather than providing a figure for the depth of shallow moonquakes, states that they occur "near the crust-mantle boundary." A diagram included in the book shows that by this he is referring to the upper region of the mantle, near where it meets the crust, thus corroborating Moore. As I see no reason to question the reliability of these sources, I feel confident stating that shallow moonquakes occur within the upper mantle. The significance of moonquakes (both shallow and deep) being particular to the mantle will become apparent in a moment. As I stated previously, twenty-eight shallow moonquakes were recorded by Apollo seismometers spanning a period of eight years. Several were quite energetic, registering 5.5 on the Richter scale—powerful enough to crack the plaster in a house and move heavy furniture around. It is shallow moonquakes, not deep ones, that set the moon "ringing like a bell" for periods of more than one hour.

If you asked a seismologist how come moonquakes can persist for many hours whereas even the biggest earthquake dies down in less than two minutes, they would tell you the key is moisture and rigidity. In the words of Clive R. Neal, associate professor of civil engineering and geological sciences at the University of Notre Dame, "Studies have demonstrated the dramatic damping effect that water has on seismic energy.

Thus, seismic energy is more efficiently propagated through the Moon, which is incredibly dry."[22] According to the official explanation, then, the Moon is a far better medium for the propagation of seismic waves than the Earth because, unlike Earth, it is an entirely dry world, and a much more ridged one. Speaking of rigidity, in a NASA article about moonquakes the Moon is compared to "a big chunk of stone or iron." Because of this, states the article, "moonquakes set it vibrating like a tuning fork."[23]

That tuning forks vibrate extremely well is an indisputable fact. Yet, I don't see how the same can apply to "big chunk[s] of stone or iron," unless they happen to be hollow. What about the issue of moisture, then? Could the Moon be a fine conductor of seismic waves, as compared to Earth, simply on account of its being bone dry? Rather than attempt to answer this difficult question, it would serve us well to bear in mind that the mantle of the Moon—the location of both shallow and deep moonquakes—could be just as wet as the Earth's, as revealed in a recent study. And since the evidence for this is extremely strong, we're forced to come back to where we started, which is trying to explain how the Moon "could be a perfect medium for the propagation of seismic waves," to quote Lang, without being at least partly hollow.

So as to eliminate any confusion, when scientists use the term "shallow moonquake" they are not referring to seismic activity caused by the impact of meteorites on the lunar surface or by the deliberate dropping of the discarded ascent stages of lunar modules and other man-made objects; nor are they referring to anything associated with the expansion and contraction of the lunar crust as the Moon moves between sunlight and darkness. Although these events can cause shallow moonquakes, true shallow moonquakes have a completely unknown origin. Scientists know, for example, whether a particular moonquake was caused by the impact of a meteorite, because the seismograph record of one meteorite impact is much like another. In other words, events of this kind have their own distinctive seismic fingerprint.

One popular theory for the cause of natural shallow moonquakes is the sudden slumping of crater rims; however, it's now beginning to look as though something rather more mysterious is involved. In a study conducted by Cliff Frohlich and Yosio Nakamura of the Institute for Geophysics at the University of Texas, Austin, a curious link was discovered between the occurrence of moonquakes and the constellations Leo and Cancer. Specifically, of the twenty-eight shallow moonquakes recorded during an eight-year period, twenty-three of them occurred while the affected area of the Moon's surface was facing Leo and Cancer. Frohlich commented, "You might well ask, if the moonquakes occur when the Moon is pointed in a certain direction, what's this all about?"[24]

What's it all about, indeed? The best answer Frolich and Yosio could come up with is something originating from a fixed source outside our solar system, such as theoretical nuggets of high-energy particles called "quark matter," are striking the Moon's surface. The existence of quark matter was first put forward by American physicist Edward Witten in 1984. Quark matter is believed to be so dense yet possess such little volume that a single particle of it—as tiny as an atom's nucleus—could weigh as much as seven pounds, while "a ton of it would be the size of a blood cell," to quote one physicist.[25] Given its theoretical density and speed, a given quantity of quark matter could zip effortlessly through a planet and out the other side.

Now, I don't have a problem with quark matter, assuming it exists, but I feel somewhat cheated by the explanation that this weird "stuff" has something to do with shallow moonquakes. If we're going to be creative about it, let's instead pretend the Moon is inhabited by a strange breed of dinosaur—call it a Lunasaurus—and when these majestic beasts are oriented toward the constellations of Cancer and Leo they get excited and let out a magnificent roar and stamp their hooves on the surface of the Moon, causing moonquakes of great magnitude.

Sarcasm aside, scientists are clearly at an absolute loss to explain what's triggering shallow moonquakes. So desperate is their need for an explanation, in fact, they've gone so far as to suggest one phenomenon they cannot understand (and may not even exist) is behind another phenomenon they cannot understand. A bit like a cat chasing its own tail, wouldn't you say? One could theorize endlessly, but the bottom line is this: moonquakes, both deep and shallow, are far too mechanical in nature, far too precise and predictable, to be triggered entirely by natural processes. All the evidence points to the fact that the Moon's bursts of seismic activity are unlike anything one would find in nature but follow an organized pattern, as though happening in accordance with the laws of some giant machine. When a pattern stands out from the noise, as is the case with moonquakes, it strongly implies that some kind of intelligence is at work. If, while wandering in the forest one day, I came across a pentagonal arrangement of rocks, I'd be a fool not to realize an intelligent being had placed them there.

So too any scientist aware of the Moon's bizarre seismic properties would be a fool not to realize it is partly artificial. There is a fascinating anecdote in *Who Built the Moon?* that gives a clue as to just how artificial the Moon might actually be. The anecdote concerns a meeting that took place in Seattle between Knight and a NASA whistle-blower by the name of Ken Johnston, Ph.D. Johnston was the manager of the Data and Photo Control Department at NASA's Lunar Receiving Laboratory during the Apollo program. He was fired from NASA's jet propulsion laboratory (JPL) solar system ambassador (SSA) program in 2007 after coming out with allegations that the organization had covered up evidence of ancient, artificial ruins on the Moon—allegations we will address in chapter 8. But for now, here's the anecdote: "Ken told Chris that at the time of the impact created by the *Apollo 13* launch vehicle the scientists were not only saying that 'the Moon rang like a bell,' they also described how the whole structure of the moon 'wobbled' in a precise way, 'almost as though it had gigantic hydraulic damper struts under it.'"[26]

While we're on the topic of NASA secrecy, Don Wilson states there is "some indirect evidence that our space agency officials are taking the 'hollow' problem of the Moon more seriously than they are letting the public know."[27] He refers to claims supposedly made by the Egyptian-born American geologist Farouk El-Baz that NASA, suspecting the Moon to have hollow regions, carried out undisclosed experiments to determine whether such hollow regions existed. The alleged experiments were "held in the strictest secrecy," the results of them never made public.[28] Wilson quotes El-Baz from a 1974 interview in an edition of *Saga* magazine: "There are many undiscovered caverns suspected to exist beneath the surface of the Moon. Several experiments have been flown to the Moon to see if there actually were such caverns."[29] El-Baz further admitted that "not every discovery has been announced."[30]

Currently research professor and director of the Center for Remote Sensing at Boston University, El-Baz participated in the Apollo program from 1967 to 1972 as supervisor of lunar science planning. Among other things, he played an instrumental role in the selection of landing sites and assisted with the training of astronauts, who affectionately called him "the King." As an example of the kind of fame El-Baz has achieved, one of the shuttle crafts in the television show *Star Trek: The Next Generation* was named in his honor. Many conspiracy researchers feel if anyone possesses valuable inside information about what was really discovered on the Moon during the Apollo program it is "the King."

Wilson's observation (that NASA appears to be "taking the hollow problem of the Moon more seriously than they are letting the public know") is just as relevant today as it was thirty years ago, when *Secrets of Our Spaceship Moon* was published. Does it not seem somewhat suspicious that scientists from NASA, JAXA, and ISRO have currently been probing the Moon for evidence of uncollapsed and partially collapsed lava tubes? The only reason these hollow regions are of interest to them, apparently, is because they are believed to make ideal bases for future lunar colonists.

But what if NASA has known about the Moon's so-called lava tubes since the beginning of the Apollo program and has only just decided to reveal them to the public? I say "so-called" because there is the further possibility we aren't dealing with lava tubes in the proper sense of the term but vast hollow tracts running throughout the entire crust of the Moon and perhaps within its mantle as well. I'm inclined to believe the latter, because it would help explain why the Moon "is an almost perfect medium for the propagation of seismic waves." Plus it has a ring of common sense to it, if you'll pardon the pun.

There is every indication, to my mind at least, that the partially hollow nature of the Moon was confirmed by NASA scientists during the Apollo era, and they are still in the process of investigating this anomaly—albeit covertly. I would suggest the real motivation behind NASA's recent and very bizarre crater bombing experiments, involving *Prospector* in 1998 and LCROSS in 2009, were to gather additional information concerning the Moon's mysterious interior. Don't forget these two very large impacts—both of which occurred at the Moon's south pole—would have triggered moonquakes of great magnitude. Is the Moon's south pole especially hollow? Could this be why NASA has taken a special interest in this region?

There is one apparent flaw with this theory. So as to cut down on costs, we are told, NASA shut down its entire lunar seismic network on September 30, 1977. There are no active seismometers on the Moon at present, and thus no means to detect or measure moonquakes. According to Richard C. Hoagland the network is still operational; NASA lied about the shutdown. What's more, it was used to record the LCROSS lunar impact—the purpose of which had nothing to do with searching for lunar water and everything to do with investigating subsurface structures on the Moon. Hoagland has provided some compelling evidence to support his argument, which I'm afraid we haven't the space to discuss here.

In December 2011, I sent Farouk El-Baz an e-mail in which I asked if he'd mind answering several questions concerning the Moon's interior. To my surprise he replied—very promptly, too. He politely informed me he was unable to answer my questions because he'd be "travelling a lot in the next couple of months . . ." and therefore would "be out of contact."[31] He ended his e-mail by referring me to a colleague of his at Boston University.

Refusing to be brushed aside, I sent him another e-mail: "I'm attempting to understand how the body of the Moon could be 'an almost perfect medium for the propagation of seismic waves' . . . without containing vast hollow areas. . . . Since you allegedly stated in an interview that 'there are many undiscovered caverns suspected to exist beneath the surface of the Moon,' I was hoping you would be able to shed some light on the matter."[32] El-Baz's reply was brief and very tight-lipped: "This [undiscovered caverns] related to lava tubes whose top layer had not collapsed."[33] El-Baz's e-mail is interesting not so much for what it does say as for what it doesn't. First, it very strongly suggests that the Moon's hollow regions—which we've been led to believe are perfectly natural lava tubes—have been of interest to NASA since the Apollo era. Second, since El-Baz never stated otherwise, it leaves open the possibility that the Moon "rings like a bell" on account of these hollow regions and by implication suggests they are not lava tubes in the proper sense of the term.

If we're willing to accept the Moon is partially hollow, we must also be willing to accept it is not entirely natural. For, to quote Sagan again, "A natural satellite cannot be a hollow object." So if the Moon isn't natural, what is it? And how did it end up in orbit around the Earth? According to what would have to be the most unconventional lunar origin theory ever proposed, our closest celestial neighbor is not a natural satellite of the Earth but an artificial one; namely, a hollowed-out planetoid that was "put into orbit around the Earth by . . . intelligent beings unknown to ourselves."[34] So wrote Alexander Shcherbakov and Mikhail Vasin, both members of the Soviet Academy of Sciences, in a paper that appeared in the Soviet journal *Sputnik* in July 1970 titled "Is the Moon the Creation of Intelligence?" Known as the spaceship moon theory, it caught the imagination of a Detroit school teacher named Don Wilson (full name Donald K. Wilson), who wrote two very impressive books on the subject, *Our Mysterious Spaceship Moon* (1975) and *Secrets of Our Spaceship Moon* (1979).

Vasin and Schcherbakov's paper—later expanded into a book (published only in Russian)—begins with a brief assessment of the three classic theories of lunar origin. They reject the fission and coaccretion theories on the grounds that Moon rocks are so different in composition to Earth rocks and the capture theory because "a catch of this kind is virtually impossible."[35] The spaceship moon theory could be seen as a more exciting and "magical" version of the capture theory. Both theories involve a foreign planetoid that took up orbit around the Earth. While the capture theory views this planetoid as some lifeless, random drifter that happened to get snagged by Earth's gravitational field, the spaceship moon theory views it as an intelligently controlled object— part natural and part artificial—which was parked in Earth orbit. It turns the capture theory on its head, because it implies the Earth was captured by the Moon—not the other way around. "Many things so far considered to be lunar enigmas are explainable in the light of this new hypothesis," state Vasin and Schcherbakov.[36]

According to their theory, the Moon started its life as a perfectly natural planetoid from somewhere far away, possibly from outside our solar system. After being hollowed out with the use of advanced technology, its interior was installed with all manner of equipment and machinery, including "fuel for the engines" and "materials and appliances for repair work." It was designed to "serve as a kind of Noah's Ark of intelligence, perhaps even as the home of a whole civilization envisaging a prolonged (thousands of millions of years) existence and long wanderings through space. . . ."[37]

Vasin and Shcherbakov point out that the Moon, like any other spaceship, would have to be durable enough to withstand the impact of meteorites plus sharp fluctuations of extreme temperature. They therefore suggest its "hull" would consist of two layers: a dense, armoring inner layer of about twenty miles in thickness, and a thinner outer layer of about three miles in thickness. Comparing the two layers, they describe the latter as a "more loosely packed covering." Between the Moon's hull and core, or its "kernel" and "shell," is a hollow region, "doubtless filled with gases required for breathing," stretching some thirty miles.[38]

No matter how farfetched and ridiculous the spaceship moon theory seems, one cannot deny it does make sense of a vast number of lunar mysteries. It explains, most important of all, why the Moon "rings like a bell" and thus appears to be partly hollow. Indeed, it is virtually impossible to account for the Moon's apparent hollowness by natural means alone. The Moon could not have been hollowed out to the extent that it seems to be unless some form of technology was involved. We'll look into this in a moment.

The spaceship moon theory sheds a great deal of light on not only why moonquakes can persist for many hours but also why they're so bizarrely predictable and mechanical in nature, especially deep moonquakes. If, as the theory suggests, the Moon contains artificial constructs, such as a double-layered hull and who knows what else in its interior, would we not expect these constructs to consistently react the same way as the Moon and the Earth interact gravitationally? And if you're skeptical about moonquakes being mechanical, allow me to provide you with an "official" quote from the *Cambridge Guide to the Solar System:* "Nearly identical seismograph records were obtained again and again, indicating that certain regions, known as *nests,* are repeatedly generating moonquakes in the same way."[39] Substitute the word "moonquakes" with "vibrations," and Lang may as well be referring to an ocean liner or a jumbo jet.

The spaceship moon theory has an interesting perspective on the origin of the Moon's maria, in addition to the strange concentrations of mass—mascons—that lie at the center of some of them. Orthodox science would have us believe the maria were produced billions of years ago by repeated outpourings of lava, when the Moon possessed a hot and active interior. But what if, instead, the maria are actually areas of the Moon's "outer shell" that have been artificially repaired? Vasin and Shcherbakov suggest these areas, having sustained heavy damage from meteorite bombardment, were flooded with a "kind of cement."

This cement, consisting partly of lunar core material, was "piped to the surface."[40] They further suggest the heavy machinery and equipment required to carry out this extensive repair work was brought directly to where it was needed and left permanently on site—a perfect explanation for the Moon's mascons.

If some kind of cement or filler was needed to patch tears in the hull of a giant spaceship, the once molten basaltic rock that makes up the maria would be absolutely ideal for this purpose. As I mentioned previously, mare rocks are extremely rich in rare refractory elements like titanium, chromium, and zirconium, which are metals known for their excellent durability and resistance to heat and corrosion. A combination of these metals, comment Vasin and Shcherbakov, "could be used on Earth for linings for electrical furnaces." They continue: "If a material had to be devised to protect a giant artificial satellite from the unfavorable effects of temperature, from cosmic radiation and meteorite bombardment, the experts would probably have hit upon precisely these metals."[41] The presence of high quantities of titanium within the Moon's maria, ranging between 1 and a little more than 10 percent, is especially puzzling to scientists, because nothing of this kind exists on Earth. Even the most titanium-rich Earth rocks contain quantities no higher than 1 percent.

Why should the maria contain such high quantities of titanium and other rare metals, including yttrium and beryllium? Were these elements added to the molten basaltic rock that formed the maria so as to imbue the material with special properties? There is evidence that apparently processed metals, including pure iron particles as well as iron particles that do not rust, and even brass, were found in lunar samples.

Don Wilson states that the pure iron particles were detected by scientists at the General Electric Research and Development Center. It was thought that the particles might have come from meteorites, until experts pointed out that there's no such thing as pure iron in meteorites; it is always alloyed with nickel. In fact, pure iron in nature is virtually unheard of. Iron, in its natural state, is never isolated. It exists in minerals (such as iron ores) formed by the combination of iron and other elements. Thus, pure iron can only be obtained by means of artificial extraction. It is a manufactured substance.

As for the iron particles that do not rust, these were apparently found in samples of lunar soil brought back by unmanned Soviet probes. One sample was collected from Mare Crisium in 1976 by *Luna 24,* the last Soviet spacecraft to have ventured to the Moon. The discovery was announced in the official Soviet government news journal *Pravda,* and is therefore difficult to verify. Pure iron that does not rust is unknown on Earth, and it's difficult to imagine how such a metal could exist without having been manufactured. The closest thing we have on Earth to iron that does not rust is stainless steel—an alloy made from steel and chromium. (To increase corrosion resistance, metals such as nickel and titanium are sometimes added as well.)

What about the alleged discovery of brass in lunar samples? This, according to Wilson, was found by Cambridge University scientists, which makes it another piece of information that's difficult to verify. Brass, an alloy of copper and zinc, is most definitely a manufactured metal.

There is a great deal about the lunar maria that seems awfully unnatural, and they could represent the strongest evidence we have that the Moon is an artificially modified world. Whenever I look at these vast, dark splotches that cover nearly a third of the lunar near side, I am very much reminded of scars or bloodstains—evidence the Moon was "wounded" in the past and has carried those wounds ever since. I can't help but feel the maria "don't belong," in the same way scars don't belong on a person's skin or freeways don't belong on the surface of the Earth.

These patches of tough, metal-rich basalt are located exactly where they ought to be in terms of providing the Moon with reinforcement for its crust. Most of the maria, as we know, can be found on the Moon's thin-crusted near side, directly covering impact basins. Without them, sections of the Moon's "hull" would be very thin indeed. Given that the far side crust of the Moon is a good deal thicker than the near side crust, it's perhaps no wonder only a tiny portion of the far side surface is covered by maria, despite being so heavily cratered. After all, what would be the purpose of filling in these "dents" with "cement" when the "hull" itself is very thick and therefore capable of withstanding just about any impact? This especially tough half of the Moon would require little maintenance.

The spaceship moon theory states that material was pumped from the Moon's interior to the surface, not only as a means of repairing cracks in the hull but also as a means of hollowing out its interior. We know volcanic activity was responsible for the formation of the maria and this activity originated in the mantle after it became hot and partially melted as a result of radioactive decay. But who's to say the volcanic activity in question wasn't wholly due to natural causes?

Scientists can't explain how the Moon managed to generate the massive amount of heat required to produce molten rock containing abundant quantities of titanium and other refractory metals (which melt at a very high temperature); nor can they explain how the Moon generated the heat required to spew forth such a great quantity of lava. In fact, it has been estimated that nearly 2.5 million cubic miles of lava erupted to form the maria, which is enough to fill ten billion football stadiums. Nor can scientists explain the mechanism by which this lava was driven to the surface from such great depths. Referring to this problem in his book *The Moon in the Post Apollo Era* (1974), the lunar expert Zdenek Kopal says it "calls for an expenditure of energy whose source is not readily apparent."[42]

That technologically advanced beings placed radioactive elements inside the Moon so as to generate the necessary heat to cause large-scale volcanic activity is not outside the realm of possibility. No matter how outlandish this explanation seems, or how difficult it is to accept, it is one that ought to be given serious consideration, for it helps make sense of all the major mysteries associated with the maria. One of the biggest mysteries of all is the unequal distribution of maria across the near and far sides of the lunar surface, a matter we've already touched upon. While it's true crustal thickness played some part in this process, in that the lava was able to penetrate to the surface more easily in those areas where the crust is thin, it could not have been the only factor involved.

For example, the Moon's most extensive mare, Oceanus Procellarum, does not correspond to any known impact basin, whereas all the others do. The oldest, deepest, and largest lunar impact basin is the South Pole–Aitkin basin, which is the region of lowest elevation on the Moon and an area where the crust is very thin. One unusual fact regarding this extensive basin is that it is filled with only a modest amount of basaltic lava. Located on the Moon's far side, between the crater Aitkin and the south pole, it is one of the largest impact basins in the entire solar system, with a diameter of roughly 1,550 miles and a depth of more than 7.5 miles. As well as being one of the coldest regions of the Moon, featuring numerous pockets bathed in perpetual darkness, it is one of the most anomalous in terms of composition.

Perhaps the hypothetical inhabitants and engineers of "spaceship Moon" saw no reason to extensively repair the South Pole–Aitkin basin, filling it with only a modest portion of "cement." As for Oceanus Procellarum, maybe it represents nothing more than an ancient "dumping ground" for material. After all, if material was extracted from the lunar interior for hollowing-out purposes (rather than for repair-related purposes) it would have had to have been dumped on the surface somewhere (presumably wherever was most convenient).

So far we've discussed two distinct types of lunar terrain: the dark maria and the pale highlands. Because of valuable data obtained by *Lunar Prospector* in the late 1990s, scientists have managed to identify a third type of lunar terrain: the Procellarum KREEP Terrane, or PKT for short. This is yet another of those weird and overly complicated names that our lab-coat-wearing friends find highly appealing. The PKT is an immense "hot spot" encompassing much of the maria on the lunar near side, so named because of its high concentrations of thorium and other radioactive, heat-producing elements, including potassium (chemical symbol K), and uranium. The "P" stands for phosphorus and the *REE* stands for rare-earth elements. Thus you get the acronym KREEP.

The prevailing theory states that as the Moon's global magma ocean began to cool, these KREEP elements refused to fit into the crystal lattices of the most common lunar minerals, which include pyroxene, olivine, and plagioclase. (For this reason they're called "incompatible" elements.) This led to the formation of pockets of KREEP-rich magma sandwiched between the crust and mantle. As the thorium, potassium, and uranium decayed, releasing heat, the surrounding mantle melted and the result was volcanic activity, or so the theory goes. That there is a high concentration of heat-producing elements beneath the PKT obviously has something to do with the intense mare volcanism that occurred within this region. That in itself is clear. What scientists don't understand is the mechanism by which KREEP became concentrated in one large area of the lunar near side. Which brings us back to the mystery of why the near side of the Moon has many maria and the far side so few.

The PKT isn't the only region of the Moon where the radioactive, heat-producing element thorium has been detected. The South Pole–Aitkin basin is another place, although here the concentration of thorium is much lower. Another thorium "hot spot" is the region in which lies the mysterious Compton-Belkovich volcanic feature. The Compton-Belkovich Thorium Anomaly, as it's become known, is situated about 560 miles from the northeastern extent of the PKT. Scientists are struggling to explain the presence of this isolated and, presumably, quite young silicic volcanic complex. Commented Bradley Jolliff, research associate professor in the department of Earth and Planetary Sciences at Washington University, "To find evidence of this unusual composition located where it is, and appearing to be relatively recent volcanic activity, is a fundamentally new result and will make us think again about the Moon's thermal and volcanic evolution."[43]

If the Compton-Belkovich volcanic complex is as young as its estimated age of eight hundred million years, then it's extremely unlikely to have formed as a result of radioactive decay, because, in the words of Jolliff, "those heat sources diminish with time and it gets harder and harder to get lavas to the surface."[40] The only alternative explanation, according to Jolliff and other experts, is that the necessary heat originated from the Moon's outer molten core. But of course we still don't know if the Moon's outer core is indeed still molten, or if it's capable of generating the kind of heat required to cause volcanic activity. What we have is yet another lunar conundrum, whereby many pieces of the puzzle are missing and what few pieces we do possess refuse to fit together.

They don't fit together, that is, unless we're willing to accept that the Moon is not a wholly natural world, but one that has been subjected to an extensive amount of tampering. Consider the possibility that radioactive elements, such as thorium, were deposited within certain, carefully selected regions of its interior, causing sections of the mantle to melt and "bleed out" onto the surface in addition to other effects. On account of its apparent young age, it's difficult to explain the formation of the Compton-Belkovich volcanic feature in terms of natural radioactive processes. What about the possibility of artificial radioactive processes? Artificially created or not, the feature is so anomalous in nature that only a highly unconventional explanation can account for its origin. A number of conspiracy theorists have suggested the feature is a thorium mine or nuclear reactor—evidence of an alien presence on the far side of the Moon. This possibility shouldn't be ruled out either.

Potassium and thorium aren't the only radioactive elements that have been found to exist on the Moon. During its twenty-month lunar orbit, the Japanese spacecraft *Kaguya* detected uranium on the surface of the Moon by using its gamma-ray spectrometer. The discovery caught the attention of those with an interest in mining the Moon's resources. For, as we know, uranium is an extremely valuable nuclear fuel, and terrestrial deposits of uranium are already running low. Later, scientists discovered the Moon had fairly low concentrations of uranium—not the commercial levels needed to warrant mining.

It's interesting to note that thorium, like uranium, can be used as a nuclear fuel. In pure form, it is a soft, silvery-white metal that darkens upon exposure to air. Three times more abundant in the Earth's crust than uranium, the vast majority of thorium that exists in nature is the isotope thorium-232. Although not directly fissile itself, meaning it cannot sustain a nuclear fission reaction, it can be converted to uranium-233, which is fissile. Because of its high abundance there has been considerable interest of late in using thorium-232 as a nuclear fuel source. Countries like India and China are already looking to build thorium-based nuclear reactors; however, the catch is thorium-232 is costly to extract and convert—problems that scientists hope to overcome in the foreseeable future. One wonders if the supposed inhabitants of the Moon are already well ahead of us in this endeavor.

Amazingly, traces of the rare radioactive isotopes uranium-236 and neptunium-237 were discovered in Moon rocks by scientists at the Atomic Energy Commission's Argonne National Laboratory in Illinois. The story was reported in 1972 in *New Scientist* and other publications, but has been largely forgotten since. Prior to the discovery, it was assumed that neither isotope occurred naturally on the Earth but could only be produced in a nuclear reactor. While both isotopes have been found on Earth, they are very rare. Traces of uranium-236, commonly known as a radioactive waste found in spent nuclear fuel, were detected in Earth ore samples by the Argonne team soon after they detected it in Moon rocks. The lunar samples were found to contain the isotope in much greater abundance than the Earth samples, by a factor of five to 350 times.

We now come to neptunium, a very strange element to find on the Moon. In pure form it exists as a silvery, ductile metal. It was discovered in 1940 by a pair of scientists at the University of California, Berkeley; they managed to produce neptunium-239 after bombarding uranium with slow-moving neutrons from a cyclotron. Neptunium belongs to the quasi-synthetic category. These are elements that occur naturally on Earth in trace quantities but generally have to be artificially produced.

The most stable of all neptunium isotopes—the one with the longest half-life—is neptunium-237. Minuscule concentrations of the isotope are found naturally in uranium ores. It forms as a by-product—in quite significant amounts—when plutonium is produced in nuclear reactors.

What were traces of uranium-236 and neptunium-237, two very rare radioactive isotopes, both generally considered artificial, doing in rocks brought back from the Moon? In a 1972 article that appeared in *The Deseret News,* titled "Moon Rock Analysis Uncovers 'Unearthly' Pair," we find the following explanation: "Scientists said the presence of neptunium and the unusual amounts of uranium-236 are probably due to solar flares which bombard the Moon with atomic particles, causing the necessary nuclear reaction."[45] So far as I can tell, no other explanations have been put forward, leaving the matter unsolved. In view of all the seemingly artificial substances found on the Moon, such as pure iron particles and ones that do not rust, could the traces of uranium-236 and neptunium-237 also belong to this category? Could they be products not of natural nuclear activity but of some form of engineered activity, possibly involving the use of nuclear reactors and such?

Vasin and Shcherbakov argue that lunar craters, though they vary greatly in diameter, are abnormally shallow and thus indicative of an artificially reinforced object. One lunar crater that could be described as abnormally shallow is the aforementioned South Pole–Aitkin basin.

During the Clementine mission in 1994, the basin was mapped topographically and found to be as deep as 7.5 miles. Scientists hoped the basin would extend well into the mantle of the Moon, leaving parts of it exposed and available for study. After all, the mantle of a planet contains important information as to the nature of that planet's total composition. Nothing about the Moon is straightforward, and the South Pole–Aitkin basin is no exception.

Simulations have shown the impact that produced the basin ought to have penetrated the crust and dug up a vast amount of mantle materials. However, all of the available evidence indicates that no such penetration occurred as plenty of crust remains beneath the basin floor. In an attempt to explain this anomaly, it has been suggested that the basin was the result of a special kind of collision, whereby the impactor struck from a very oblique angle (rather than directly) and was traveling at low velocity. It could just as well be argued that the basin is unusual because the lunar crust itself is unusual, having been artificially reinforced. This would lend support to the spaceship moon theory.

Before jumping to such a conclusion, it would serve us well to remember the Moon isn't the only body in the solar system to have sustained some extraordinarily huge impacts. Allow me to offer an example. If you've seen the original Star Wars series you'd be familiar with the Galactic Empire's "ultimate weapon"—the Death Star. A moon-size space station, the Death Star is capable of destroying entire planets with its powerful energy beam, which looks like a concave disc. The first Star Wars movie to feature the Death Star is *Episode IV: A New Hope,* which was written and directed by George Lucas and released in 1977. Another unusual "moon synchronicity," came to light years after the release of the movie: we had a "Death Star" in our very own solar system. The object in question is Mimas, the smallest and innermost of Saturn's major moons, discovered by William Herschel in 1789. Mimas is a heavily cratered object, believed to consist largely of ice, with a diameter of only 250 miles.

Few details were known about Mimas until during the early 1980s, when the probes *Voyager 1* and *2* took close-up photographs of its surface. Higher resolution photographs were obtained by Cassini-Huygens in 2005. These photographs revealed a very interesting feature on the surface of Mimas: a colossal crater measuring 80 miles in diameter and 6 miles deep. Named Herschel, it has a prominent central peak and is recognized as the one of the largest impact structures in the solar system relative to the size of the body on which it exists. Because its diameter is almost a third of that of Mimas, Herschel stands out like a giant eye. It is this "giant eye" that gives Mimas an uncanny resemblance to Lucas's fictitious Death Star. It is noteworthy that in *Episode IV: A New Hope* the Death Star is described as having a diameter of roughly 100 miles, which is close to the diameter of Herschel crater, at 80 miles. Also, if we divide 100 by 250 (the diameter of Mimas) we get a fraction of $^2/_5$.

I mention Mimas not only for its synchronistic significance but also because of its striking crater Herschel, which has been attributed to an asteroid impact. And what a devastating impact it must have been. A NASA website states the impact "probably came close to breaking the moon apart."[46] Fractures associated with the impact have been identified on the surface of Mimas opposite the side of Herschel. It is rather surprising, perhaps, that the impact didn't inflict greater damage on this very tiny moon. And the same could be said of the impact that produced the Moon's South Pole–Aitkin basin—the largest known impact crater in the entire solar system.

We know the Moon is tough and very resilient, its crust a good deal thicker than the crust of the Earth. But can we say it's tougher than other bodies in the solar system, to the extent that its toughness can only be accounted for by artificial means? Not necessarily, though it would be wrong of us to rule out this possibility.

So far we've discussed the Moon's external strength, but not so much its internal strength. To gain some insight into this matter, it would help to take a closer look at the Moon's peculiar, lopsided shape.

The Moon, as we know, is by no means a perfect sphere, possessing not one but two major deformities. The most obvious and expected of these is that it's slightly squashed; its equatorial width is about 2.5 miles greater than its pole-to-pole height. This feature is known as an equatorial bulge, and they can be found on all rotating celestial bodies, including Earth, whose equatorial width is 26.5 miles greater than its pole-to-pole height. The cause of this phenomenon is centrifugal force. The second major deformity we previously touched upon is its far side is higher in elevation than its near side. The far side crust, as stated earlier, is around 9.3 miles thicker than the near side crust, giving the Moon a very prominent backside.

The asymmetrical nature of the Moon's crust is one of its greatest unsolved mysteries. A number of explanations have been put forward, all of them far from conclusive. One study looked at how the gravitational pull of the Earth might have shaped the Moon billions of years ago, when it was young and partly molten, and found that the lunar crust ought to be distorted in a similar fashion to the Earth's ocean tides. The study came up with a Moon whose crust is thinnest at the poles and thickest at the equator, especially in the regions in line with the Earth. Some of these details are obviously consistent with reality. Indeed, the lunar crust is thinnest at the poles, while there is a "crustal bulge" on the far side of the Moon positioned in line with the Earth. But where, you might ask, is the second "crustal bulge," the one that ought to exist on the near side of the Moon, also in line with the Earth? Tides, after all, are symmetrical in nature.

The only explanation offered by scientists as to the absence of this hypothetical near side bulge is that it might have disappeared over time as a result of volcanic activity or other geological processes. However, this tells us nothing at all, except that the theory itself is in need of a lot of work. According to another, more recent theory, the Moon acquired its far side bulge as a result of a collision it sustained in the past with a second, smaller moon of the Earth. But instead of removing material, the impactor deposited material, smearing the lunar far side with an extra layer of crust. It would be unfair to say this theory is completely devoid of merit. I do wonder, though, if we need another giant collision to explain the formation of the Moon. Scientists obviously find them extremely appealing, coming up with "big whacks," "giant impacts," "big bangs," and "almighty knockers" at every chance they can get.

The theory that the Moon acquired its far side bulge as a result of gravitational forces early in its history when its crust still floated on a sea of magma does have a ring of plausibility to it. The only problem with this theory is it leaves the near side of the Moon unaccounted for. There is a way around this problem if we accept the Moon originated from somewhere else, as stated by the spaceship moon theory. Explains Don Wilson: "The spaceship Moon traveling throughout the cosmos undoubtedly passed (and perhaps visited) many other worlds and stars that could have placed a gigantic pull on it. Perhaps the Moon zoomed into another celestial orbit that caused it to wrench itself out of shape."[47]

In addition to the mystery of how the Moon came to acquire its far side bulge, there is the mystery of how it manages to support it. "This bulge is a non-equilibrium one and must be supported by some curious internal characteristic of the Moon, such as great strength of the interior or some variation in the density of the body of the Moon," observed Urey in 1962.[48] Although we now know there are variations in density within the body of the Moon—comprised, as it is, of a geochemically distinct crust, mantle, and possible core—Urey's words are still of relevance today. The lunar scientist Zdenak Kopal came to much the same conclusion as Urey, noting, in 1971, there "must be considerable strength in the deep interior of the Moon, and that this has been true since the Moon acquired its irregular shape." Why, asked Kopal, was the Moon unable to adjust its shape so as to relieve the pressure associated with its huge far side bulge? Because no such adjustment has occurred, he concluded, "surely indicates considerable strength in the deep interior."[49] If, as the spaceship moon theory says, the Moon does indeed possess some kind of inner hull, this would explain where it gets the necessary internal strength to maintain its far side bulge.

In terms of making sense of some of the Moon's biggest mysteries, the spaceship moon theory hasn't failed us, yet. Like any theory, it needs to be put to the test. And what better way to test it than by exploring one of the Moon's greatest mysteries of all: its magnetic field. As everybody knows, the Earth generates its own magnetic field. Pass a current through a wire and you'll find it produces a magnetic field around itself, which can be detected with the use of a simple compass. By wrapping the same length of wire around a cylinder, you have what's called an electromagnet. Each loop of wire puts out its own magnetic field, and when you bring all these loops together the result is a magnetic field pattern similar to that obtained by a bar magnet, known as a dipole field. We've all seen what happens when you place a bar magnet on a piece of paper and sprinkle it with iron filings. The iron filings arrange themselves in accordance with the magnetic lines of force, rendering visible the magnet's dipole field.

The magnetic field of the Earth is similar to that of a bar magnet's dipole field. It's as though the center of the Earth itself were occupied by a giant bar magnet, roughly aligned with the planet's axis of rotation, with the north of the magnet pointing down, toward the south geographic pole, and the south of the magnet pointing up, toward the north geographic pole. Magnetic fields "travel" from a north to south direction so that, in the case of Earth, magnetic lines of force leave the bottom of the planet, curve up, cross the equator, and then curve down, entering the top of the planet. According to the widely accepted dynamo theory, the magnetic field of the Earth is generated deep within the planet, where heat produced by means of radioactive decay in the solid inner core drives convective motion in the liquid iron outer core. This activity produces an electric current, which in turn gives rise to a magnetic field. The dynamo theory views the Earth's magnetic field as the product of a dynamo effect. (A dynamo is a machine that converts mechanical energy into electrical energy.) From the perceptive of this theory, then, the Earth possesses an internal, self-sustaining dynamo.

The Earth's magnetic field, which exists in a state of constant fluctuation, surrounds the planet like a giant, protective cocoon, extending past the atmosphere and well into space. By shielding Earth from the fury of the solar wind, it also shields us from possibly lethal solar particles and other forms of space radiation. Scientists use the term "magnetosphere" to refer to the extensive protective cavity carved out in the solar wind by the Earth's magnetic field (see plate 6). Not all planets in the solar system possess magnetospheres; only ones that generate detectable, global magnetic fields. Planets belonging to this category include Mercury, Jupiter, Saturn, Uranus, and Neptune, the two exceptions being Mars and Venus. Ganymede, the largest of Jupiter's satellites—bigger than the Moon and Mercury itself—is the only moon known to possess a magnetosphere.

As for Mars, patches of residual magnetism were detected on the red planet by the *Mars Global Surveyor*. The patches served as evidence that it possessed a global magnetic field (and thus a magnetosphere) up until four billion years ago. The means by which this presumed field got "switched off"—suddenly or gradually—remains unknown. One promising theory is a series of very large impacts "knocked out" Mars's internal dynamo, and hence its ability to generate such a field. Without a magnetosphere to protect it, a planet's atmosphere gets stripped away by the solar wind. And this, it would seem, is what happened to Mars. There is every indication the thin Martian atmosphere was previously more substantial.

It came as no great surprise to scientists when, at the beginning of the space age, magnetometers carried aboard Soviet and American spacecraft detected little or no trace of a magnetic field on the Moon. The Moon, being the size it is, was not expected to have the necessary internal equipment to generate such a field. The matter became rather more complicated when rocks brought back from the Moon by Apollo astronauts were analyzed and found to contain remnant magnetism. "We don't know what the magnetic story is, but we do know that there are some very strange magnetic properties in these rocks which are not expected," commented geophysicist Paul Gast, Ph.D., of Columbia University, at a Lunar Rock Conference held in 1969.[50]

Rocks that form in the presence of a magnetic field contain a record of the strength and orientation of that field, and this remnant magnetism, as it's called, can be accurately measured in the laboratory. In the case of the Apollo samples, some of them were found to be highly magnetized, indicating that they formed in the presence of a strong, steady field. To determine if the rocks had acquired their magnetism from the Earth's magnetic field during transportation, one lunar sample was taken back to the Moon. The experiment proved otherwise. Remnant magnetism has been found in a lunar rock dated at 4.3 billion years old, pointing to the almost undeniable conclusion that the Moon had a global magnetic field very early in its history.

During the Apollo era, satellites were used to measure the Moon's magnetism from lunar orbit while the astronauts also took measurements from the ground. Discovered across the surface of the Moon were numerous "pockets" of magnetism. Some of the localized regions measured as broad as 62 miles with magnetic fields of between one-hundredth and one-thousandth the strength of Earth's magnetic field. Decades later came *Lunar Prospector,* and with it a better understanding of the magnetic properties of the Moon. The strongest magnetic fields were detected on the Moon's far side. For reasons unknown, these fields happen to be located directly opposite to a number of impact basins on the lunar near side; namely, Imbrian, Serenitatis, Crisium, and Orientale. Because the basin rock itself is weakly magnetized, there is speculation that the impacts that carved out the basins demagnetized the crust in these locations while at the same time magnetizing the crust on the opposite side of the Moon.

Associated with some of the Moon's strongest regions of magnetism are deeply mysterious markings called "swirls," which no lunar scientist knows what to make of. "Swirls remain one of those big unanswered questions in lunar science," commented planetary scientist Catherine Neish, Ph.D.[51] They are named as such because they resemble the swirling patterns that form when milk is poured into a cup of coffee. Because of their high albedo, these rolling, sinuous markings show up best when contrasted against the dark maria.

One of the Moon's most prominent swirls is Reiner Gamma, located on the Oceanus Procellarum. It was first photographed up close in 1966 by NASA's *Lunar Orbiter II.* Before then it was believed to be nothing more than a very messy-looking crater. As with all lunar swirls, Reiner Gamma coincides with a region of high magnetic irregularity. According to one theory the swirls represent exposed silicate materials that have essentially remained preserved on account of being protected from the solar wind by the magnetism present in those areas. Another, less plausible theory states the swirls were left behind by cometary impacts. What is so puzzling about these markings is they appear to be entirely unique to the Moon.

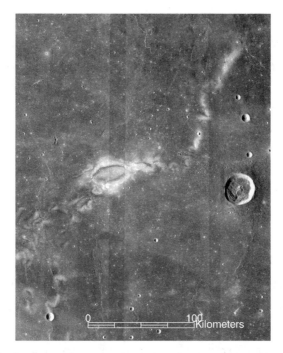

The lunar swirl Reiner Gamma. Photo courtesy of NASA.

Scientists are now fairly certain the Moon had a global magnetic field early in its history—similar to the Earth's, only a great deal weaker—generated by means of a churning molten core (dynamo) that was active for more than four hundred million years. But whereas the Earth's dynamo is self-sustaining, scientists believe the Moon's was not. For the Moon, on account of its small size, would have lacked the necessary internal temperature differences for convection to have occurred within its molten outer core. This means the Moon's dynamo could not have operated under its own steam. The question, then, is what helped power the Moon's dynamo? From where did it receive its energy?

One possible answer is Earth. We know Earth and the Moon interact gravitationally, and they did so to a greater extent in the past when they were closer together. With this in mind, it's been suggested that the Moon's molten outer core (which may still be molten today) was stirred, so to speak, by strong gravitational tugs from the Earth.

This supposedly involved a back-and-forth kind of stirring, such as the type produced by a coffee stirrer, whereby the Moon's solid mantle was rotated against its core. Another theory states the stirring in question resulted from large impact events. Time will tell if either or both of these theories are valid. The undeniable fact is the Moon appears to have had a global magnetic field early in its history (at least four billion years ago), which operated for a while and then suddenly quit, and which must have been generated by means of a dynamo—one dependent on outside help.

There is a theory that states the Moon acquired its magnetism from the magnetic field of the Earth when the two bodies were closer together; meaning the Moon had neither a global magnetic field nor, for that matter, a dynamo. Although there is plenty of evidence to counter this theory, it hasn't been ruled out entirely. Experts have pointed out that for the Moon to have been magnetized by Earth in this way, the two bodies must have been extremely close—so close that the Moon would have likely broken up under the Earth's gravitational field, having come near or within the Roche limit. The *Roche limit,* a very important principle in astronomy, refers to "the minimum distance to which a large satellite can approach its primary body without being torn apart by tidal forces."[52] The Roche limit applies to natural satellites. But if we're dealing, instead, with an artificially reinforced satellite, a different set of rules applies. A spaceship Moon with a double-layered hull could theoretically orbit the Earth well within the Roche limit without being torn apart by Earth's gravity.

Perhaps, billions of years ago, when the Moon first "arrived," it was parked extremely close to Earth, well within the Roche limit, with its far side facing Earth. Not only would this explain how the Moon became magnetized—its far side, especially—it would explain how it acquired its far side bulge. Assuming this feature to be a distortion produced by the strong gravitational pull of the Earth, what is it doing on the far side of the Moon? Common sense would suggest the near side of the Moon ought to be the side with the bulge, for here the Earth's gravity is strongest. Could the Moon have been "flipped" at some point in the past so that what was once its near side is now its far side?

The spaceship moon theory has numerous strengths, but it is not without its weaknesses either. It falls short the same way the capture theory falls short: by failing to take into account that the Earth and the Moon have a shared ancestry. Let us not forget the two are extremely similar in composition, and their oxygen isotope signatures are practically identical. If the spaceship Moon theory were true in every respect, it would mean the Moon is an object with the same capabilities as a spaceship, and, in the manner of a spaceship, traveled a vast distance to get here. And yet, we can be fairly sure the Moon has been here all along, keeping the Earth company. The theory, though difficult to accept in its entirety, seems to be at least partly accurate.

The theory is accurate in so far as it states the Moon is not a wholly natural world, but rather, to borrow a term from Don Wilson, a "hybrid world," possessing both natural and artificial characteristics. Relying solely on the evidence we've come across so far, we have no way to determine if the Moon is artificial to the extent that its interior is packed with machinery and equipment, such as one would find in the interior of a Boeing 747. What we do know for sure is the Moon has been tampered with, or terraformed, if you will. Part of this process clearly involved an extensive amount of hollowing out, no doubt within the crust, but possibly within the mantle as well. I use the word "extensive" because only an extensively hollow object would "ring like a bell" as the Moon does. Next we come to the maria, which are natural in the sense that they formed as a result of volcanic activity and artificial in the sense that the volcanic activity in question was engineered (possibly with the use of radioactive substances). Either the maria were created to patch up sections of the Moon's damaged crust, or they have no such purpose but merely represent huge deposits of "excavated" material from when the Moon was artificially hollowed out.

Let's imagine for a moment the Earth and Moon constitute a double-planet system. Go back in time, about 3.8 billion years, to when the two planets were very young. For argument's sake, let's say they orbit each other at a distance of approximately 12,500 miles (about nineteen times closer than they are today), and the Moon, although heavily scarred by meteorite impacts, is entirely free of maria, having not yet undergone any volcanic activity. What's more, the Moon has a self-generated, global magnetic field (and thus magnetosphere), preventing its atmosphere (thin though not as thin as it is today) from being stripped away by the harsh solar wind.

Let's further imagine that because the two planets grew up together, as stated by the coaccretion theory, the Moon orbits the Earth's equatorial plane—a far more logical position than the current one. I won't begin to speculate on complicated matters such as tidal friction, for that would only undermine the purpose of this exercise. We will assume the Earth-Moon system exists in a state of perfect harmony, and the Earth is home to some very basic organisms as its climate is well suited to life. (It is possible life appeared on Earth as early as 3.8 billion years ago, scientists believe.)

An ancient alien intelligence, while journeying through the darkness of space, happens to catch sight of the Earth-Moon system and is instantly drawn to its presence. Parasitic in nature, it seeks out a suitable host and feeds off it, but in such a manner as to remain undetected. Of the two planets—which, being so closely linked, are essentially one—the Earth is of greater interest to the parasite, for it is here life has emerged and will eventually blossom. It is a fertile world where water exists in abundance, whereas the Moon is much the opposite. If the parasite were to make a direct connection with its host—in this case, the Earth—it would run the risk of being discovered. So, instead, it decides to attach itself to the Earth's companion, the Moon, thereby gaining indirect and covert access to the Earth.

The parasite, having chosen the Moon as its domain, sets about transforming it into a suitable habitat. Little by little, it "takes over" the Moon. Or, perhaps more suitably, it infests it. Like a tick burrowing into the ear of a dog, it starts at the surface and works its way inside, through the crust, then the mantle, then the core, carving out a labyrinth of tunnels as it goes. Here, in the very heart of the Moon, the parasite wreaks such sickening destruction the planet itself almost dies. With its dynamo no longer functioning, the Moon's magnetic field—its "energy field"—shuts down. All that remains are a few patches of magnetism, of energy—a sign that life still clings to the Moon, albeit barely. As with any living thing that has been "wounded" so deeply, the Moon "bleeds." Vast quantities of hot molten rock gush from its interior, covering sections of its surface.

By the time the Moon has been completely taken over by the parasite, it has become almost an extension of the parasite's body. The parasite can do what it wants with the Moon, including altering the way it functions. To have control of the Moon means to have control of the Earth. Realizing this, the parasite makes subtle changes to the orbital characteristics of the Moon, gradually shifting it away from the Earth's equator. Meanwhile, as time passes and life on Earth continues to flourish and evolve, the parasite does what parasites do: it feeds on its host to the detriment of its host. The parasite remains alive to this day, keeping a careful eye on humanity and all that exists on Earth from its huge, celestial watchtower.

I wish to emphasize that what you just read was less of a theory and more of a thought experiment. The purpose of this experiment was to guide us in the right direction in our search for the truth about the Moon. I find it just as hard to accept that the Moon is literally possessed by an ancient alien parasite as I do to accept that the Moon is literally a giant spaceship. From a metaphorical perspective, bizarre and unorthodox ideas such as these often make a great deal of sense and deserve to be explored.

We've come to a point where, based on the evidence we've looked at, we know with some degree of certainty the Moon is a hijacked and artificially modified world—and this I mean literally. Further lunar discoveries await us.

5

OUR RELATIONSHIP WITH THE MOON

It is an indisputable fact that the Moon was extremely important to the ancients—even more so than the Sun. Beacause the Moon was intimately bound with vital matters such as birth and death, fertility, conception, growth, the afterlife, and time, it was central to their very reality. They kept careful track of the lunar cycle and organized their lives in accordance with this cycle. The Moon, after all, was the only chronometer they had—the only means by which they were able to reckon time for periods longer than a day.

Some of the oldest paintings, tools, and engravings ever found have a distinctly lunar theme. There is a well-known paleolithic artifact consisting of a small, flat piece of bone engraved with strokes and notches that was discovered in a rock shelter called Abri Blanchard in the Dordogne region of France. Carved by a Cro-Magnon man about 25,000 BCE, the artifact has been identified by experts as a very primitive lunar calendar. It features a total of sixty-nine marks, some of which look like full Moons, others like half-Moons. The marks are arranged in the form of a serpent, as if to capture the very rhythm of the Moon's

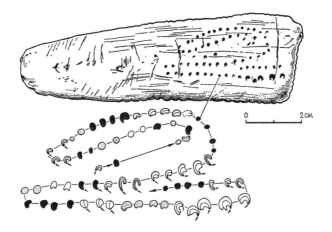

A sketch of the Abri Blanchard bone, with the various phases of the Moon clearly evident. Drawing after A. Marshack, 1970.

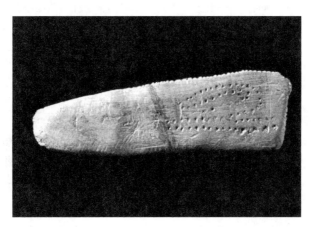

The Abri Blanchard bone. Photo courtesy of Harvard University, Peabody Museum.

waxing and waning. The marks make up two complete "Moons," showing two periods of waxing, two periods of waning, and two periods of darkness or invisibility.

There are hundreds of caves in the Dordogne region of southwestern France; some of them are famous for the paleolithic artworks they contain. The most famous of all is Lascaux, discovered in 1940 by four teenage boys. The cave is decorated with thousands of paintings and engravings of animals and symbols. Lascaux Cave includes horses, deer, and bison, in addition to some very strange mythical creatures. There is little doubt whatsoever that some of the illustrations have profound shamanistic significance—significance that cannot be decoded by the minds of contemporary humanity as we are trapped in a limited left-brain mode of consciousness. After being opened to the public in 1948, the condition of the cave and its precious artworks quickly deteriorated. Today, public access to the cave is prohibited.

In 1911 a French physician happened to be exploring the area near Lascaux (at the time undiscovered) when he came across the carving of a female figure on the wall of a limestone rock shelter. Known as the Venus of Laussel, the paleolithic carving dated between 18,000 and 22,000 BCE depicts a naked, pregnant woman with large breasts and vulva, her left hand resting on her swelling belly and her right hand holding up the crescent horn of the bison, which is inscribed with thirteen notches (see plate 7). She is entirely faceless, perhaps to give greater emphasis to her body. The carving, measuring seventeen inches in height, was once entirely painted with red ochre—the color of blood. It now lies in a museum in Bordeaux, France.

What "message" were our prehistoric ancestors trying to communicate when they carved the Venus of Laussel? No one knows precisely; however, we can be fairly confident she represents a perceived connection between the Moon and female fertility. First of all, the horn is clearly suggestive of the crescent Moon. There is obvious significance in the fact that its notches add up to thirteen—a number with strong lunar connotations. A woman menstruates, on average, every twenty-eight days and approximately thirteen times a year. So this could also relate to the human menstrual cycle. Then we come to the woman's bulging and clearly pregnant belly, which couldn't be more symbolic of fertility.

Our prehistoric ancestors, having recognized the Moon follows a 29.5 day cycle while menstruation occurs an average of every 28 days, were probably convinced of a deep connection between the two cycles. Supporting this belief, no doubt, was the further recognition that the average gestation period for humans (the time between conception and birth) is 266 to 270 days, which is almost nine complete synodic months. It's worth noting that gestation periods vary from mammal to mammal. For a Virginian opossum it's only twelve days; for an Indian elephant, twenty-two months.

"Women especially are affected by the Moon and their menstrual cycle is intimately linked to this celestial body," declares one popular women's health website.[1] Interestingly, the word "menstrual" derives from the Latin *mensis,* which simply means "month." Moreover, the words "month" and "Moon" are cognates, for, in the past, a month was equal to a "Moon" (a synodic month). "The moon is the first chronometer . . . practically everywhere the month is denoted by the same word as moon," writes Martin Nilsson.[2]

From an orthodox scientific perspective, the cycle of the Moon has no bearing whatsoever on the human female reproductive cycle; their similarity in length is merely coincidental. Scientists argue, for example, that if the two cycles were in fact synchronized, then one would expect the same to apply to other mammals as well, but it doesn't. For example, rats have an estrus cycle of four to five days, while cats have an estrus cycle of fourteen to twenty-one days. It's important to recognize the difference between menstruation and estruation. Menustration is particular to higher primates (though not all higher primates), and estruation applies to all other mammals. Explained simply, estrus coincides with ovulation. A female mammal in estrus is receptive to copulation. She is "in heat."

Why is menstruation particular to higher primates? Belonging to this category are human beings, apes, and monkeys. As is common knowledge, the closest living relatives to humans are the great apes, which include gorillas, chimpanzees, orangutans, and bonobos (also called the pygmy chimpanzee). Great apes are highly intelligent and certainly more so than monkeys. Chimpanzees, with which human beings share 98 percent of their DNA, are able to use simple tools in a variety of different ways. They can even be taught to communicate using sign language because their language, capability is so highly developed. Scientists believe chimpanzees are related to humans more closely than any other species of great ape and that the two species diverged from a common ancestor about five to seven million years ago. A regular menstrual cycle occurs in great apes with durations similar to that of human beings. For a chimpanzee it is 36.5 days; a gorilla, 30.5; and an orangutan, 30.5.

We now come to monkeys, with which we have less in common than apes. Monkeys belong to two main groups: New World and Old World. New World monkeys are found in South and Central America, whereas Old World Monkeys are found in Asia, Africa, and Europe. The two groups have been separated genetically and geographically for a very long time, and there are some very significant differences between them, such as appearance. One such difference, as pointed out by the late Stan Gooch in his remarkable book *The Dream Culture of the Neanderthals,* is that menstruation occurs in Old World monkeys, which are related to humans and apes, but not in New World monkeys. Yet the menstrual cycles of Old World monkeys are completely dissimilar to that of human beings. The capuchin, for example, follows an 18-day cycle; and the howler, 23.

You're probably wondering at this point why there is an obvious correlation between the menstrual cycles of great apes and humans yet none whatsoever in the case of humans and Old World monkeys. Could this having something to do with the evolutionary split that occurred some thirty million years ago between the Old World monkeys and the ape-human lineage? According to Gooch, the answer is yes. "The inference is that apes—of which we are one—evolved as a group because its females abandoned the normal mammalian estrus cycle and developed instead a menstrual cycle governed and regulated directly by the action of the Moon."[3]

The importance of Gooch's argument cannot be denied. It does have one minor hiccup though. Considering chimpanzees are supposed to be our closest living relative, why, then, isn't the chimpanzee menstrual cycle closer to that of a human being? Why is it, in fact, the least closely matched cycle in comparison to the other species of great ape? How do we explain the 8.5-day gap? In *The Dream Culture of the Neanderthals,* Gooch points out that the closest menstrual cycle connection is between that of a human being and a gibbon, which menstruates, on average, every 30 days. He goes on to argue that "from a large number of psychological and physiological points of view, modern western man seems closer to the gibbon and orangutan than he does to the chimpanzee and the gorilla."[4] A fact Gooch seems to ignore is gibbons don't belong to the great ape family; they are classified as lesser apes.

As for the human-orangutan connection, recent genetic studies have revealed these fascinating Asian primates are more closely related to humans than previously thought; the two species share 97 percent of their DNA. In his controversial book *The Red Ape,* physical anthropologist Jeffery Schwartz challenges scientific convention by arguing that orangutans, not chimpanzees, are our closest living relatives. Schwartz bases his argument more on morphological than molecular (DNA) evidence. He brings up the fact that human and orangutan teeth are remarkably similar, to the extent that they are commonly misidentified for each other in the fossil record. Then there's the fact that orangutans have a hole in the roof of their mouth, a feature once considered unique to humans. Schwartz adds, "Humans and orangs have the widest-separated mammary glands, and they grow the longest hair. Humans and orangs actually have a hairline, in contrast to virtually all primates, where the hair comes down to the top of the eyes."[5]

Alternative researchers like Gooch, who see a direct correlation between the human female menstrual cycle and the cycle of the Moon, believe the phenomenon originated at a time when our ancestors slept out in the open, their bodies exposed to the varying light of the Moon's waxing and waning. Attached to this is the further idea that the development of civilization, including the invention of indoor electric lighting, has resulted in a disruption of the female menstrual cycle so that it no longer follows the cycle of the Moon as closely as it once did. Gooch cites the work of Louise Lacey, author of the book *Lunaception* (1974), who claims it's possible to bring the menstrual cycle in synch with the Moon's cycle by sleeping with the light on during certain nights of the month as if to simulate the natural light of the Moon.

While some studies indicate a link between the human menstrual cycle and the cycle of the Moon, many others do not; the differences make the matter a very confusing one. In my opinion there must be a connection between the two. I do wonder, though, if the phenomenon has less to do with biology and more to do with psychology and the power of belief. It's no mystery that a person's mental state can have a profound effect on their physiology—as shown, for example, by the placebo effect.

Perhaps our female prehistoric ancestors—convinced, as they were, of the Moon's complete governance over fertility—developed over time and solely through the power of belief a menstrual rhythm in harmony with the cyclic activity of the Moon. The Moon was so central to the lives of our prehistoric ancestors (both men and women), they would have thought about it almost constantly—if not consciously then subconsciously. Whereas the people of today spend many hours each day absorbed in the world of television, film, and Internet, so it must have been for ancient humanity in relation to the Moon. So, then, were the women of the past so attuned to the Moon that they underwent menstruation precisely in accordance with its 29.5-day cycle? I wouldn't be surprised if they did. It figures too that because humanity has gradually lost touch with the Moon in a spiritual and symbolic sense, this would no longer be the case today.

This theory makes sense in relation to human females, but what about in relation to female great apes? We must not forget that great apes, like humans, menstruate according to a cycle with apparent significance to the Moon. In the case of female orangutans, which menstruate on average every 30.5 days, it's difficult to imagine how the power of belief could have any bearing on the matter. I doubt orangutans think about the Moon or pay any attention to it whatsoever; however, considering orangutans do possess a very primitive form of culture, we cannot rule out this possibility altogether. Perhaps orangutans, gorillas, and chimpanzees have some kind of inner "relationship" with the Moon, albeit an extremely basic one.

The Moon has been associated with women and fertility since the very earliest times—an association that extends not only to menstruation but also to conception and childbirth as well. Like most people, I tend to think of the Moon in entirely feminine terms, and the lunar deities who stand out strongest in my mind are predominately female. We find, though, in most early cultures the Moon was personified as male. This generally stemmed from the belief women underwent menstruation as a result of being "ravaged" by a wily, lustful, and usually well-endowed Moon god. According, for example, to the Uaupes of the Upper Amazon, when a girl first undergoes menstruation it's because she's been "deflowered by the Moon." For the Japanese, menstruation was attributed to sexual intercourse with the Moon god, a belief that also existed in India, Siberia, Greenland, and Alaska, among other regions.

From the Matacos of northern Argentina comes a bizarre and erotic story explaining how the Moon is accountable for bringing menstruation and the female fertility cycle into being. The story goes as follows: A long time ago, when the Moon was a young and good-looking fellow, he had a reputation as a very fine hunter. His eating habits, though, were most irregular. Following a successful hunting venture he'd gorge himself on game, becoming so fat and full that he could eat no more. He'd then subject himself to a fourteen-day period of starvation, at the end of which his body, utterly emaciated, would consist of virtually nothing. After this, the Moon would go out hunting again, and on and on the process went.

The handsome, athletic Moon selected a girl from the local village to be his wife. She died only five days after the wedding. He soon took another wife, but she died too—again only five days after the wedding. The same thing happened with his next wife. All these mysterious deaths made the people very suspicious, so they decided to interrogate the Moon. He told them he couldn't remember what had happened, except that, on each occasion, he'd gone to bed with his wife and woken up to find her dead. Despite this, the Moon couldn't resist choosing another wife.

The girl's family came up with a plan to try and keep her safe, advising her to stay awake all night so as to keep a close eye on her husband. This she did, sleeping at her family's house during the day. All went well at first, and the Moon came across as a kind and decent fellow. Not until it was time to "get physical" did she notice something unusual about her husband. He was extremely well endowed, his organ big enough to split her in two. No wonder, she realized, that his previous wives had died. She was forced to leave her husband, naturally enough, whose reputation kept the other women away.

Years later the Moon paid his fifteen-year-old granddaughter a visit. (No explanation is given as to where she came from.) She and the others in her village were unaware of the Moon's lethal effect on women. And so, when the Moon asked that she return home with him, her answer was yes. One night, during the journey, the Moon came on to the girl while she was asleep, penetrating her not fully but only partway and instantly waking her up. This left her torn, bleeding, and in pain, but nonetheless still alive. Ever since then women have menstruated—all because of the Moon.

Despite all this talk of a male Moon, it's important to recognize the longheld and universally accepted idea that women and the Moon are connected. We could sum this up by saying the Moon embodies the feminine principle. Speaking of matters of a feminine nature, there is a strong body of evidence, based on artifacts like the *Venus of Laussel,* that suggest the prehistoric era belonged to the "Mother Goddess," also known as the "Great Mother Goddess" or "The Great Goddess." As the dominant cultures and religions of today are male oriented, those of prehistoric times were female oriented, and the Mother Goddess reigned supreme. Anthropologist E. O. James defines the Mother Goddess as "the great symbol of the Earth's fertility." He adds, "[e]ssentially she was represented as the creative force in all nature, the mother of all things, responsible particularly for the periodic renewal of life."[6] The main attributes of the Mother Goddess include creativity, birth, fertility, sexual union, and the sequences of growth.

Those who worshipped the Mother Goddess—no matter what name they gave her—saw essentially no distinction between the Earth and the Moon. From their holistic point of view, the two were composed of the same substance, each being an identical expression of the Mother Goddess. "Earth goddesses were Moon goddesses, and Moon gods were also forms of the Great Mother, either as sons or as consorts, and often both," explain Jules Cashford.[7]

So important was the Moon to our prehistoric ancestors that they went to the trouble of creating basic lunar calendars, in addition to lunar-associated carvings like the *Venus of Laussel*. But what else did they leave behind? A map of the Moon, perhaps? A similar question crossed the mind of astronomer Philip Stooke, Ph.D., of the University of Western Ontario, Canada. Why, he asked himself, was there no recorded map or drawing of the Moon older than about five hundred years? Older, that is, than the one drawn by Leonardo da Vinci in about AD 1505. Convinced "there just had to be an older map somewhere," Dr. Stooke began searching for one, targeting old manuscripts, history books, and the excavation records of Neolithic sites on the British Isles.[8]

At a five-thousand-year-old Neolithic passage mound called Knowth, located in County Meath, Ireland, Stooke found exactly what he was looking for. Knowth contains two rock-lined passages, one facing west, the other facing east. Each is more than one hundred feet in length. Forming the end of the passage facing east is a carved stone named Orthostat 47. It has an arrangement of lines and dots carved into it—markings that, to the casual observer, mean nothing at all. It was Stooke who recognized these seemingly random markings perfectly match the lunar maria. "I was amazed when I saw it. Place the markings over a picture of the full Moon and you will see that they line up," commented Stooke in a 1999 BBC News article. "It is without a doubt a map of the Moon, the most ancient one ever found. . . . You can see the overall pattern of the lunar features, from features such as Mare Humorum through to Mare Crisium."[9]

The lunar map has been dated at 4,800 years, making it about ten times older than the one drawn by da Vinci. Such is the alignment of Knowth's eastern passage that, originally, light from the rising Sun or Moon would have been able to penetrate all the way to the end, illuminating the lunar map. The map actually consists of not one but three images. The first image is clear and easy to recognize; it shows the maria as seen on a full Moon a little after midnight. The two remaining images also show the maria, though oriented at different angles from the first. The reason for this, in the opinion of Stooke, was to "depict the apparent rotation of the maria on the disc of the full Moon as it crosses the sky in the course of the night."[10] What's unusual about these "rotated" images is, unlike the first, they don't show the maria in full, nor are they as structured and accurate. One of the images shows the maria "upside down," as though imagining how the Moon would appear if positioned beneath the horizon or from the perspective of someone living in the Southern Hemisphere. "The people who carved the Moon map were the first scientists," said Stooke. "They knew a great deal about the motion of the Moon. They were not primitive at all."[11]

During the Neolithic period, there occurred in many parts of the world a transition from hunter-gatherer societies to agricultural societies. Prior to the emergence of agriculture and animal domestication about twelve thousand to eleven thousand years ago, hunting and gathering was the global norm. Lacking the knowledge required to grow and store their own food, people went in search of it. They foraged, to use the correct term, and this entailed a largely nomadic existence within relatively small communities. Survival meant spending vast amounts of time hunting or trapping small or large animals, fishing, and gathering shellfish, wild fruits, vegetables, seeds, nuts, and so on. There was little time for anything else, especially fun or leisure.

In pre-agricultural times, especially before the development of large-scale agriculture, the Moon was given precedence over the Sun concerning matters of growth and the sustainment of life. We can be fairly sure the Sun's life-giving properties were well recognized by early humanity, such as the fact that plants grow toward the Sun and quickly perish if entirely deprived of sunlight. Nonetheless, there was a very deep and persistent belief that the spirit of growth emanated from the Moon, and all life was regulated by the Moon. Intrinsic to this belief, as we've already seen, was the perceived connection between the Moon and female fertility. I'll elaborate further on this again in a moment.

In the case of hunting and gathering, one obtains whatever food is available from nature at the time. No attempt is made to control or dominate nature. If your intention is to collect shellfish from the beach, for example, you simply grab whatever happens to be lying around. And if these are scarce, you venture out and collect something else. In the case of agriculture, on the other hand, you force nature to comply with your will. You might keep goats and breed them, for example. Or you might plant crops of barley or wheat, clearing the land and plowing the soil to do so. Anyone who has practiced some form of agriculture—as basic as keeping three tomato plants—would be aware of how much agricultural processes are dependent on the seasons. Tomatoes are a warm-season vegetable, which grow best throughout the summer months. If you try growing them in winter, they won't stand a chance.

As soon as our prehistoric ancestors started practicing agriculture, they would have recognized very quickly that the seasons relate not to the Moon but the Sun. They no doubt would have believed the seasons are controlled by the Sun, not yet understanding seasons are in fact governed by the tilt of the Earth's axis in space as it journeys around the Sun every year. The result was a shift of focus from lunar to solar, whereby it was realized the light and heat of the Sun are responsible for growth, not the Moon's waxing and waning. With all the attention on the seasons and hence the solar year, the lunar calendar was superceded by the lunisolar calendar, which consists of lunar months structured around a solar year. Today most of the world follows an entirely solar calendar.

To learn more about these matters, we must travel back in time many thousands of years to an ancient region called Mesopotamia—what historians call "the cradle of civilization," which roughly corresponds to modern-day Iraq. The name *Mesopotamia* is Greek for "the land between two rivers"—the Tigris and the Euphrates.

As hunting and gathering gave way to farming and animal domestication, fertile lands were sought, attracting the first settlers to Mesopotamia at about 10,000 BC. It was in southern Mesopotamia—later known as Sumer—where the Ubaidians, a non-Semitic people, laid the groundwork for the first true civilization, settling within the region between 4500 and 4000 BC. They lived in large villages in houses constructed from mud-brick. Their agricultural practices were highly developed and involved growing crops of wheat, barley, and peas, for which they constructed simple irrigation systems. They developed trade and established industries, such as weaving, leatherwork, pottery, metalwork, and masonry. Adding to the population and cultural diversity of the region was a large influx of various Semitic peoples.

The year 3300 BC marked the arrival of the mysterious Sumerians, a non-Semitic people whose origins remain unknown. It's been suggested that they traveled from Anatolia, a region now part of Turkey. They brought with them a language unrelated to any other, and it became the prevailing language of the region. Intelligent and innovative, the Sumerians constructed vast and complex irrigation systems, which allowed them to practice intensive, year-round agriculture. Their inventions include the wheel and wheeled vehicles, such as carts and chariots; the plow; the potter's wheel; the sailboat; and hand tools such as axes, chisels, and saws; not to mention the cuneiform writing system; the sexagecimal (base sixty) number system; and the lunisolar calendar. Before long the Sumerians were the dominant group in the region. Their civilization would prevail until roughly 2000 BC.

The Sumerians were hardly the most unified people. By the

third millennium BC their population was divided among twelve independent city-states: Kish, Erech, Ur, Sippar, Akshak, Larak, Nippur, Adab, Umma, Lagash, Bad-tibira, and Larsa. Each city-state was built around a temple with a grand pyramidal structure called a Ziggurat, which was surrounded by steps on all sides. The temple was the heart and center of the community, and each temple was dedicated to a particular deity, existing in the form of a statue. Surrounding the temple was the city itself, which was enclosed within a protective wall. On the outskirts of the city lay farmland and villages. It was believed these gods and goddesses watched over the city-states in which they "dwelled," and their powers could be harnessed for the benefit of those city-states. Nippur, for example, was the home of Enlil, a god of wind and agriculture, while in Erech resided two important divinities: the sky god An and the goddess Inanna, known as "Queen of the Sky."

The Sumerians saw themselves as subservient to the gods, whom they tried to please and serve as best they could. They provided the gods, who were forever hungry and in need of appeasement, with endless offerings of food, incense, alcohol, and animal sacrifices. Superstitious to their absolute core, the Sumerians remained ever fearful of the gods, whose actions and intentions they tried to fathom and predict with the use of ritual and divination. Of course matters such as these fell to the priests—those specially appointed to mediate between the gods and humanity. Of the four main social classes in Sumer, the priests occupied the top position and exercised the most authority. They, through consultation with the gods, were the ones who made the important political decisions, such as those regarding land and property, agriculture, trade, commerce, and war. One can only imagine how much they must have abused their power. At the very bottom of the social ladder were the slaves, some of them prisoners of war; others born or sold into slavery; others forced into slavery by debt. Children could be sold into slavery if their parents so wished.

It's been claimed in books, articles, and so on that Sumer was a matriarchal society ruled primarily by women; however, such was not the case. Although women had a number of basic rights, they were not treated as equal to men. They were allowed to run businesses, own property, and buy and sell slaves. And, if their husbands happened to be absent, they could partake in legal matters. Women of high status could become priestesses. The life of the average woman, though, was one of endless domestic duties, such as cooking, cleaning, and raising children. Exploring this issue at a deeper level we find that within Sumerian religion the role of the goddess was just as significant as that of the god. In this respect, at least, women had equality.

In contrast to orthodox Christianity, for example, which could be accused of disregarding the feminine principle, the polytheism of the Sumerians is rife with important female deities, some of them overtly sexual in nature. Take the aforementioned Inanna, for example, a goddess of sexual love, fertility, and warfare. The most prominent female deity in ancient Mesopotamia, Inanna was portrayed as a strong, seductive, and beautiful warrioress with an unbridled sexual appetite. It's tempting to compare her to the famous 1980s pornographic actress Nina Hartley, if only for humorous reasons. Some rather "naughty" incidents are known to have occurred at temples where Inanna was worshipped. Her priestesses would give their bodies willingly to male visitors, as though channeling the energies of the goddess herself.

Inanna's parents were Nanna, god of the Moon, and Ningal, a reed goddess. Nanna (whose Babylonian counterpart is Sin) was considered "a great lord" and "a light shining." His primary dwelling place was the city of Ur, located near the mouth of the Euphrates River. There, his temple, the Great Ziggurat of Ur, still stands today after having been reconstructed to some extent (see plate 8).* Not only did Nanna establish the month, bring the year to completion, and illuminate the darkness of the night, he revitalized the world. Of special importance was his relationship with the cowherds, who were crucial to the livelihood of the people. By governing

---

*It underwent partial restoration during the 1980s under the rule of Saddam Hussein, for whom the site held special significance. No ordinary bricks were used for the restoration effort; each was stamped with the name of the dictator.

the rise of the waters, the growth of reeds, and all matters associated with life and fertility, Nanna ensured that the cowherds prospered, thereby producing an abundance of dairy products. His emblem, the crescent, was sometimes represented as a pair of bull horns.

One famous tale describes how Nanna and Ningal came to be united in love, and it couldn't be sweeter and more poetic. In it, Nanna is required to earn the affection of his ladylove by carrying out numerous tasks of renewal. This includes refilling the rivers, restocking the fish in the marsh and the game in the forest, ensuring the ripening of grains and vegetables, providing honey and wine for the table, and so on. At the end of all this toiling, Ningal agrees to join him. "O Lord Nanna, in your citadel, I will come to live," she declares joyfully. "Where cows have multiplied, calves have multiplied, I will come to live."[12] They also express their love physically.

As rivalry escalated between the Sumerian city-states, a system of monarchy was introduced whereby each city-state adopted its own king. City-states fought one another for control of the river valleys, developing better and better weapons with which to kill each other. Each struggled to gain ultimate supremacy. Distracted by ceaseless internal conflict, the Sumerians rendered themselves vulnerable to external conquerors. First to gain control were the Elamites, who reigned only briefly. Next were the Akkadians, a Semitic people led by a king named Sargon, also known as Sargon the Great. He founded the city of Agade (Akkad)—the exact location remains unknown to this day—and it became the capital of his mighty empire. During Sargon's one-hundred-year dynasty, from 2334–2279 BC, unity existed among the city-states and a new model of government was introduced.

By around 1900 BC the civilization known as Sumer was no more: the whole of Mesopotamia was conquered by a Semitic tribe called the Amorites, who are thought to have originated from Arabia. Their occupation led to the establishment of the mighty empire of Babylonia, led by King Hammurabi, with Babylon as its capital. Under Hammurabi's reign, which lasted from 1792–1750 BC, science and scholarship were promoted and the citizens were required to follow his strict code of laws—all 282. Legend states that Hammurabi received his laws from the Sun god Shamash via a process of divine inspiration (in a manner similar to Moses and his Ten Commandments). Shamash (or Utu to the Sumerians) was the son of the Moon god Sin. He was a deity of justice and equality, a judge of both gods and men, who, with his blinding light, conquered darkness and evil.

Ancient clay tablets and other objects inscribed with Hammurabi's code of harsh and unforgiving "solar" laws (known simply as the Code of Hammurabi) have been uncovered, preserved, and the text deciphered, providing us with a fascinating glimpse into the values of Babylonian society. The laws adhere to the principle of retribution ("an eye for an eye, a tooth for a tooth") and pertain to trade, commerce, marriage, divorce, assault, theft, slavery, and debt. The laws reveal a society that afforded few rights to women and punished women harshly for only the slightest of incidents. A husband could have his wife drowned, for example, simply for being a "gadabout. . . . [N]eglecting her house [and] humiliating her husband."[13] The same punishment applied to a wife found guilty of committing adultery. Children, not just wives, were expected to obey the man of the house in the strictest sense possible. "If a son strikes his father, his hands shall be hewn off," states the code.[14]

The Babylonians were more patriarchal and more "solar oriented" than their predecessors, the Sumerians. Nowhere is this better exemplified than when we compare the religions and mythologies of the two cultures. In Sumerian mythology we find Nammu, the Great Mother Goddess, the source from whom all life sprang. A personification of the primeval sea, Nammu gave birth to both An (heaven) and Ki (earth), in addition to all the other gods. Some sources state she instructed her son, the aforementioned Enki, to create humankind out of clay and blood so we could be of service to the gods. A god of water, commonly represented as a half-goat, half-fish creature, Enki evolved into a major deity within the Sumerian pantheon. He adopted many of his mother's functions and eventually surpassed her in importance.

The Babylonian equivalent to Nammu is Tiamat, also a primeval goddess of the ocean, but with the added characteristics of a bloated, hideous monster. Tiamat makes an appearance in the famous Babylonian epic of creation, a poem called the *Enuma Elish,* which dates back to the era of Hammurabi. The epic, replete with descriptions of violence and gore, relates the defeat and massacre of Tiamat by her great-great grandson, Marduk, earning him the title of Lord of the Gods of Heaven and Earth. During their battle Tiamat gives birth to an army of horrible, venom-filled creatures, which includes giant sea serpents, demons, horned snakes, bull men, and fish men. None of this fazes Marduk. Having shot Tiamat in the belly with an arrow, which "cut[s] through her insides, splitting the heart," he stomps on her legs, smashes her skull with a club, and severs her arteries.[15] He then splits her body down the middle. One half he uses to make the sky, the other half, the earth.

Marduk continues the process of bringing forth order out of chaos by arranging the constellations, establishing the calendar, and appointing the Sun and the Moon to their rightful positions in the sky. Later in the *Enuma Elish,* he creates humankind out of the blood of Kingu, the consort and son of Tiamat, as well as her collaborator in battle. This creation acts as servants to the gods. Without describing the epic in further detail we can see its sole purpose was to elevate Marduk above all the other gods; to make him head of the entire Babylonian pantheon. Having annihilated Tiamat the "queen," Marduk the "king," had the right to the entire kingdom and then some. A goddess was replaced by a god. Marduk assumed the throne of the most important (though not highest) god in the Sumerian pantheon, Enlil.

So who was this Marduk character anyway?* Originally a deity of thunderstorms and later fertility and vegetation, Marduk was both the

---

*There is an interesting connection between the Lunar Society and the god Marduk. As I mentioned in chapter 3, the Scottish inventor William Murdock (1754–1839), who is credited with the development of practical gas lighting, was a member of the Lunar Society. In Duncan A. Bruce's *The Mark of the Scots,* it is revealed that Murdock's mastery of gas lighting led to his being "proclaimed a deity by Nassr-ed-din, shah of Persia, who believed him to be a reincarnation of Merodach, or Marduk, the god of light."[16]

chief god of the city of Babylon and the national god of Babylonia—
an empire founded, don't forget, by the solar-loving warrior-king
Hammurabi. Marduk had not several titles but a maximum of
fifty. Each title was that of a deity or divine attribute. Marduk's
star was Jupiter—Jupiter being, of course, the largest planet in the
solar system. In Western astrology (which makes use of the zodiac, a
Babylonian invention) Jupiter is understood to be a planet of expansion,
optimism, morality, and the law as well as excess, overindulgence, and
irresponsibility. There's little doubt many of these attributes apply to
the Babylonian empire itself.

Horses and dogs were sacred animals to Marduk. Dragons were
also sacred and the reason why his city, Babylon, had its walls adorned
with images of dragons. In later times Marduk became known as Bel,
which comes from the Semitic word *baal,* meaning "lord." Speaking
of the "Lord," there are countless similarities between Marduk and
Yahweh, the god of the Israelites, leading some researchers to conclude
that they're the same deity worshipped under different names. The
Babylonians celebrated Marduk's victory over Tiamat on an annual
basis by reciting the *Enuma Elish* during the New Year Festival. Clearly
Marduk occupied an important place within their culture and religion.

Although the Babylonians were defeated in 331 BC and their empire
crushed, few can deny the enormity of the mark they left on history,
or the significance of their contributions in the areas of art, science,
mathematics, medicine, philosophy, and literature. A predominately left-
brained people, they are remembered especially for their astronomical
and mathematical contributions. For example, their sexagecimal
(base sixty) number system is still used today (albeit in a somewhat
modified form) when it comes to measuring angles and calculating the
time. Despite being a Sumerian invention, it was vastly improved by
Babylonian mathematicians, who also gave us the precursor to zero in
addition to the abacus—the ancestor of the modern computer.

The Babylonians deserve some credit for the invention of the first
working calendar. Theirs was derived from the Sumerian calendar but

included a number of important refinements. It was based on a year of twelve lunar months, beginning in spring. A month could be either 29 or 30 days, giving a total of roughly 354 days for the entire year. The months began with the first visibility of the new Moon at evening—the appearance of the first lunar crescent. So as to bring the calendar in line with the solar year of approximately 365 days, intercalary months were inserted on a semi-regular basis. Realizing the calendar was beginning to lose pace with the seasons, the Babylonians would decide to lengthen the year with the addition of a thirteenth month, making it 384 days, in which case some years would consist of twelve months, others thirteen, and others perhaps fourteen; it all depended on the degree to which the calendar was out of synch with the solar year. The procedure was rather a haphazard one.

As the Babylonians became more adept at charting the heavens, they were able to apply this knowledge toward improving the accuracy of their calendar. Around 500 BC, they implemented a standardized system of intercalation: every nineteen-year period, seven intercalary months were inserted at precise times. Specifically, the years three, six, eight, eleven, fourteen, seventeen, and nineteen of the period were all extended to thirteen months. What the Babylonians had discovered, essentially, is that nineteen solar years corresponds almost precisely to 235 lunar months. If you do the math, you will find the difference between the two is only a matter of hours. This fascinating correspondence is known as the Metonic cycle, after the Athenian astronomer who discovered it, named Meton (432 BC). (Clearly, though, the Babylonians got there first.)

The lunisolar calendar of the Babylonians never truly vanished from use; a very similar one is used by the Jewish people of today, called the Hebrew calendar. Like the Babylonian calendar, it's based on a year of twelve lunar months, each month consisting of twenty-nine or thirty days. As a further similarity, it follows the Metonic cycle, keeping it in line with the solar year. We also find the Jewish names of the month are practically identical to the Babylonian names of the month. The Jewish Nisan (which corresponds to March-April of the Gregorian calendar) was called Nisanu by the Babylonians, for example. These similarities seem unusual, but the Jews lived in exile in Babylonia from 587–538 BC, during which time they adopted certain aspects of the surrounding culture.

The Hebrew calendar has little use besides keeping track of religious observances. In contemporary Israel, for example, it's used alongside the Gregorian calendar. Generally speaking, lunisolar calendars like the Chinese, Tibetan, and Hindu have no purpose other than religious and cultural. The lunisolar calendar is like the manual typewriter—an antique. Although the manual typewriter has been rendered archaic by the invention and widespread use of the personal computer, some authors such as Fredrick Forsyth haven't replaced their typewriters. The reason for this isn't (in most cases) because they can't afford a computer, but because they've grown strongly accustomed to using the typewriter and because it is something to which they attach sentimental value. The situation is much the same with the lunisolar calendar. Eventually, perhaps, they'll no longer be used whatsoever.

The Islamic (or Hijri) calendar is the only purely lunar calendar in widespread usage today. It is based on a year of twelve lunar months, which alternate between twenty-nine and thirty days in length. The only exception is the twelfth month, Dhū al-Ḥijja, the length of which varies in accordance with a thirty-year cycle so as to keep the calendar synchronized with the true phases of the Moon. For the first eleven years of the cycle, Dhū al-Ḥijja consists of thirty days, for the remaining nineteen years, twenty-nine. A year, then, can be either 354 or 355 days. And since this is slightly shorter than a solar year, the months gradually regress through the seasons every 32.5 solar years. The holy month of Ramadan, which is the ninth month of the Muslim calendar, might fall on spring one year, and in a decade's time, winter.

Ramadan begins, like all months of the Muslim calendar, with the first appearance of the lunar crescent after a new Moon. The Moon's appearance is confirmed visually by more than one witness rather than simply calculated. It is the duty of Muslims to follow such a calendar, for as is stated in the Koran: "It is He [Allah] who made the sun to be a shining glory and the moon to be a light of beauty, and measured out stages for her; that you might know the number of years and the count of time."[17] With Islam being rather a fanatical religion, it should come as no surprise that their calendar is used in an official capacity throughout much of the Muslim world. The calendar is especially important in Saudi Arabia, a fiercely Muslim nation. Other Muslim countries, such as Turkey, have adopted the Gregorian calendar, using the Islamic calendar for religious purposes only.

There are some curious connections between Islam and matters of a lunar nature. For example, Mohammed and the Moon appear together in one of the best known stories in the Islamic tradition called "the splitting of the Moon." According to the story, Mohammed was asked to prove himself by splitting the Moon in two. And so, raising his hands to the sky, he proceeded to perform the miracle. Obeying his command, the Moon first descended to the top of the Kaaba* and circled it seven times. The Moon then entered his right sleeve and emerged from his left. Next it entered the collar of his robe, dropping to the skirt and splitting into two. One half of the Moon appeared in the east of the skies, the other in the west, before finally reuniting.

Only because of the Islamic religion has the purely lunar calendar survived to this day. Dominating the world is the Gregorian calendar, which has nothing to do with the Moon and everything to do with the Sun. Its rise to preeminence is entirely understandable since the invention of agriculture, when close observations of the seasons became essential to our very survival. As a result of becoming highly attuned to the Sun, we've become less attuned to the Moon. The question remains: What drove humanity to murder the Earth/Moon Goddess and replace her with a solar-based masculine deity, such as in the case of Tiamat and Marduk?

---

*The Kaaba is a small, stone, cube-shaped shrine located at the center of the Great Mosque in Mecca, Saudi Arabia. Muslims make sure to face the Kaaba while conducting their daily prayers, believing it to be the most sacred spot on Earth.

A number of theories have been put forward to explain this incident, one theory placing the blame on urbanization. (And let us not forget that urbanization is a product of civilization, the latter made possible by agriculture.) The theory states that as soon as our ancestors started living in noisy, stressful, and crowded urban environments, their more aggressive and competitive traits were accentuated. And since these traits stem primarily from the male, men became the dominant sex, their dominance pervading all aspects of life, including, of course, religion and spirituality.

Another theory—as put forward by the late Leonard Shlain in his popular book *The Alphabet Versus the Goddess*—states that the invention of literacy was responsible for the "fall" of the goddess. Once, on an archaeological site tour of the Mediterranean, Shlain was amazed to learn that almost every shrine he visited had once been dedicated to a goddess but then, for reasons unknown, reconsecrated to a god. The fact dawned on him "that all early peoples worshipped some manifestation of the goddess." A physician and surgeon with a deep appreciation of the arts and humanities, Shlain was left wondering: "What event in culture could have been so immense and so pervasive as to change the sex of God?"[18] The question led Shlain some five thousand years back in time and to the ancient civilization of Sumer, where reading and writing first originated. For it was during this period in history, says Shlain, the goddess began to lose power.

The Sumerians are known to have invented the first system of writing. One of their earliest forms of script was pictographic in nature; crude pictures were used to represent various objects. By around 3000 BC, their script had developed to the point where it consisted of intricate wedge-shaped (cuneiform) symbols. Gradually their cuneiform began to represent the syllables of the Sumerian language itself. They used a blunt reed, called a stylus, in much the same manner as we would use a pen, writing not on paper but on wet clay surfaces. When the Akkadians took over Sumer, they adopted and modified the Sumerian system of writing, creating their own style of cuneiform. Written in

Akkadian cuneiform, on a series of ancient clay tablets, is one of the oldest recorded stories in the world—the famous "Epic of Gilgamesh." Soon after the invention of Sumerian script came the hieroglyphic writing system of the ancient Egyptians. These early writing systems were so highly complex as to be practically impossible for the average person to master; literacy was reserved for few. Not until the invention of the alphabet, in about 1500 BC, did literacy become accessible to the general population.

The human brain looks something like a walnut, consisting of two hemispheres joined at the middle by a thick bundle of nerve fibres called the corpus callosum, without which they would be unable to communicate with each other. Each hemisphere is responsible for a different set of functions. The left hemisphere controls the right side of the body, and the right hemisphere controls the left side of the body. They could be described as two different personalities. The left brain is the logical, analytical, factual, objective, and practical side of our being. Science, mathematics, words, and language—all of these belong to the domain of the left brain. The right brain, on the other hand, is the imaginative, creative, intuitive, subjective, and emotion-based side of our being. It appreciates symbols, images, philosophy, and religion. Whereas the left brain gets caught in a worm's-eye view of the world, the right brain sees the "bigger picture." As opposed to the two being "balanced," the left brain tends to dominate our existence. We are very much a left-brained society.

But this, argues Shlain, was not always the case. According to him, our distant ancestors were a predominately right-brained, goddess-worshipping people, their society a generally peaceful and feminine-centered one. But with the development of widespread literacy—made possible by the invention of the alphabet—came greater use of the practical left brain, which caused society to become more aggressive and patriarchal and hence, suppressed the goddess. Shlain goes on to argue that the invention of photography and, later, television gave rise to the feminist movement of the twentieth century. Women started to reclaim some of their power due to both photography and television engaging the right hemisphere of the brain—watching television in particular. "As our culture becomes more image-based, we're balancing our hemispheres. Through this new re-wiring, we're becoming a much gentler and kinder society," writes Shlain.[19] We can only hope that Shlain is correct.

Even though the goddess was under attack in the ancient world (whether due to the invention of writing it's difficult to say), she nonetheless endured for quite some time. Out of ancient Egypt came the infamous Isis, wife and sister of Osiris and mother of Horus. The amount of reverence she attracted is astonishing. She had worshippers throughout the entire Roman Empire, even in places as far as England and Afghanistan. In the words of Elen Hawke, a British author and practicing witch, Isis is "the queen of all the goddesses of all time," and "the most popular female deity ever."[20] Very much the ultimate representation of motherhood, one popular depiction shows her nursing an infant Horus. Her other roles include a healer, sorceress, and aide to the dead. Isis was the deity to call on if you were ill, needed comfort or protection, or simply a receptive ear. While it's true she belongs to the lunar goddess category, she is far more than that.

Almost every goddess that exists is bound to be associated with the Moon to some extent. It is those goddesses specifically associated with the Moon that chiefly concern us here. Of these, many are Greco-Roman in origin, such as Artemis, Selene, Hecate, and Diana.

Certainly no book on the Moon would be complete without giving mention to the Greek Selene. *Selene* is the Greek word for "Moon," while *Luna,* her Roman counterpart, is the Latin word for "Moon." Selene was the sister to Helios, the Sun god. Both were said to traverse the sky in horse-drawn chariots; Helios made his journey during the day, and Selene made her journey at night. She was typically depicted as a young, beautiful, pale-faced woman with a crescent Moon adorning her head and a torch in her hand.

According to myth, Selene was a passionate and romantic soul who engaged in numerous love affairs. She even gave herself to the vigorous and lustful Pan on one occasion. He managed to seduce her by wrapping himself in a white fleece so as to conceal his bestial, goatish appearance. But Selene's true love was a mortal by the name of Endymion. She was attracted to the young and handsome shepherd as soon as she set eyes on him. Realizing, though, she would eventually lose him to old age and death, she did what was necessary to prevent this from happening: she induced in him a state of perpetual sleep. The story goes on to state that Selene made nightly jaunts to a certain mountain cave, for it was here that her dreaming lover resided, and that she bore him fifty daughters. (How this was possible, given that Endymion was stone-cold asleep, is never explained.) In another version of the myth, it was Endymion's own wish that he be placed in a state of perpetual sleep—a wish bestowed by Zeus, the almighty god of thunder and lightning.

The Greek poet Theocritus (300–260 BC) did a splendid job of capturing the story in verse.

> *Endymion the shepherd,*
> *As his flock he guarded,*
> *She, the Moon, Selene,*
> *Saw him, loved him, sought him,*
> *Coming down from heaven,*
> *To the Glade at Latmus,*
> *Kissed him, lay beside him,*
> *Blessed is his fortune,*
> *Evermore he slumbers,*
> *Tossing not nor turning,*
> *Endymion the shepherd*[21]

Why would Theocritus describe Endymion's fortune as a "blessed" one? While it's true that, in one sense, he's blessed to be immortal and to have the love of a beautiful goddess, in another sense he's terribly cursed. He's cursed because he's lost in his dreams. He's trapped in a state of slumber from which he can never awaken, while Selene, like some kind of hungry vampire, visits him night after night to feast on his youth and good looks. Most of us, I'm sure, would prefer wakefulness and clarity over endless dreams and illusion, no matter how desirable the lunar goddess involved. In this tale, Selene reveals herself to be deeply flawed; so great was her attachment toward Endymion that she essentially resorted to imprisoning him.

In Greek mythology Selene is mentioned in connection with the fearsome Nemean lion. A beast of great size whose claws could pierce any armor and whose hide was impervious to weapons, it plagued the district of Nemea before eventually being defeated by Hercules. He ended up strangling the animal with his own bare hands. In one version of the story, it was Selene who gave birth to this frightening creature. Selene, in the words of Robert Graves, "dropped it to Earth on Mount Tretus . . . in punishment for an unfulfilled sacrifice, she set it to prey upon her own people."[22] Sacrificing animals to the gods, might I add, was a common practice among the ancient Greeks and Romans. In the case of Selene, most of the animals selected for this purpose were of the cloven-hoof variety, due to the resemblance between the symbol for Selene in the earliest Greek script, which looks like two back-to-back "C's," and the shape of the hooves themselves. It was customary for the animal to have the sign of Selene branded on its flank prior to the sacrifice.

Selene is closely linked to the Greek Artemis, who is closely linked to the Roman Diana, resulting in much confusion. I won't attempt to untangle this intricate mythological web, for that is beyond the scope of this book, but I will clarify a few basic facts. First, Selene and Artemis are not the same deity, though they often get confused with each other. Selene was the Moon personified as a goddess in the most basic and pure sense—the Moon incarnate, so to speak. She was closely associated with the full Moon but also with the darkness of the new Moon. Artemis, on the other hand, was a goddess of the hunt, vegetation, wild animals, chastity, and childbirth and was associated with the waxing Moon.

When we try to make a comparison between Artemis and Diana the matter becomes rather more blurry. Basically, an early identification was made between the two, such that Diana became the Roman counterpart of Artemis. Like Artemis, Diana was a goddess of wild animals and the hunt. She probably started off as an indigenous woodland goddess with no particular association to the Moon, having acquired this association as a result of being linked to Artemis. To add to this list of names is the frequently overlooked Hecate, also a lunar goddess, whom I'll say more about in a moment.

Artemis was the daughter of Zeus and Leto and the twin sister of Apollo. Artistic depictions of her show a strong, athletic-looking woman equipped with a bow, a pouch full of arrows, and a stag or hunting dog by her side. Many things about her seem contradictory, such as the fact she is described as both vicious and fond of hunting, as well as a great protector of young children and suckling animals. One of her titles was Mistress of Animals. "She has a ruthless, cruel aspect that shows itself when those things dear to her are threatened," explains Elen Hawke.[23] When Artemis was a child of three, we are told, she requested her father, Zeus, to grant her—among other things—eternal virginity, a bow and arrow with which to hunt, and a troop of nymphs to assist her. These she was given.

In terms of sexuality and femininity, Selene and Artemis contrast markedly: the former soft, dainty, and womanly with an appetite for romantic liaisons with men, and the latter boyish, virginal, and tough, with little or no interest in the opposite sex. To be a virgin, in the classical Greek sense, is to be self-possessed, dignified, and independent. It's not that Artemis was devoid of sexual desire, but she didn't need a man to make her complete. In fact, it was the company of females that she favored—hence her faithful nymphs. When not busy hunting or fishing, we are told, Artemis was either dancing or bathing with her nymphs. And if a man happened to catch sight of her naked body, she would command her hounds to pursue him and rip him apart. Once a hunter named Actaeon made the mistake of spying on Artemis while she bathed. This so greatly enraged her that she turned him into a stag, and he was torn to pieces by his own hunting dogs—all fifty of them.

One gets the impression Artemis despised men and was possibly of the lesbian persuasion. This didn't prevent her from falling in love with the giant and masculine Orion, who was known for his good looks and hunting prowess. The story states that Apollo didn't approve of their relationship, either because he was jealous, or because he felt that Orion, being a mere mortal, was unworthy of his sister's love. In an attempt to kill Orion, Apollo had him targeted by a giant scorpion. Orion fought the scorpion as best he could, but, unable to defeat it, he fled into the sea. Moments later, along came Artemis. By this stage Orion was so far away that Artemis was unable to recognize him. All she could perceive was the tiny figure of a man. Apollo tricked her into thinking that the apparent stranger in the distance had insulted one of her priestesses. And so, filled with anger, she took aim with her bow and shot the apparent stranger dead. Realizing her mistake, she turned Orion into a constellation eternally pursued by a scorpion—a permanent reminder to the world of Apollo's terrible crime. Another version of the story states that Orion was killed by Artemis for attempting to rape her.

There are some dark and morbid stories in relation to the worship of Artemis at Sparta. According to one story a quarrel broke out among a group of rival devotees who were sacrificing together at an altar, resulting in the deaths of several of them. Shortly thereafter the rest were struck down by plague. Worried, the locals sought the advice of an oracle, who declared the only way to end the situation was to placate Artemis with offerings of human blood. The victim to be sacrificed was chosen by a vote, and the victim's blood poured over the altar. Apparently the ritual was performed every year, until it was eventually banished by the lawgiver Lycurgus. From then on it was ritual floggings rather than human sacrifices that were conducted as a means to appease the lunar goddess. Boys were used in these rituals, and they were flogged so violently that they bled.

We now come to the deeply enigmatic Hecate, who, according

to the Greek poet Hesiod, was daughter of the Titan Perses and the nymph Asteria. A triple-natured deity, she was believed to rule over the kingdoms of Earth, heaven, and sea. Her rulership also encompassed birth, death, and growth, as well as all three phases of the Moon: waxing, waning, and full. Mainly, though, she was a goddess of the dark Moon, and thus, the underworld.

A common depiction of her shows three women in one, joined back-to-back, each woman carrying a burning torch. The image makes sense when we realize Hecate was associated with crossroads, where two or more roads intersect; or, in the metaphysical sense, a place where two or more realms intersect, such as the physical and spiritual realms. Statues of Hecate were erected at crossroads and offerings were made to her. Given her ability to see into different realms simultaneously and to permit or deny access between those realms, Hecate was the chief goddess residing over magic, spells, and witchcraft. In addition, she was thought to have dominion over spirits, ghosts, and other creatures of the night. Like Artemis, she had a special affinity with hounds.

Having taken a look at Hecate, Artemis, Selene, and Diana, we're now equipped with a better understanding of what the Moon represents mythologically, spiritually, and indeed overall. If one crucial fact has emerged so far it's that lunar deities are, generally speaking, ruthless, territorial, possessive in nature, and capable of great acts of vengeance. In almost every case in which the Moon is personified, one characteristic stands out: severity. The Moon is not soft, nor gentle, nor compassionate, but a very "harsh mistress" indeed.

Leaving behind Greco-Roman goddesses, we now come to Ixchel, or "Lady Rainbow," an important mother goddess of the Maya. The wife of the sky god and supreme lord of creation, Itzamná, she presided over the Earth and the Moon and was also patroness of womanly crafts. A triune deity, she was represented in three different forms, each associated with a particular phase of the Moon. As maiden (waxing Moon) she has large breasts and is holding a rabbit, a sign of fertility; as mother (full Moon) she has a pregnant belly; and, lastly, as crone (waning Moon) she is a wrinkled and hunched old woman pouring water out of an upturned vessel. On her head rests a coiled serpent, which serves as a symbol of rebirth and transformation. (The snake is an animal that sheds its skin and so appears to be reborn each time it does so.) This last image suggests the Maya saw Ixchel as responsible for rain and flooding.

Indeed, there is an ancient and universal belief that the Moon is related to all things watery, from the dew on the grass to the fluids inside the human body. No doubt this belief originated in part from the simple recognition that the Moon governs the tides as well as in part from the Moon's association with growth and fertility. Particularly in hot and arid environments, such as Egypt, the Moon was regarded as an agent whose effects were contrary to those of the parching, hostile Sun. As the Sun dried up the drinking pools and scorched vegetation, the Moon brought life, nourishment, and moisture. "They reason that the Moon, because it has a light that is generative and productive of moisture, is kindly towards the young of animal and the burgeoning plants," wrote the Greek biographer Plutarch in regard to the Egyptians.[24]

In China it was believed that "the vital essence of the Moon governs water."[25] Among the Native Americans, there were numerous watery beliefs about the Moon. The Pueblos called the Moon "Water maiden," while the Sioux believed the Moon held a pitcher of water. Pliny the Elder, the brilliant Roman intellectual, described the Moon in watery, life-giving terms, even going so far as to declare "the blood of man is increased or diminished in proportion to the quantity of her light."[26] In the writings of Cornelius Agrippa (1486–1535), the German occult philosopher, water is referred to as the "lunar element." Reference to the Moon's watery reputation can even be found scattered throughout the works of William Shakespeare. In a *Midsummer Night's Dream,* for example, the character Titania, queen of the fairies, calls the Moon "the governess of floods," who, "pale in her anger, washes all the air."[27]

The ocean, the Moon, and abduction all come together in the well-known Maori story of Rona, the "Tide Controller." Rona was the daughter of Tangaroa, a god who controlled the sea. One moonlit night, Rona made a trip to the local stream to fetch some water in a bucket. As she was carrying the water back home to her children, the Moon disappeared behind some clouds. No longer able to see where she was going, Rona tripped over the root of a tree. "You cooked-head Moon," she shouted, directing her anger at the Moon. This Moon, like other Moons in other stories, was as touchy and precious as a spoiled house cat. Insulted and annoyed, it took her in its clutches, pulling her up toward the heavens, along with the tree to which she clung. Rona, the tree, and her pail of water can still be seen in the Moon to this day. The power to control the tides is something she brought with her when she was taken to the Moon.

Stories similar to this one—in which we are told that the markings on the Moon represent an abducted woman holding a bucket—can be found throughout the world. In many of these stories it is explained that rain occurs as a result of the woman emptying her bucket.

Water and the Moon are closely allied in the case of the Hindu deity Chandra. A god of the Moon, he is depicted as a handsome, youthful, and very regal individual. He looks, perhaps, like a prince. Like the Greek Selene, he rides a chariot across the sky at night, only his is pulled by ten white horses. The animal most sacred to him is the hare—an obvious reference to the "hare in the Moon." Chandra is accountable for fertility, rain, dew, the growth of plants, and the governing of the tides. To make matters somewhat confusing, Chandra is virtually synonymous with the Vedic Soma, which is both a deity and a sacred, intoxicating drink.

Soma was used in the Vedic religion for ritualistic purposes, particularly when offerings were made to the gods. Not just anyone was permitted to drink this divine substance, but usually only priests or members of the higher castes. It apparently left the drinker feeling highly exhilarated and may have induced hallucinations. The soma juice was obtained by crushing the stalks of a particular mountain plant, which, also called soma, has never been properly identified. The closest guess is hemp.

It would be a mistake to think of soma in purely prosaic terms. From a symbolic and spiritual perspective, soma was the life force itself. It was the "elixir of life" and the "nectar of the gods." Drinking it brought communion with the heavenly realm, great ecstasy, and even immortality. A number of different cultures have their own versions of Soma. In the Bible it's referred to as Manna. This "bread from heaven" is described as white in color, with a taste similar to that of honey. And so the Moon was thought to be a receptacle for not just the rain that falls on the Earth but for the very "waters of eternal life."

Since the Moon appears to die and come to life again each month, it has long been considered a symbol of immortality. Of all the myths that explore this connection, one of the finest is the story of Heng-O, the Chinese goddess of the Moon. Her husband was the brilliant archer Shen I, who, in exchange for building a palace for the Queen Mother of the West, was granted a pill of immortality. Intending to take the pill at a later date, he took it home and hid it in the rafters. While he was out one day attending to business, Heng-O noticed a delicious smell and saw a glow emanating from the ceiling. Curious to determine the source of the glow, she soon found the pill and, unable to resist it, popped it in her mouth. She instantly became weightless and found herself floating off the ground. Having arrived home at that very moment, Shen I was less than pleased to discover that not only was his precious pill missing, but that his wife was floating out the window.

Frightened, Heng-O continued her ascent, eventually reaching the Moon. Here, on this cold, glassy, luminous world, she found she wasn't entirely alone; nearby was a huge cassia tree. Feeling a little ill, she coughed up the pill's capsule coating, which immediately transformed into a white rabbit. The Chinese are a resourceful people and so was this particular bunny. It made itself comfortable under the cassia tree and started to grind the elixir of immortality. Meanwhile, Heng-O, as a result of sipping some lunar dew and swallowing some lunar cinnamon, transformed into a toad. She resides on the Moon to this day, just like her rabbit companion. (Here we have the aforementioned "hare in the

Moon." The toad on the other hand is more difficult to spot, being an image in negative space.)

The conclusion of the story is a more or less happy one. After being granted immortality by the King Father of the East as a reward for his heroic acts of service, Shen I was assigned to live on the Sun. Having long forgiven his wife, he decided to visit her, riding to the Moon by means of a sunbeam. Once there, he built her a palace in which to live, called the Palace of Boundless Cold. He was unable to stay forever, of course, on account of his important solar duties. Every month, though, when the Moon is full, he pays his wife a visit. On these occasions the two are united: he being the active force of yang, she being the receptive force of yin, causing the Moon to shine with the brilliance that it does. While it's easy for him to visit her, it's not possible for her to visit him. For, as we know, the Moon shines by reflected light.

Is there any mythological evidence for the hypothesis we established earlier? Namely, that the Moon is a part-natural, part-artificial world? Amazingly, there is. The evidence to which I refer can be found in myths that concern how the Moon acquired its maria. In many of these myths is an indication that the maria are an unnatural feature of the Moon; they ought not to be there. We are told, for example, that the maria represent a woman or a man who remains imprisoned on the Moon, perhaps because they offended the Moon and so were kidnapped by it, or perhaps on account of a crime they committed.

Stories involving a man in the Moon are largely European in origin, and they tend to deal with the issue of Christian morality. One German folktale tells of an old man who went out into the woods to collect sticks on a Sunday morning—Sunday, of course, being the day of the Sabbath for Christians. Having collected a large bundle of sticks, the man slung it over his shoulder and started for home. On the way he happened to run into a man dressed for church, who demanded to know why he was working on a Sunday, the holy day of rest and worship. "Sunday on Earth, or Monday in heaven, it makes no difference to me!" he laughed. The man was less than happy with this response. "Since you don't value Sunday on Earth," he declared, "you shall have a perpetual Moon-day in heaven. You will stand in the Moon for eternity, bearing your bundle, as a warning to all Sabbath-breakers." A moment later the stranger was gone, while the old man, together with his staff and bundle of sticks, found himself trapped in the Moon. He remains there to this day. In a similar story from Luxembourg, a man is placed in the Moon as punishment for stealing turnips from his neighbor's yard.

And then there is the Polish story of Mr. Twardowski—another sinner. According to one version of the story, Mr. Twardowski was a sixteenth-century nobleman who was lazy and lived purely for the attainment of pleasure. Eventually, lacking the funds required to continue his hedonistic lifestyle, he decided to make a deal with the devil, selling his soul in exchange for great wealth. But before doing so, he told the devil that he was planning to take a trip to Rome soon, and that he didn't want his soul to be taken until he arrived there. After the devil had agreed to these conditions, Mr. Twardowski quipped that he wasn't planning to go to Rome after all. He had deceived the devil!

Although far from pleased about this, the devil was obliged to keep his end of the bargain. Years later, he managed to trick Mr. Twardowski into dining at an inn called Rome. Mr. Twardowski knew better than to persist with his antics, allowing the devil to escort him to hell. During the journey there, just as they were flying over the Moon, Mr. Twardowski began to reflect on his wasted life and was overcome by feelings of remorse. He started to sing a hymn that his mother had taught him. Before he knew it, the devil had vanished and he found himself plummeting toward the Moon. Still to this day he resides on the Moon, though his life there is rather a bleak one. No longer does he joke of how he once deceived the devil.

In another version of the story, Mr. Twardowski sells his soul to the devil not for monetary gain but in exchange for great knowledge and magical powers. His associate, whom he turned into a spider earlier, ends up joining him on the Moon. Because the spider has the ability to travel to the Earth and back via a thread, he uses it to keep track of what's occurring below.

In these "man in the Moon" stories, the Moon is seen as a kind of purgatorial realm where souls are forced to serve time for the sins they committed on Earth. Usually they have no chance of ever getting off the Moon; they're trapped there for eternity—a reference, perhaps, to the Moon's perceived deathlessness. The Moon is very much associated with purgatory in the Fourth Way teaching of Gurdjieff—more on this later.

The Chinese have a tale very similar to the ones just described, although in this instance the sinner is a boy. One day the boy (let's call him Wang) saw another boy in his neighborhood take care of an injured bird. Wang noticed that as soon as the bird had recovered it gave the boy a magic pumpkin seed, which, once planted, grew into a huge vine bearing pumpkins filled inside with gold and silver pieces. Wanting the same for himself, Wang deliberately injured a bird and took care of it, receiving from it a pumpkin seed just like the other boy. The vine that grew from the seed was huge, its runners reaching all the way to the Moon. Wang's heart sank when he noticed there were no riches inside the pumpkins that sprouted from his vine. Instead, from one of the pumpkins emerged an old man, bearing a bill written in red ink. Wang's "karmic debt" was no small amount.

Following the old man's instructions, he boarded the pumpkin vine elevator, which carried him straight up to the Moon. He was immediately escorted to the Palace of Boundless Cold (this being the residence of the immortal Moon goddess), where he was handed an ax and told to chop down the cassia tree that grew in the nearby courtyard, on the agreement that if he succeeded in his task he would be free to return to Earth. But here's the catch: the cassia tree is a symbol of immortality; it cannot be destroyed. Thus, poor Wang is still on the Moon to this day, trying in vain to cut down the cassia tree. Every time his ax takes a chip from the tree, it immediately repairs itself. To make matters worse, each swing of the ax earns him a painful peck in the back by a white rooster.

The hare is an important animal in folklore, mythology, and religion. In many parts of the world today, both the hare and the rabbit are considered unappealing and attract such labels as "vermin," "oversexed," and "stupid." Hence, we find such phrases as "breed like rabbits" as well as others. The Australians in particular despise rabbits and hares with a passion. After they were introduced into the country from Europe during the mid-nineteenth century, their numbers quickly rose to plague proportions, and their destructive effect on the ecology is still a major problem today.

In Britain, as in Australia, rabbits and hares are seen as a threat to agriculture and the environment and so are hunted without mercy. When we look back through history we find the hare was once sacred in Britain, especially to the Saxons. A Germanic people who arrived in Britain during the fifth century, the Saxons originally followed a pagan religion. One goddess of theirs (whose name remains lost) was depicted as a woman with the head or mask of a hare, holding a Moon-disc across her belly. The disc contains the image of a woman's face, similar to that of the goddess herself. Much later, after paganism had been eradicated and Christianity firmly established, the hare became associated with witches and black magic—not just in Britain, but throughout Europe. In England, even up until the nineteenth century, hares were sometimes killed in the belief that they were witches in disguise, or the so-called witch's familiar. In fact, the hare was "the most common disguise of a witch in all the Northern countries of Europe," according to the folklorist William Henderson.[28]

Farmers have long noted the unusual behavior displayed by hares when caught in a burning field. Rather than make a run for it, the hare tends to remain in its hideout. Frozen, it would seem, in a state of hesitation until the flames come along and roast it alive. As a result, the hare has long been symbolic of sacrificial death. Realizing this, the significance of the Easter Bunny (formerly the Easter Hare) makes a good deal of sense; especially when we consider that Easter, the date

Lunar rabbit deity of the Saxons, about which little is known.

of which is regulated by the full Moon, celebrates the resurrection of Christ, who is said to have sacrificed himself on behalf of humanity. It is well recognized that Easter is a festival with pagan roots, and so I'll refrain from elaborating on the matter. As for the Easter egg—a gift from the Easter Bunny—this is a symbol of rebirth. Because hares (and rabbits) give birth to very large litters, we find the hare is also a symbol of fertility. To be precise, gestation for hares is about thirty days, which is remarkably close to one synodic month. And so here we find an additional hare-Moon connection.

Whereas man in the Moon stories are predominately European and deal mostly with issues of sin and crime, hare in the Moon stories are predominately Asian and deal mostly with issues of sacrifice and immortality. It is not to suggest the hare in the Moon is restricted to Asian culture. The Native American tribes, the Maya, the Aztec, and the Africans all have their own distinct myths about the hare in the

Moon. (And by this I am referring to both hares and rabbits. They belong to the same family, after all, while the terms hare and rabbit are frequently used interchangeably.)

In Sanskrit, the word for "hare" and the word for "Moon" are practically identical; the former is *sasa* and the latter is *sasin,* meaning "that which is marked by the hare." In a well-known folktale of Indian origin (there are numerous versions), the hare is portrayed as a kind, noble, and entirely selfless creature. The story states that Buddha was wandering through the woods on one occasion when, having become lost, he encountered a hare. The Buddha told the hare he was poor and hungry, and so the hare, overcome with compassion, instructed the Buddha to make a fire so that he could cook and eat him. Thanking the hare, the Buddha set about making a fire. It was decided that, at the appropriate time and for the sake of simplicity, the hare was to throw himself into the flames. This the hare did. Only instead of meeting a fiery death, he found himself safely in the hands of Buddha, who had caught him in mid-flight. So as to honor the hare for his act of self-sacrifice, the Buddha placed him in the Moon for the whole world to see.

The hare as he appears in African mythology is altogether different in nature from the noble hare of the Buddhists, being very much a selfish trickster. In a tale that originated among the Hottentots of southern Africa, it is the hare's fault why humans aren't immortal. According to the story, the Moon sent a hare to Earth with the task of imparting an important message to humankind: because the Moon dies and rises to life again, so would humanity. The hare approached the people and told them that because the Moon dies and rises to life no more, so too would humanity. When the Moon learned that the hare had bungled the message, she was more than a little angry. She tried to crack his head open with a hatchet but missed and split his lip instead—an explanation for the birth defect in humans known as a hare-lip (also called a cleft lip).

There are two different endings to the story. One ending states that the hare retaliated, scratching the face of the Moon with its claws and so leaving behind dark scars, which the Moon still bears to this day. The other ending states the hare simply fled and is still fleeing the Moon. The San (or Bushmen) of southern Africa, a people closely related to the Hottentots, have a story very similar to the one just described in which the Moon's messenger starts off as a man but is later turned into a hare as a result of a curse placed on him by the Moon. This is his punishment for bungling the Moon's message. In this and other stories we've looked at, the hare plays the role of an intermediary between the Moon and humankind, as though able to pass freely between the heavenly and Earthly realms.

An Aztec hare in the Moon story says that the gods had to willingly sacrifice themselves by jumping into the fiery abyss in order to create both the Sun and the Moon. At first the Moon was as bright and as hot as the Sun. One of the gods decided to change this by hurling a rabbit at the face of the Moon. The resultant dark blotch dimmed the brilliance of the Moon, leaving it cold and pale. Is this story trying to tell us something important? Namely, that the Moon was originally a beautiful, glowing world, but for reasons somehow related to the creation of the maria it was rendered virtually dead, or was somehow corrupted? Jules Cashford observes, "There is a memory in many disparate cultures of a time when the Moon was as bright as the Sun, not only in light but also in importance."[29]

It's remarkable how many sources declare the Moon was once as bright as the Sun. The idea can be found, for example, in a legend that originated among the Pueblos, which states the Moon gave up part of her light so people could sleep at night. According to a certain Brahmanical text, the Moon and the Sun were comparable in brightness until the Sun "took to himself the Moon's shine; although the two are similar, the Moon shines much less, for its shine has been taken away from it."[30] In the Old Testament in the book of Genesis, the Sun is described as the "greater light" and the Moon the "lesser light." In contrast, the talmudic literature of the Jews declares that the Moon and the Sun were created equal in greatness and luminescence, but God "diminished" the Moon as punishment for being jealous of the Sun. This is analogous to an old

Islamic legend that says God created the Sun and the Moon equally bright. But because the creatures of Earth were unable to distinguish between day and night, he commanded the angel Gabriel to soften the light of the Moon. This Gabriel did, by brushing his wings against it.

Some of these legends and such are probably metaphorical references to the gradual dominance of Sun worship over Moon worship. But we must also be prepared to consider a literal interpretation of the matter. Could there have been a time in the distant past when people gazed up at the heavens and saw a Moon far brighter than the one we see today?

With this question in mind, consider how much more effectively the Moon would reflect the light of the Sun—and therefore how much brighter it would be—if it had no maria. Of course, the maria are supposed to be billions of years old, as determined by radiometric dating. But what if they're actually far younger than this—say, thousands of years old? For indeed there are numerous indications the Moon acquired its maria and was rendered dimmer within the history of humanity. Plus, if the maria are indeed artificial in origin, there's a good chance they haven't been dated correctly. Clearly, in this instance, a different set of rules would apply as far as dating techniques are concerned.

6

THE MOON MATRIX

*The power of the Moon often touches the darker side of
the human soul.*

ARNOLD L. LEIBER, *THE LUNAR EFFECT*

Can the Moon bring out the worst in human behavior, particularly
when it's full? Can it drive some people to commit acts of violence, even
murder? In the case of David Berkowitz, the notorious "Son of Sam"
serial killer, questions such as these are worth examining. Born Richard
David Falco on June 1, 1953, in Brooklyn, New York, he was adopted as
an infant by hardware store owners Pearl and Nathan Berkowitz. Since
childhood, Berkowitz showed signs of being a troubled loner, eventually
developing a penchant for larceny and pyromania. His aggressive and
antisocial tendencies became especially pronounced when, at the age of
thirteen, his adoptive mother died from breast cancer. To make matters
worse, he didn't get along with his adoptive father's new wife. In 1971,
when he was eighteen, the couple moved to Florida without him.

That same year, Berkowitz joined the army. He left after three years
of service, but not before developing into an excellent marksman. Shortly

thereafter he was reunited with his birth mother, Betty Falco. He was, she revealed, an illegitimate child; his father was a man with whom she'd had an extramarital affair. This greatly disturbed Berkowitz, and before long he drifted out of contact with his mother.

Believing his mind to be under the influence of demons, Berkowitz attempted to murder fifteen-year-old Michelle Forman on Christmas Eve of 1975, stabbing her repeatedly with a hunting knife. Fortunately, she survived. Not long after the incident, Berkowitz relocated to the city of Yonkers, New York. His new neighbor, Sam Carr, owned a black Labrador retriever named Harvey, and Berkowitz came to believe the animal was possessed by an ancient demon that commanded him to slaughter people.

Berkowitz performed his first successful murder on July 29, 1976, about an hour after midnight. His victim was eighteen-year-old Donna Laurie, whom he shot in the chest. At the time of the incident, Laurie was seated in a parked car accompanied by her friend Jody Valente. Valente was shot in the thigh but survived. Roughly a month later Berkowitz felt the need to kill again, firing several shots through the windows of a parked car. Its occupants were Carl Denaro and his girlfriend, Rosemary Keenan. Though both survived, Denaro sustained a serious gunshot wound to the head.

Over the next nine months Berkowitz carried out an additional six attacks, mostly targeting young brunette women and most of them seated at the time in parked cars with their boyfriends. At the scene of one of the murders Berkowitz left behind a handwritten letter signed "Son of Sam." It was a reference to the demon he believed resided in Sam Carr's dog. Part of the note read, "I feel like an outsider. I am on a different wave length then [sic] everybody else—programmed too [sic] kill."[1]

Berkowitz was arrested on August 10, 1977, after someone reported him loitering near the scene of his last shooting. After pleading guilty to his crimes, which included the death of six people, he was sentenced to 365 years in prison.

He is currently housed in a New York State maximum security prison and, since being "born again" in 1987, is a very devout Christian. He has a good prison record and no intention of seeking parole, stating in 2002, "I believe I deserve to be in prison for the rest of my life. . . . I have accepted my punishment."[2] Berkowitz cited his resentment toward women, his mother especially, as one of the reasons that led him to kill, adding he found the acts to be sexually arousing.

There is a very bizarre twist to the story. Berkowitz claims to have conducted the crimes not alone but in participation with members of a violent satanic cult. Apparently the cult consisted of some two dozen core members in New York with ties across the United States and was involved in illegal activities, including drug smuggling and child pornography. A letter of his published in the *New York Post,* dated September 19, 1977, ominously states, "There are other Sons out there, God help the world."[3]

Berkowitz's satanic cult claims are difficult to dismiss. He has stated he cannot reveal everything he knows about the cult because doing so could jeopardize the safety of his family members. As for his own safety, an attempt was made on his life in 1979 when a fellow inmate (whom he refused to identify) attacked him with a knife. The wound required a total of fifty-two stitches. It was around this time when Berkowitz named John Carr and John Carr's brother, Michael, as fellow cult members, along with their father, the aforementioned Sam Carr. Both John and Michael Carr died under extremely suspicious circumstances: John died in February 1978 from a rifle shot to the face (an apparent suicide), and Michael died in October 1979 as a result of an apparent drunk-driving accident, despite his clinical aversion to alcohol.

Even if Berkowitz did act alone and the cult part of his story was a fabrication, there remains an occult motivation to the killings. Berkowitz had a deep and obsessive interest in magic and the occult. At the time of his arrest, police searched his apartment to find its walls covered in "satanic graffiti." And here's where the lunar element creeps into the story: of the eight attacks Berkowitz committed, five of them occurred on a night when the Moon was either new or full. Some would call this merely a coincidence. Others would say Berkowitz acted, to some extent, under the Moon's maleficent influence; it triggered his violent tendencies. Others—myself included—would say Berkowitz performed the killings in a ritualistic fashion, whereby he chose to murder on nights when the Moon was new or full because of the occult and magical significance of those occasions.

After all, the spells and rituals of witchcraft and ceremonial magic are often conducted in accordance with the phases of the Moon. According to those who partake in such matters, the waxing Moon is a time for benevolent magic and the waning Moon is a time for malevolent magic. As for the significance of the full Moon, this is thought to be a time when one's psychic powers are operating at their optimum level. As Elen Hawke clarifies, "At full Moon, the point of maximum strength has been reached in terms of occult force and power. . . . It is as though the full Moon is a gateway to other realms of being, the energies of which flood our plane and interact with us."[4] Those particularly attuned to the Moon and its phases tend to agree the full Moon carries an uncomfortable feeling of accumulated tension with it, which can only be relieved through engaging in mental and physical activity.

The association between the full Moon and madness—not to mention phenomena of an eerie nature—is so deeply ingrained in our collective psyche as to require little explanation. For many of us, thinking about the full Moon immediately conjures up images of werewolves, witches' covens, and the mentally unbalanced in the thrall of madness. Despite scientists arguing to the contrary, many subscribe to the belief that the full Moon has the ability to influence us somehow, if not spiritually then at least physiologically. There are more traffic accidents during the full Moon; more hospital admissions; more suicides; more crime—we've all heard such claims, and few of us question their validity.

The belief that the Moon, particularly when full, can induce strange or abnormal behavior has been around for centuries, even longer. Need I mention that the word *lunacy* comes from the Latin *luna,* meaning "Moon"? Many superstitions, some very old, warn of the supposed dangers of prolonged exposure to moonlight. Pregnant women in particular were told to avoid the light of the Moon, in the belief that they would either have a miscarriage or that their child would be rendered deranged. A similar result was thought to occur in the case of a child conceived by the light of the Moon. Sleeping under the Moon's rays was considered just as harmful. Even the Talmud warns against it, the supposed effect being madness. Today, when we say that someone is "moonstruck" we mean they are either mentally unbalanced or romantically sentimental. Originally the term had a very literal meaning. It referred to a supposed condition whereby a person's vision and sanity were weakened as a consequence of falling asleep with the Moon shining on their face.

It was the widespread opinion of early physicians that the amount of moisture in the brain varied in accordance with the waxing and waning of the Moon; therefore, the Moon was also the cause of various mental conditions, including bouts of insanity and epileptic attacks. Paracelsus (1493–1541), the remarkable Swiss medical pioneer and alchemist, was convinced the Moon had the "power to tear reason out of a man's head by depriving him of humors and cerebral virtues."[5] The Moon was most powerful when full, he believed. Sir William Hale, a chief justice of England, was thinking along similar lines when he wrote in the 1600s that the "Moon hath a great influence in all diseases of the brain . . . especially dementia."[6] In England during the nineteenth century, saying the Moon could cause madness was equivalent to saying rain fell from the sky; it was accepted as fact. For example, in *Peter Parley's Almanac for Old and Young,* published in 1836, it is written, "When the Moon is near the full, or new, people are more irritable than at other times, and headaches and diseases of various kinds are worse. Insanity at these times has its worse paroxysms . . ."[7]

Sir William Blackstone (1723–1780), the esteemed English jurist, defined "a lunatic, or *non compos mentis*," as "one who hath . . . lost the use of his reason and who hath lucid intervals, sometimes enjoying his senses and sometimes not, and that frequently depending upon the changes of the Moon."[8] In England, the concept of being a lunatic reached a new level of credibility with the introduction of the Lunacy Act of 1842. This held that lunacy was a condition in its own right, distinct from mere insanity. To be *insane* was to be chronically and incurably psychotic whereas to be a *lunatic* was to be "afflicted with a period of fatuity in the period following after the full Moon."[9] There is the story of a certain British soldier who was charged with the murder of his comrade in 1940. He used the excuse of "Moon madness" as his defense, adding his supposed condition was trigged by the full Moon. (It's unknown whether his defense was taken seriously.)

Many things once deemed factual by officialdom are today considered false by officialdom (and vice versa), and such is the case in regard to the Moon's supposed ability to influence human behavior. I recall asking my tenth-grade science teacher, and later my psychology teacher, if there was any evidence to support the belief of a lunar effect on human behavior. In both cases I was told there is no evidence that the Moon can influence human behavior. Although some studies seemed to show that traffic accidents, hospital admissions, and so on increased slightly whenever the Moon was full, these studies were proved to be flawed and untrustworthy.

Seeking a different perspective on the matter, I approached an acquaintance of mine—a gentleman with a spiritual bent. I wanted his opinion regarding the lunar cycle and human behavior, and he said the two are most certainly related.

He began by explaining that at those times when the Moon is new or full its gravity is combined with that of the Sun, producing especially high tides (spring tides). It figures, he continued, that if the Moon can affect the Earth's ocean water in such a way, it must have a similar effect on the human body, which is, after all, around 70 percent water. The effect would no doubt be strongest when the Moon is new or full.

This idea seems sensible enough, and indeed there are many who believe it to be true; however, when we examine it more closely, we find that it doesn't add up. There are two main reasons why. First, the Moon can only affect unbounded bodies of water, and the water in the human body is bounded—contained within our tissues—and remains shielded from the Moon's tidal forces. Second, although the Moon produces a significant effect in terms of gravitational attraction when it comes to the Earth's ocean water, the same doesn't apply to the human body on account of the vast difference in scale. In fact, the gravitational force exerted on the human body by the Moon is so minute as to be virtually nonexistent. The astronomer George O. Abell calculated a mosquito resting on one's arm would exert a greater gravitational pull than would the Moon.

The idea that our behavior is influenced by the Moon—specifically, by the gravitational force it exerts on our watery bodies—gained much support in the late 1970s with the publication of Arnold L. Lieber's hugely influential *The Lunar Effect*. Lieber is a psychiatrist (now retired from active practice) from Miami, Florida, whose controversial lunar ideas have been the target of much derision by skeptics around the globe.

As a medical student in training at Jackson Memorial Hospital in Miami, where he worked in the mental wards, Lieber became aware of "a peculiar pattern of events." Namely, that there "were recurring periods when behavior of [mental] patients was more disturbed than usual." Years later, having taken up psychiatry as a career, Lieber again noticed this "peculiar pattern of events" concerning the behavior of patients in mental wards.[10] He found his work colleagues noticed it too, and they overwhelmingly blamed it on the appearance of the full Moon.

Convinced there might be a link between lunar activity and human aggression, Lieber began to investigate the matter. In perhaps his most famous study, conducted with the help of clinical psychologist Carolyn Sherin, Lieber looked at 1,887 homicide cases that occurred in Dade County, Florida, between 1956 and 1970. They also looked at an additional 2,008 homicide cases that occurred in Cuyahoga County, Ohio, from 1958 to 1917. To ensure accuracy, the only cases used in his study were those for which the time of injury had been recorded. He ignored those cases that only recorded the time of death. This is because homicide victims don't always die straight away; it might be days, perhaps weeks, after the incident.

For both sets of data, graphs were constructed showing the frequency of homicides relative to the lunar cycle. The Florida graph was the more impressive of the two. "The results were astounding," explains Lieber. "Homicides in Dade County showed a striking correlation with the lunar phase cycle. Homicides peaked at full Moon . . . with a secondary peak immediately after new Moon." Specifically, there were eighty-six homicides at full Moon, with sixty-three being the number of homicides expected by chance. Lieber continues, "[o]ur results indicated that murders became more frequent with this increase in the Moon's gravitational force."[11]

The Ohio graph exhibited a similar pattern to the Florida graph, only the peaks did not correspond to the new and full Moon. Rather, they were displaced to the right, occurring about three days subsequent to both a new and full Moon. Commenting on this anomaly in *The Lunar Effect*, Lieber suggests it might represent "a lag" in lunar influence owing to variations in geographical location, particularly in relation to latitude. He points out the fact that ocean tides vary in height with respect to latitude, and the highest tides occur near the poles.

It was by making observations such as these that Lieber came to develop his controversial "biological tides" theory, which places great emphasis on the fact that human beings consist of about 70 percent water. Explains Lieber:

I believe the gravitation force of the Moon, acting in concert with other major forces of the universe, exerts an influence on the water of the human body—in you and in me—as it does on the oceans of the planet. Life has, I believe, biological high tides and low tides governed by the Moon. At new and full Moon these tides are at their highest—and the Moon's effect on our behavior is strongest.[12]

From a mainstream scientific perspective (and for reasons already discussed) there is much to criticize about Lieber's biological tides theory. According to the theory, most of our biological processes, including fluid balance and hormonal activity in the brain, ebb and flow in response to the activity of the Moon in the same manner as the tides on Earth. Lieber's theory proposes that when biological tides are high, it "sets the human machine on edge," not necessarily causing strange behavior, but making it more likely to manifest.[13] Especially susceptible in this respect are people who already suffer from poor impulse control and low stress tolerance—those who only need the tiniest nudge to send them over the edge.

Lieber's biological tides theory has no credibility in so far as mainstream science is concerned, nor does his research in general. Others have tried to replicate the results of his study into homicide rates and the lunar cycle. With no success, it suggests murders do not increase around the time of the new and full Moon. In fact, countless studies of this kind have been conducted over the decades. Some of the researchers have a genuinely open-mind, while others are zealous "debunkers" with a frighteningly mechanistic view of reality. The countless studies were conducted in an attempt to determine if there is any correlation between the cycle of the Moon and such phenomena as traffic accidents, emergency room admissions, violence in prisons, domestic violence, crisis calls to police or fire stations, alcoholism, and suicide. Rather than waste time by going into specifics, allow me to state here that the vast majority of these studies were generally unimpressive, showing little or no sign of any such correlation.

In reviewing the results of these studies I was struck by how similar they are to the results of studies conducted in relation to psi phenomena. ("Psi" refers to a wide range of paranormal phenomena or abilities, such as telepathy, clairvoyance, and psychokinesis.) Why similar, you ask?

Because they both adhere to the principle of what Colin Wilson calls "James's law," in reference to an observation made by the philosopher William James concerning psychical research. This states, in Wilson's words, "that there seems to be just enough evidence to convince the believers, and never quite enough to convince the sceptics."[14] The famous extrasensory perception (ESP) experiments conducted by the psychologist J. B. Rhine at Duke University during the earlier half of the twentieth century are some of the best examples to which James's law applies.

It's easy to see James's law at work in a number of studies concerning the Moon and human behavior. For example, a study conducted in 1979 by psychologists Frey, Rotton, and Barry looked at fourteen types of calls made to police, fire, and ambulance departments over a two-year period in a large Midwestern city. A total of fifty-six tests were conducted, whereby each type of call was analyzed in relation to the lunar cycle. Of these, six achieved a "very small" but statistically significant result. The researchers concluded, based on the large number of tests conducted, the outcome of the study was consistent with chance expectation. In a 1985 meta-analysis, conducted by Rotton and Kelly, the results of nearly forty previous lunar effect studies were combined. The authors discovered "although a few statistically significant relations emerged . . . phases of the Moon accounted for no more than 1 percent of the variance in activities usually termed lunacy."[15]

They next divided the lunar cycle into four equal sections and discovered the activities in question took up 25.7 percent—ever so slightly higher than the 25 percent expected by chance. They were "not impressed by a difference that would require 74,477 cases to attain significance in a conventional . . . analysis."[16] Referring to this assessment in his book *The Conscious Universe,* the parapsychologist Dean Radin points out: "Although a 0.7 percent increase may not sound like much, in a city with a population of, say, one million, it could translate into hundreds of additional 911 crisis calls per day during 'adverse' lunar phases."[17] It's worth noting that Radin, whose research we'll examine in a moment, is a believer in psi, having "concluded that some psychic abilities are genuine."[18] The situation has James's law written all over it.

It's fair to say the Moon-human connection is a complicated and controversial matter that cannot be grasped merely with the use of statistics. Conversely, the Moon-animal connection is rather less difficult to comprehend, particularly in the case of marine creatures, such as the Californian grunion. A small, slender, sardine-size fish, the grunion is remarkably attuned to the cycle of the Moon, making use of the spring tide for reproductive purposes. Twice a year, during the first three to four nights after a new or full Moon and beginning soon after high tide, thousands of grunion journey to the beaches of southern California to spawn. Once there, the females dig themselves into the sand, tail first, to deposit their eggs beneath the surface. Meanwhile, males come to mate with them, fertilizing the eggs with their milt. The process is over in minutes, sometimes less, and the grunion return to the ocean. The eggs incubate while protected beneath the wet sand, hatching about twelve days later with the arrival of the next spring tide. The young grunions are carried into the ocean by retreating waves.

As well as being very tasty when grilled (so I'm told), the grunion is something of an enigma. How are they able to determine when it's the right time to spawn, and to such a precise degree? What is the biological mechanism involved? The obvious answer would be that it has something to do with detecting variations in the intensity of Moon light, but this possibility has been ruled out. Rather, scientists think the fish "are able to detect minute changes in water pressure caused by the rising tides."[19]

The oyster is another sea creature that's highly attuned to the cycle of the Moon. Frank A. Brown, Ph.D., a specialist in the field of biological rhythms, conducted a series of experiments throughout the fifties, sixties, and seventies, whereby he tested the effects of remote environmental factors on a variety of plant and animal life, ranging from seaweed to salamanders to potatoes. An overview of his research is given in Lyall Watson's classic book *Supernature* (1973). In perhaps his most remarkable experiment, Brown gathered some oysters from the sea near Long Island Sound, Connecticut, transporting them more than one thousand miles to his laboratory at Northwest University in Evanston, Illinois, where he placed them in tanks.

Oysters follow a tidal rhythm, whereby they open their shells to feed at high tide and close them when the tide begins to recede. Brown noticed that, for the first two weeks, the oysters continued to adhere to the tidal rhythm of their original location; however, on the fifteenth day their rhythm began to slip until they eventually settled into a new one. It soon dawned on Brown that the new rhythm corresponded to the passage of the Moon overhead. Meaning, in other words, they were opening their shells during what would have been high tide in Evanston even though they were inland. Since the oysters were kept in a darkened, isolated environment under constant temperature and were unable to detect the rising and setting of the Sun, it's sensible to suggest they might have been responding to the subtle gravitational pull of the Moon. We'll discuss this possibility further in a moment.

In a similar kind of experiment conducted by Brown, four hamsters were placed in a cage and their metabolic activity recorded over a two-year period by means of a hamster wheel. Artificial lighting was used to keep the animals under carefully controlled conditions of "night" and "day." They soon learned to adapt to these conditions, such that they exhibited a decreased amount of activity just prior to the light being switched on and a greater amount of activity just prior to the light being switched off. The amazing thing about the experiment is the hamsters showed an apparent lunar periodicity in the metabolic sense, demonstrating greater activity for four days following the full Moon and for several days following the new Moon.

In another experiment Brown used a different kind of rodent—a rat. Placed in a cage under unchanging conditions of light and temperature, the animal was found to be half as active when the Moon was above the horizon as it was when the Moon was below the horizon.

The results of Brown's experiments seem to support the age-old notion that life on Earth is attuned to the Moon; that all things dance, if you will, to a lunar rhythm. Apparently even the spud is no exception to this rule. Brown painstakingly gathered about a decade's worth of metabolic (photosynthesis, and so on) data for potatoes, which were kept under constant conditions. He found they were able to "detect" whether the Moon had just appeared over the horizon, whether it was setting, or whether positioned at zenith. "[T]he similarity of changes such as these in metabolic rate with the time of lunar day can be plausibly explained only by saying that all are responding to a common physical fluctuation having a lunar period," concluded Brown.[20]

Could this "common physical fluctuation" have something to do with the gravitational pull of the Moon? Many researchers dismiss this possibility on the basis that the effect involved is very weak, especially in relation to objects with little mass. We know the gravitational force exerted on the human body by the Moon is practically nothing, rendering Lieber's "biological tides" theory extremely untenable. There is another, more sensible theory, which looks at the phenomenon in terms of magnetism rather than gravity. This recognizes the Moon produces regular changes in the Earth's magnetic field, and, further, life on Earth is sensitive to these changes, from single-celled organisms to human beings.

Indeed, a complex and little-understood relationship exists between the lunar cycle and the state of Earth's magnetic field. This relationship also extends to the Sun, which has its own cyclical activity. Expressed simply, the Earth, Moon, and Sun form an interconnected system—a trinity—where energy is exchanged from one body to another and to which life on Earth is connected. These sound like purely astrological notions, yet they're entirely in agreement with scientific opinion. Besides, much of what astrologists have been saying all along is gradually being confirmed by science. Even the biologist Richard Dawkins, the high priest of scientism, would be hard pressed to deny that life on Earth is shaped by "cosmic influences."

The energy output of the Sun isn't constant but fluctuates in accordance with variations in sunspot and solar flare activity in much the same way the flow of a stream fluctuates in accordance with variations in rainfall and other factors. These fluctuations have an impact, of course, on the magnetic field of the Earth, which in turn has an impact on us and other living things. A solar flare is a sudden eruption of intense high-energy radiation from the Sun's surface. They are capable of causing significant disturbances in the Earth's magnetic field—known as a geomagnetic storm. It figures an increase in solar activity is accompanied by an increase in the solar wind, while the more intense the solar wind the greater the buffeting sustained by the Earth's magnetosphere. It helps to think of the Earth's magnetosphere as a giant windsock that changes shape depending on the intensity of the solar wind.

Like a windsock, the magnetosphere of the Earth has a tail. Known as the magnetotail, it points away from the Sun, extending out into space for some considerable distance, well beyond the orbit of the Moon. In fact, it has been observed that the full Moon passes through it. "The Moon enters the magnetotail about three days before it is full and takes about six days to cross and exit on the other side," according to Tim Stubbs, a University of Maryland scientist.[21] Running down the middle of the magnetotail is a region called the "plasma sheet," consisting of trapped charged particles, most notably electrons. Thanks to data gathered by the Kaguya probe, scientists have determined that as the Moon intersects the plasma sheet it absorbs these electrons, acquiring a strong negative charge. Yuki Harada, a graduate student at Kyoto University in Japan, commented, "We are suggesting the presence of a relatively strong electric field around the moon when it is inside the Earth's magnetosphere, but the origin of this electric field remains a mystery."[22]

During this process there is a big difference in the amount of charge that builds up on the day side of the Moon (its near side in this situation) as compared to the amount of charge that builds up on the night side of the Moon (its far side in this situation). A bombardment of UV photons on the day side of the Moon means that electrons have difficulty accumulating; the photons knock the electrons away. The opposite situation occurs on the night side of the Moon, where electrons accumulate with ease, bringing the surface charge up to hundreds, perhaps thousands, of volts. I should add that none of the Apollo Moon landings occurred while the Moon was full, that is, while the Moon was passing through the Earth's magnetotail. Thus, scientists can only speculate as to what conditions would be like on the surface of the Moon at this time.

One thing for sure is electrostatic activity would be at an all-time high, perhaps causing problems for astronauts and equipment—the same issue mentioned earlier in relation to lunar craters. NASA is concerned that astronauts might find themselves "crackling with electricity like a sock pulled out of a hot dryer."[23] Another concern is the very real possibility of lunar "dust storms." Scientists have long suspected that very fine particles of lunar dust, once charged, are capable of floating above ground by means of electrostatic repulsion. The process creates a thin, dusty atmosphere of sorts. If true, the dust would flow in a wind-type manner from the Moon's highly charged night side over to its weakly charged day side. Most of this activity would occur along the terminator. (Energy flows from areas of high concentration to areas of low concentration.)

This observation of the Moon passing through the Earth's magnetotail is an important and far-reaching one in that it may provide a basis for many of the strange and "superstitious" beliefs surrounding the full Moon, particularly in relation to its effect on human behavior. The full Moon seems to be a time when the Earth and the Moon establish a powerful connection, involving the transference of energy from the former to the latter. That most of this energy ends up on the Moon's far side may or may not be significant. If we imagine energy is transferred in both directions, Hawke's earlier given description of the full Moon as a "gateway to other realms of being, the energies of which flood on to our plane and interact with us," makes a good deal of sense. Is it these "energies" that cause certain people to succumb to "full Moon madness?"

The fact that the Moon, when full, passes through the Earth's magnetosphere no doubt has some bearing on the relationship I mentioned earlier between the lunar cycle and the state of Earth's magnetic field. How these two tie together exactly is something that awaits further scientific investigation. In fact, this whole area is a scientific work in progress. This leaves us with no other choice but to make as much sense as we can out of the few fragments of information that we do possess. We know the Sun is a source of electromagnetic radiation and some of this energy, consisting of ionized particles of gas, behaves in the manner of a wind—a solar wind. Situated within this ever-changing stream of electromagnetic radiation is Earth, which is orbited by the Moon.

Picture the sun's energy as a river and the Earth and Moon as two stones situated within this river, positioned slightly apart from each other. Let's say the Earth is the larger of the stones, and the Moon is the smaller one. If you were to position the smaller stone slightly upstream from the larger one—in the case of a new Moon configuration—it would obviously act as a barrier of sorts, deflecting water away from its larger companion, and so reducing the amount of water that can reach it. In this way, the water would be made to flow in a different manner around the larger stone. If you were to then position the smaller stone off to the side of the larger one—in the case of a first-quarter or third-quarter Moon configuration—the result would be different still, in terms of how the water flows around the larger stone. A variety of different configurations would yield a variety of different results.

I realize how limited this example is, but hopefully it illustrates my point. You see, by orbiting the Earth, the Moon alters the properties of the electromagnetic environment in which the Earth is immersed and by extension affects the Earth's magnetic field. Lieber explains this by saying the Moon, during its journey around the Earth, "continually intersects the earthward flow of electromagnetic radiations from the

[s]un," resulting in "daily and monthly variations in the Earth's magnetic field."[24]

The K-index is a scale ranging from zero to nine that is used to measure geomagnetic turbulence: one being calm and five or more indicating a geomagnetic storm. Related to this is the Ap-index, which measures the general level of geomagnetic activity over the globe for a given day. Simply by taking note of the Ap-index during the course of a certain period—say, one hundred synodic months—a number of researchers have observed a subtle but evident relationship between the lunar cycle and geomagnetic activity.

In a study by Radin, he looked at geomagnetic field (GMF) data recorded in the 1980s over a period of more than 120 lunar cycles. He found geomagnetic activity was quietest around the time of the full Moon. In a considerably older study, published in a 1964 issue of the *Journal of Geophysical Research,* NASA scientists Stolov and Cameron analyzed thirty-one years of GMF data in an attempt to determine "whether genuine influences of the Moon on geomagnetic activity may be present."[25] The study showed a decrease in geomagnetic activity by approximately 4 percent for the seven days leading up to the full Moon and an increase in geomagnetic activity for the seven days preceding the full Moon, also by approximately 4 percent.

The results of these studies would be insignificant if it weren't for the fact that life in general is sensitive to changes in the Earth's magnetic field. Experiments conducted to test the sensitivity of magnetic fields on organisms—from very simple organisms to the very complex organisms—have yielded some extremely remarkable results. One notable experiment of this kind, conducted by J. D. Palmer, an associate of Frank Brown, involved the green algae *Volvox.* These pinhead-size balls dwell in ponds, lagoons, and other freshwater habitats. Each tiny ball is a colony of cells, numbering from five hundred to sixty thousand. The cells in each colony work together in a harmonious, coordinated fashion, as though possessed of one mind, enabling the colony to propel itself through the water in a distinctive rolling motion. (*Volvox* is Latin for "rolling.")

A colonial organism is one that consists of a colony of single-celled organisms, whereas a multicellular organism is one that consists of more than one cell. Curiously enough, *Volvox* fall into neither category. Rather, they represent the link, in the evolutionary sense, between the two categories. It would not be incorrect to describe them, as many scientists do, as the simplest multicellular organism on Earth. They've existed in their present form for at least two hundred million years.

In his experiment, Palmer placed a total of 6,916 *Volvox* into a special glass vessel with a long thin neck that pointed toward magnetic south, with the intention of observing their direction of travel upon emerging from the neck of the vessel. He divided the colonies into three even groups, testing the first group under natural conditions. The second group was tested under the influence of a bar magnet, which was placed in line with the magnetic field of the Earth to produce an augmentative effect. He also used a bar magnet with the third group, this time positioned east to west, at right angles to the magnetic field of the Earth.

The results were very compelling. With the second group, 43 percent more *Volvox* than normal turned to the west. The same phenomenon was observed in the case of the third group to a heightened extent, with 75 percent more *Volvox* than normal turning to the west. The obvious conclusion to draw from this experiment is *Volvox* are equipped with the ability to sense a magnetic field to the point of being aware of the directions of force within that field. Referring to this experiment in his book *Supernature,* Watson adds: "The fact that *Volvox* is an archaic form shows that life's awareness of magnetism goes back a long way and is probably very deep-seated."[26]

The ability to sense a magnetic field is one not unique to *Volvox*. Many life-forms on Earth possess such a sense, using it to perceive direction, altitude, or location in order to find their way around. Scientists call this magnetoception. Life-forms known to possess magnetoception include bacteria, turtles, lobsters, sharks, rodents, stingrays, honey bees, frogs, salamanders, and fruit flies, not to mention migratory birds. The sense is especially pronounced in blind mole rats, one of nature's most efficient tunneling machines.

The blind mole rat, an odd-looking cylindrical-shaped creature with short-limbs, protruding incisor teeth, and eyes and ears so minor as to appear virtually nonexistent, can be found primarily in the Middle East in countries such as Israel. Hardly the most attractive of God's creations, it's perhaps a good thing that the blind mole rat spends the vast majority of its time underground, hidden away within its vast, mazelike network of burrows. Its existence is a solitary one.

Once thought to be entirely blind, new research has shown the blind mole rat can see to a very limited extent, using its tiny, vestigial eyes to detect light to tell when a predator has breached one of its tunnels. Their eyes are no use to them when it comes to subterranean navigation. Indeed, there are times when they have to dig considerable distances in search of bulbs and roots to feed on and then find their way back home again. Fortunately they are equipped with a highly developed ability to sense the magnetic field of the Earth, as proved by extensive scientific research.

In a study conducted in early 2000 by Tali Kimchi of Tel Aviv University and his colleagues, wild mole rats were placed in two types of laboratory maze, and their magnetoception abilities were put to the test. The first maze was shaped like a wheel with eight spokes extending out from a central hub. The animals were encouraged to find their way, via the most direct route, from a feeding place located in one part of the maze to a nesting box located in another. This they managed to do well, becoming better with experience. Next, with the use of a device consisting of two electromagnets, one placed in either side of the maze, the scientists were able to alter the surrounding magnetic field, tilting it 90 degrees from what it would normally be. It was found that under these conditions the animals were less likely to navigate the maze correctly. The second maze was rectangular and considerably more intricate than the first. The animals were encouraged to find their way through the maze via the most direct route to a food reward placed at the end. They once again proved competent at the task, becoming better with experience. As soon as the electromagnets were switched on, their performance plummeted; they were less likely to find shortcuts.

The animals, while subjected to the artificial magnetic field, made fewer errors in the case of the first maze (which involved traveling a short distance) than they did in the case of the second maze (which involved traveling a long distance). The results of the experiment indicated to Kimchi that "the mole rat can use internal cues to orient short distances, while for long distance orientation it also uses a magnetic compass," enabling it to get around "at least as efficiently as sighted animals aboveground."[27]

When birds and other life-forms navigate by means of the Earth's magnetic field, they do so by using it to assess direction at the initial stage of their journey. The blind mole rat, on the other hand, possesses a more advanced form of magnetoception; it's able to monitor and maintain its course as it goes along. It would appear that no other animal can do this. Commented zoologist Pavel Nemec of Charles University in Prague, "It is amazing that mole rats live in a permanently dark environment and are able to navigate so well. Other animals use sight to correct their mistakes. Mole rats use the Earth's magnetic field."[28]

Moving the discussion above ground, who hasn't marveled at the beauty of a monarch butterfly? Large in size, with striking orange and black wings, the monarch is extremely easy to recognize. They are common in North, Central, and South America but can also be found in countries such as Australia and New Zealand.

What makes the monarch such an amazing insect is not so much its appearance but its long annual migrations. Every autumn in North America millions of monarchs gather in preparation for the almost two-thousand-mile migration south. Their journey takes them to overwintering sites in either southern California or the mountains of Mexico. With the appearance of spring comes mating and the monarchs begin their journey north, laying their eggs along the way on the leaves of milkweed plants. From these eggs emerge a new generation of monarchs.

Plate 1. Earth and Moon shown to scale. Image courtesy of NASA.

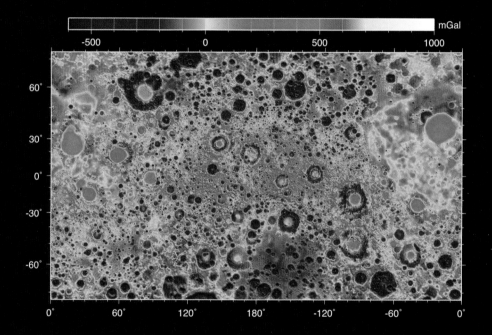

Plate 2. A map of the Moon's gravitational field, with the far side in the center and the near side on either side. Areas of lower than average gravity are shown in blue, while areas of higher than average gravity are shown in red. Generally, the red parts of the map (higher than average gravity) correspond to regions of high elevation—for example, mountains; while the blue parts of the map (lower than average gravity) correspond to regions of low elevation—for example, craters. The large red dots are mascons. The map was constructed from data obtained by NASA's Gravity Recovery and Interior Laboratory (GRAIL) mission. Image courtesy of NASA/JPL-Caltech/GSFC/MIT.

Plate 3. Thoth. Photo courtesy of Jon Bodsworth.

Plate 4. *Apollo 13* mission patch. Image courtesy of NASA.

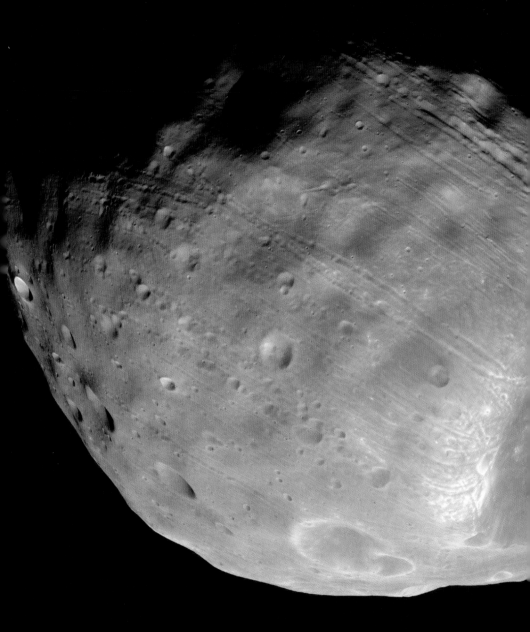

Plate 5. Phobos, the larger of Mars' two moons, photographed in 2008 by NASA's Mars Reconnaissance Orbiter (MRO). Image courtesy of NASA.

Plate 6. The Earth's magnetosphere. Image courtesy of NASA.

Plate 7. The Venus
of Laussel.

Plate 8. The Ziggurat of Ur, Iraq. Photo courtesy of Rich Lilly.

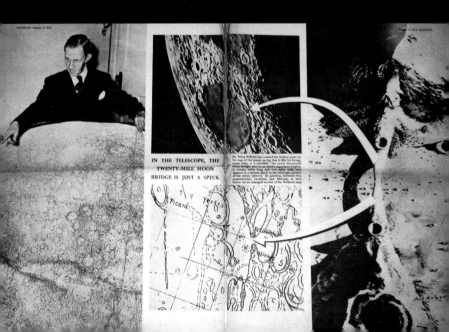

Plate 10.
Ingo Swann.

Plate 11. *Lilith.* Painted by John
Collier in 1892.

Plate 12. A map of the Moon's gravitational field with the influence of surface topography removed (known as a Bouguer anomaly map). The map reveals variations in crustal thickness as well as variations in crust or mantle density. Areas of lower than average gravity are shown in blue, while areas of higher than average gravity are shown in red. The large red dots are mascons. The map was constructed from data obtained by NASA's Gravity Recovery and Interior Laboratory (GRAIL) mission. Image courtesy of NASA/JPL-Caltech/CSM.

Bouguer gravity
(mGal)

600

0

−600

Plate 13. Three astronauts exploring a lunar lava tube. The illustration was done in the mid-1980s by space artist Pat Rawlings on behalf of NASA. G. Jeffrey Taylor, Planetary Scientist at the Hawaii Institute of Geophysics and Planetology (HIGP) in the School of Ocean and Earth Science and Technology at the University of Hawaii at Manoa, commented to me in an e-mail: "[The image was] part of a project I worked on with [geologist and lunar scientist] Dr. Paul Spudis. . . . The idea was to illustrate future geological field work on the Moon. . . . The picture is a metaphor for a future in which people live and work and explore on the Moon. It shows a lava tube because Spudis and I like volcanic features and because they are recognizable to people." Image courtesy of NASA.

After maturing, they continue heading northward. In fact, not one but several generations of monarchs are required to complete one annual migratory trip. In terms of life span, there are two different types of monarchs: ones that live only a month or so, and ones that live for seven to eight months. The latter—the "Methuselah generation"—are the ones that make the long migration south.

Scientists and nature enthusiasts have long marveled at the ability of these tiny, fragile creatures to migrate such long distances, year after year, and to the exact same locations as the ones before them. For a long time it was believed that they navigate solely by means of the Sun, despite there being one crucial flaw with this theory: monarchs continue to travel in the right direction even when the Sun is completely obscured by clouds. While scientists are fairly sure monarchs do use the Sun for navigational purposes, they now think the insects also make use of the Earth's magnetic field; meaning monarchs are equipped with an inbuilt magnetic compass similar to that of birds and other creatures. This magnetic compass "may serve as a backup system to keep the butterflies on track when they can't see the sun," speculates neurobiologist Professor Steven Reppert of the University of Massachusetts Medical School, whose research is focused on trying to unravel the mysteries of monarch migration.[29]

Reppert and his colleagues discovered that monarchs possess two types of photoreceptor (light sensitive) proteins called cryptochromes, which enable them to sense both UV light (which humans are unable to perceive since it lies outside of the visible spectrum) and the Earth's magnetic field. The scientists conducted experiments in which they inserted monarch cryptochrome genes into the genomes of fruit flies. These fruit flies were specially engineered to lack their own cryptochrome gene. The scientists found that these genetically "rescued" flies were able to sense a magnetic field. In comparison, fruit flies devoid of cryptochrome exhibited no such sense. Cryptochromes can be found in a wide variety of plant and animal life and are known to play an important role in the generation and maintenance of circadian rhythms (biological cycles). Especially compelling is the role of cryptochrome in magnetoception.

Scientists have discovered that the ability of birds to sense the magnetic field of the Earth is explainable, in part, by the presence of cryptochromes in the photoreceptor neurones of their eyes. Furthermore, a recent experiment has shown magnetoception in birds is dependent not only on light but also clarity of vision, suggesting birds can literally see the magnetic field of the Earth, possibly in the form of patterns of light and shade superimposed over their "standard" vision.

The experiment was undertaken by Katrin Stapput of Goethe University in Germany and involved robins. When the robins were fitted with translucent frosted goggles, which still permitted light to get through, they became disoriented and unable to navigate. This happened only when the right eye (controlled by the left side of the brain) was covered by the goggles. If the left eye (controlled by the right side of the brain) was covered, the robin could navigate perfectly well. Complicating the issue of magnetoception in birds is the fact that pigeons and other migratory species are believed to have small crystals of the mineral magnetite (a magnetic oxide of iron) embedded in their beaks.

What, you might ask, has this got to do with humans? A great deal, it would seem, for it's been discovered that our eyes, like those of migratory birds, contain cryptochrome, so that we too might be able to sense the Earth's magnetic field. In an experiment similar to the one involving monarch butterfly genes, Professor Reppert and his team inserted the cryptochrome gene of humans into the genomes of fruit flies. The experiment proved successful; the human gene restored the fruit flies' ability to sense a magnetic field. According to Professor Reppert, these results constitute strong evidence that humans have the capacity to sense the magnetic field of the Earth, perhaps without even being aware of it. In his own words, "I would be very surprised if we don't have this sense; it's used in a variety of other animals. I think that the issue is to figure out *how* we use it" [emphasis in original].[30]

Of course, the experiment proved little beyond the fact that human cryptochrome can act as a magnetic sensor when inserted into fruit flies. Whether the protein acts as a magnetic sensor in humans is another question entirely. Roswitha Wiltschko, who conducted pioneer research in magnetoception in birds, commented on the experiment: "To sense the magnetic field, one does not only need a molecule like cryptochrome, but also an apparatus that picks up the changes in that molecule and mediates it to the brain. [The fruit fly] obviously has this apparatus, but humans? I have my doubts."[31] Researchers have speculated on the possibility that our distant ancestors did make use of cryptochrome for magnetoception but that the protein no longer serves this function today.

Of relevance to the question of magnetoception in humans is the strong possibility that our bodies, like those of homing pigeons, dolphins, honey bees, and bats, contain crystals of magnetite. In 1992 a California Institute of Technology (Caltech) geobiologist by the name of Dr. Joseph Kirschvink announced that he'd found crystals of magnetite in human brain cells, publishing his findings in an issue of *Proceedings of the National Academy of Sciences*.

Furthermore, in a paper published in *Nature* in 1983, a team of scientists at the University of Manchester, England—among them Robin Baker, Ph.D.—claimed to have found deposits of ferric iron (most likely magnetite) in human sinus bones, particularly in the ethmoid bone. The ethmoid bone is located at the roof of the nose, between the brain cavity and the nasal cavity. Presumably referring to the University of Manchester finding, the science journalist B. Blake Levitt states that the ethmoid bone "has a high concentration" of magnetite crystals.[32] She adds (without specifying from where she obtained the information) that crystals of magnetite have also been found in the blood-brain barrier, a physiological mechanism whose function is to prevent certain substances in the blood from entering brain tissue while allowing other substances to enter freely.

No scientist has conducted more research on the issue of human magnetoception than the aforementioned biologist Robin Baker, Ph.D. In the mid-1970s, while employed at the University of Manchester, Baker became intrigued by the possibility of a magnetic sense in human beings and was curious to know how people would perform "when treated as homing pigeons."[33] Baker conducted a wide range of experiments in human magnetoception and spent nearly a decade doing so, in many cases using University of Manchester zoology students as study participants.

In one type of experiment participants were blindfolded and taken on long and winding bus trips through the countryside. Once the bus had reached its destination, the subjects, while remaining blindfolded, were individually led off the bus and asked to point in the direction of "home" (the direction from which they'd originated). Overall, the participants were remarkably accurate with their estimates of home direction. Yet, as soon as the blindfold was removed and the participants were able to view their surroundings, performance plummeted; they gave poor estimates of home direction. This indicated to Baker that the participants were making use of a subconscious magnetic sense to keep track of their journey.

Baker was curious to see if attaching bar magnets to the heads of the participants would interfere with their ability to sense the Earth's magnetic field, as shown to be the case in experiments involving homing pigeons. As expected, there was a noticeable difference in performance between participants fitted with magnets and participants not fitted with magnets; the latter group performed much better in the experiment.

Taken as a whole, the results of Baker's human magnetoception studies are very persuasive, leaving us, in his words, with "no alternative but to take seriously the possibility that man has a magnetic sense of direction."[34] In Baker's view, human magnetoception is an innate subconscious sense whose origins date back to hunter-gatherer times, when the very survival of our species depended on our ability to find our way from place to place without becoming lost; and while we no longer rely on this sense to the extent that we once did, due in part to the invention of the compass and other navigational tools, it still plays a role in our everyday lives.

Such an ability—or "sixth sense"—would have been extremely valuable to early humans. It's possible that our distant ancestors were able to literally see the magnetic field of the Earth, using it to find their way back to camp after long expeditions in search of game. And what about dowsing? Anyone who's experimented with dowsing for a sufficient period of time would know that the phenomenon is genuine. Those who dismiss it as bunk have clearly never held a pair of dowsing rods. Although dowsing is generally thought to make use of "earth energies," "ley lines," and so forth, and not the Earth's magnetic field per se, I would be very surprised if the phenomenon has nothing to do with magnetism.

Another paranormal phenomenon that might have some bearing on magnetoception in humans is the ability to perceive auras, or energy fields, around living and nonliving things. That a small percentage of the population possess this ability is extremely difficult to deny. Could it be that those able to perceive auras are actually utilizing some form of visual-based magnetoception, such as what a robin uses?

It would be ignorant to think that humans, contrary to virtually every other living thing, lack the capacity to sense the Earth's magnetic field. Although we may not be as highly attuned to geomagnetic processes as our forebears perhaps were—or to the profound extent that animals and insects are—our connection to the Earth's magnetic field is a very deep one. We are electromagnetic beings living in an electromagnetic universe.

The "powers that be" continue to suppress and deny the negative impact that cell phones, smart meters, WiFi, and other such devices have on human health, both mental and physical, even though all the evidence screams to the contrary. I've seen plants sicken and die as a consequence of being in close proximity to these devices and their powerful electromagnetic fields. I used to have a WiFi modem in my own household until the headaches and restlessness became unbearable and I decided to get it switched off. My health improved almost immediately. It's been suggested, by the way, that the excessive amount of electromagnetic pollution in our environment may be interfering with our ability to sense the Earth's magnetic field.

Of course, not all of the electromagnetic energies to which we're subjected in the environment are artificial. Some of them are perfectly natural and unavoidable. One such example is the Earth's magnetic field, which is governed, to some extent, by the Moon. More and more evidence is accumulating to suggest we are deeply affected by changes in the GMF, particularly with regard to mental health. In a study conducted by Oleg Shumilov of the Institute of North Industrial Ecology Problems (INEP) in Russia, a correlation was found between geomagnetic activity and suicide rates. By examining GMF data from 1948 to 1997, Shmilov identified three seasonal peaks in geomagnetic activity for every year. These peaks occur from March to May, in July, and in October. He then looked at the suicide rate pertaining to the same forty-nine-year period for the small Russian city of Kirovsk. Amazingly, he discovered that peaks in the suicide rate coincided with those peaks in geomagnetic activity.

Shmilov's findings are supported by a 2006 Australian study, published in *Bioelectromagnetics*, that detected a correlation between geomagnetic storm activity and the incidence of suicide. The study made use of national suicide statistics for the period 1968 to 2002 and included a total of 51,845 incidents of male suicide and 16,327 incidents of female suicide. The study found "suicides among females increased significantly in autumn [which, in Australia, begins March 1 and ends May 31] during concurrent periods of geomagnetic storm activity."[35] No such effect was observed in the case of males. According to the authors of the study, Berk, Dodd, and Henry, these results suggest "perturbations in ambient electromagnetic field activity impact behavior in a clinically meaningful way."[36]

Since suicide is generally caused by depression, could it be that depression increases during times of high geomagnetic activity? A 1994 study published in *The British Journal of Psychiatry* addressed this very question. The study found a 36.2 percent increase in male hospital admissions related to depression in the second week following geomagnetic storms. Commented Michael Rycroft, former head of the European Geosciences Society, "The intriguing correlation between geomagnetism and suicide justifies more research into its mechanism."[37]

Many researchers point to the pineal gland as being instrumental in this "mechanism." An endocrine gland is one that secretes chemical substances, called hormones, directly into the bloodstream. There are seven such major glands in the human body, and by far the most mysterious of these is the pineal—the so-called third eye. Rene Descartes (1596–1650), the famous French philosopher-mathematician, undertook an in-depth study of the pineal gland and attributed to it metaphysical properties, calling it "the principal seat of the soul."[38] It is at the pineal gland, he believed, that the mind (or soul) interacts with the body. In humans, the pineal gland is about the size of a grain of rice, while its shape has been compared to that of a pinecone (hence its name). It's located almost in the very center of the brain, exactly between the two hemispheres.

The pineal gland is the body's source of melatonin, a hormone that regulates the circadian cycle. The secretion of melatonin by the pineal gland is a process that depends on the presence or absence of light. At night, when we turn off the bedroom light and climb into bed, the pineal gland is stimulated via photosensitive cells in the retina and produces more melatonin, inducing drowsiness. When we wake up in the morning and our eyes detect light, melatonin production is inhibited. "Cyclical fluctuations of melatonin are vital for maintaining a normal circadian rhythm," states one source.[39]

The pineal gland has been shown to exhibit great sensitivity with regard to sources of magnetism, and there is a strong possibility that it responds to changes in the Earth's magnetic field. There have been experiments conducted in which individuals have been placed in underground bunkers to shield them from the ambient electromagnetic field, and their circadian rhythms were observed to gradually desynchronize and lengthen. They were brought back to normal again with the use of artificial magnetic fields. Could significant fluctuations in the GMF disrupt a person's circadian rhythms to the extent that it negatively affects their mental health, perhaps tipping them over the edge just enough to commit suicide? According to psychiatrist Kelly Posner of Columbia University in the U.S., the answer to this question is a definite maybe. She explains:

> The most plausible explanation for the association between geomagnetic activity and depression and suicide is that geomagnetic storms can desynchronize melatonin production and circadian rhythms. . . . The circadian regulatory system depends upon repeated environmental cues to [synchronize] internal clocks. Magnetic fields may be one of these environmental cues.[40]

Based on the evidence we've looked at, it would appear only significant disturbances in the GMF—geomagnetic storms and the like—are capable of affecting one's mental health by desynchronizing melatonin production and circadian rhythms. Now, as far as mainstream science is concerned, geomagnetic disturbances have much to do with the Sun, yet little, if nothing, to do with the Moon. Thus, as we've seen, scientists are willing to consider the possibility of a solar influence on human behavior but not so willing to consider the possibility of a lunar influence on human behavior.

The important role played by the Moon is its altering the electromagnetic environment in which the Earth is immersed, and so producing regular changes in the Earth's magnetic field. This role is unfortunately overlooked by most scientists. Previously we saw that geomagnetic activity appears to be quietest around the time of the full Moon, when the Moon happens to pass through the Earth's magnetotail. It strikes me as rather obvious that any creature equipped with some form of magnetoception, such as the blind mole rat or the monarch butterfly, would be aware of these lunar-induced alterations in the GMF, subtle though they are. I see little reason why humans would be unable to detect these changes as well, in light of the findings discussed in this chapter.

Researchers T. Nishimura and M. Fukushima at the Graduate School of Medicine at Kyoto University, Japan, have taken seriously the fact that GMF conditions alter in connection with lunar activity, and suggest this might explain why animals respond to the full Moon. In a 2009 paper published in *Bioscience Hypotheses,* the scientists state that "animals can clearly detect the changes in magnetic field intensity that occur at full Moon."[41] They continue: "We think that moonlight increases the sensitivity of animals' magnetoreception. . . . We propose a hypothesis that animals respond to the full Moon because of changes in geomagnetic fields, and that the sensitivity of animals' magnetoreception increases at this time."[42] Of obvious relevance to their theory is how GMF fluctuations can disrupt melatonin production in the pineal gland and affect circadian rhythms. (And let us not forget melatonin production is dependent on light, while full Moon nights are less dark than other nights.)

I consider the issue of whether magnetoception exists in humans to be intimately bound with the issue of whether our behavior is affected by the phases of the Moon. The existence of the former entails the existence of the latter, and vice versa. If it's true that humans are able to sense the magnetic field of the Earth, which, in my opinion, is practically certain, it means we're receptive to every single change, no matter how minute, that takes place within that field. It implies too that our bodies would respond to these changes even on an emotional level. And of course, studies have shown that we do seem to be affected by GMF fluctuations via the magnetically sensitive pineal gland. It seems to me that the pineal gland, the cryptochrome in our eyes, and the magnetite crystals believed to exist in the ethmoid bone and elsewhere in the body constitute one system through which we remain linked to the Earth's magnetic field—and to the Moon and the Sun.

We now return to Dean Radin, whose research indicated geomagnetic activity is quietest around the time of the full Moon. Taking this research further, Radin conducted a study into the use of psi in the casino, with the objective of determining whether the performance of gamblers varies with respect to the lunar cycle. He was familiar with the notion that magical forces are augmented by the full Moon. "Magic," according to Radin, is "the primeval origins of what we now call psi."[43] A description of the study is given in Radin's groundbreaking book *The Conscious Universe* (1997), in which he also refers to a 1965 telepathy test conducted by Andrija Puharich (1918–1995). Puharich was a neurologist, parapsychologist, and author with a military and intelligence background whose involvement with the Israeli psychic Uri Geller is well documented.

The objective of Puharich's experiment was to determine if psi performance varies with the lunar cycle. Interestingly, he found the highest telepathy scores were achieved around the time of the full Moon as well as around the time of the new Moon. The former result was more pronounced than the latter. His experiment was based on the idea that psi performance might be related to gravity—an idea Radin essentially dismisses. Rather, he says the outcome of Puharich's experiment is better understood in terms of fluctuations in the GMF. According to Radin, there is evidence to suggest psi performance improves on days when the GMF is quiet; the theory behind this is that the brain is less distracted at this time and better detects psi.

Given the extremely private (and obviously corrupt) nature of the casino industry, casinos are extremely reluctant to allow people access to their daily profit-and-loss records. As luck would have it, an executive at the Continental Casino in Las Vegas, who happened to have an interest in parapsychology, was happy to provide Radin with four years of gaming data for his study. This consisted of daily payout percentages for five casino games, among them roulette, keno, and craps. Not surprisingly, for each of the five games the data showed the casino gained vastly more money than they lost, while gamblers lost vastly more money than they won. For each dollar player in roulette, an average of seventy-five cents was returned to the gambler—the mean payout percentage for that game. As a further example, the mean payout percentage for slots was fifty-three cents. The higher the payout percentage was for a game, the better the performance of the gamblers in that game.

First, Radin wanted to see if a relationship was present between the lunar cycle and the average payout percentage for the five games combined. He found such was indeed the case; payout percentages were highest around the time of the full Moon. Specifically, payouts peaked on the day of the full Moon, at 78.5 percent, while the lowest payouts occurred about a week before and after the new Moon, dropping to a low of 76.5 percent. The results were certainly compelling with odds against chance of twenty-five to one. Elaborates Radin, "This finding suggests that by gambling on or near days of the full Moon, and by avoiding the casino on or near days of the new Moon, over the long term gamblers may be able to boost their payout percentage by about 2 percent."[44] Next, Radin took a look at GMF conditions during the course of this one lunar cycle and discovered, as expected, the payouts peaked when GMF conditions were quietest.

During the next stage of his study Radin was curious to see if slot-machine payouts had anything to do with the lunar cycle. Once again, payouts were found to be highest around the time of the full Moon. Radin decided to take a closer look at the four years of slot-machine payout data to which he'd been given access. He noticed within this period there were six major jackpots recorded, and, remarkably, four of these jackpots occurred within one day of the full Moon. "The odds against chance of seeing four out of six jackpots this close to the full Moon, when jackpots presumably occur at random, is sixteen thousand to one," notes Radin.[45]

Also remarkable was Radin's discovery that the peak average payout rates for each of the other games all clustered around the full Moon: for blackjack, three days before; for craps, three days after; for keno, one day before; and for roulette, one day after. In summary, then, three of the five games—slots, keno, and roulette—exhibited peak payout rates within one day of the full Moon, producing odds against chance of just over two thousand to one. Radin, whose reputation as a scientist is utterly impeccable, has established that fluctuations in casino payout rates are consistent with the cycle of the Moon, and, further, payout rates are "reliably *predictable* using mathematical models" [emphasis in original].[46]

Finding that payout rates at a casino follow a lunar cycle is admittedly quite amazing. If Radin's reasoning is correct, it means psi ability waxes and wanes with the Moon, becoming most potent when the Moon is full due to changes produced in the Earth's magnetic field. On a side note, there could be some significance in the fact that the peak average payout rates for all five games occurred either during the three days prior to the full Moon or during the three days subsequent to the full Moon; the time period constitutes a period of seven days, including the day of the full Moon, while the Moon passes through the Earth's magnetotail for approximately six days. This, I'm inclined to believe, is not merely a coincidence. Having examined Radin's research, one is left wondering if the lunar influence to which our minds are subjected operates on a level so deep that it encompasses psi processes.

The British conspiracy theorist David Icke, while hardly the most cautious and scientific of researchers, has some fascinating conclusions to offer regarding how the Moon affects human consciousness. Some of his conclusions deserve mention here, if only because they encapsulate the sheer mystery and intrigue that has built up around the topic of the Moon. And so, with an open mind, let's take a look at Icke's lunar research.

Icke was born in Leicester, England, April 29, 1952, to parents Barbara and Beric. In his autobiography, *In the Light of Experience,* he describes his father as a "complex man" who, unable to afford to study medicine as he'd wanted, was forced to endure a life of hardship and virtual poverty. The result was "a large chip on his shoulder" and a very argumentative personality. He received little income from his job in the drawing office of the Gents Clock Factory, and so the Icke family struggled financially, having to live in a council house estate in the suburbs. Says Icke, "To say we were skint is like saying it is a little chilly at the North Pole."[47] His mother, a strong yet gentle woman, did her best to contribute to the family budget by getting jobs as a factory worker and a cleaner.

As a child Icke was shy and nervous, preferring to spend his days alone. His anxiety was so great during morning school assembly that there were times when he had to excuse himself, feeling as though he might faint, and he ran home on one occasion.

Icke's confidence and self-esteem rose at the age of nine when he started to play for the junior school soccer team and realized he had a passion and talent for goalkeeping. Rather than pay attention to his schoolwork—although very intelligent, he lacked motivation when it came to academic matters—he spent his days playing and thinking about soccer, making it his ambition to become a professional goalkeeper. This he achieved at the age of fifteen, first playing for Coventry City and later Hereford United. He was forced to retire from sport at the age of twenty-one, after his body became riddled with chronic rheumatoid arthritis, which caused excruciating pain.

Saddened but undefeated, he decided to become a journalist and began his new career in 1973 as a news reporter with *The Leicester Advertiser.* "I got the job because no one else wanted it," he says.[48] Icke progressed through the journalistic ranks very quickly—a testament to his talent as a communicator. By the early 1980s he was presenting sports on television for the BBC and even appeared on programs such as *Grandstand.* It wasn't long before feelings of disillusionment began to settle in. "I found television to be a deeply insecure world full of insecure, often shallow, and sometimes vicious characters."[49] Icke drifted away from journalism to become involved in the environmental movement, which culminated in his election as a national speaker for the

British Green Party. Shortly after he joined the party it quickly rose to success. At the European Parliament election of 1989 the party won a record 15 percent of the vote.

Before long, due to infighting within the party and other issues, Icke became disillusioned with the Green Party and decided politics was a waste of time and not his true calling in life. The party, he felt, was a lie, claiming to be the "new politics" when in reality it was "just another version of the old politics with the same old methods, manipulations and reactions."[50] When it came time to renew his party membership, he didn't bother to do so.

Beginning in 1989, while working on a book about environmental issues called *It Doesn't Have to Be Like This*, Icke found himself undergoing a "spiritual awakening." Strange and synchronistic events began to manifest in his life, and he felt as though his awareness was expanding. He was amazed by how the book seemed to write itself, often producing an entire chapter in one day—a phenomenon that sounds a lot like "automatic writing." While working on the book there were moments when he found himself wondering if he'd written the material on the page in front of him, unable to remember having done so. "It was like I was writing it all in a dream," he says.[51]

Also around this time, he began to sense a presence around him—a presence that lingered for months. During one of the most significant events of his "awakening," which took place in March 1990, he was instructed by a voice in his head to look at some books in a local newsagency. On the way there he suddenly felt the soles of his feet become "incredibly hot." This was accompanied by what felt like magnets pulling his feet to the ground. Once inside the newsagency, his attention was drawn to a copy of *Mind to Mind* by the famous psychic and spiritual healer Betty Shine, which he purchased and read immediately.

He soon got in contact with Shine. During their third consultation together, she closed her eyes and said she could see (in her mind) a "Chinese-looking" figure named Wang Yee Lee. Icke, who was lying down on a couch at the time, felt an odd sensation on his face, as though it were covered by a cobweb. Meanwhile Shine, playing the role of intermediary, passed on to Icke a number of messages from the "Chinese-looking" figure. One message included the following: "He [Icke] is a healer who is here to heal the Earth and he will be world famous; he was chosen as a youngster for his courage. He has been tested and has passed all the tests; sometimes he will say things and wonder where they came from. They will be our words."[52]

These messages and other similar messages received through other psychics left Icke convinced he'd been chosen by benevolent forces to fulfill an important spiritual mission on behalf of the human race. He describes how he felt at the time: "There was a shadow across the world that had to be lifted, a story that had to be told, and I, for whatever reason, was going to tell it."[53]

Driven by strong spiritual urgings, Icke took a trip to Peru at the beginning of February 1991, not long after writing the book *Truth Vibrations*. At one point during the trip he heard about an ancient Inca site called the Sillustani Ruins and felt an overwhelming desire to visit the location. Having paid a guide and driver to take him there, he explored the ruins for an hour or so, all the time wondering what had compelled him to visit the site. It was a sunny, quiet day, and he was the only tourist in the area. Feeling underwhelmed, he boarded the bus to go back to the hotel. Just as he was leaving the site his attention was drawn to a nearby mound, and he heard a voice in his head say "come to me." He told the driver to stop so he could exit the bus. Once at the top of the mound, he noticed a circle of standing stones. While standing in the middle of the stone circle he felt as though magnets were pulling his feet to the ground—the same sensation he experienced the previous year. Icke picks up the story:

My arms then stretched out above my head without any conscious decision from me. . . . My arms were like that for over an hour. I felt nothing until it was over, and then my shoulders were in agony. . . . I felt a drill-sensation in the top of my head and I could feel a flow of energy going the other way, too, up from the ground through my feet and out through the top of my head. . . . I heard a voice in my mind that said: 'They will be talking about this moment a hundred years from now.' This was followed by: 'It will be over when you feel the rain.' I stood there, unable to move as the energy increased to the point where my body was shaking as if plugged into an electrical socket. . . . I kept moving in and out of conscious awareness.[54]

The "voice" had promised rain, and very soon it rained. Icke claims the storm clouds that brought the rain appeared almost suddenly, as though a curtain had been drawn across the previously clear and sunny sky. As soon as he felt the rain strike his body, the experience came to an end and he managed to stagger back to the bus, his muscles sore and stiff. Icke then attempted to "diffuse" some of the energy that remained surging through his body by placing his hands on a crystal. That night, his feet "continued to burn and vibrate," preventing him from getting to sleep.[55] The sensation lasted some twenty-four hours.

He later came to believe—as he still does today—the experience involved the awakening of Kundalini energy. While the experience was positive in the sense that it expanded his awareness and caused him to "suddenly see myself and the world in a totally different way," it also left him somewhat "burnt out." Icke continued, "Looking back, I can liken the experience to pressing too many keys on a keyboard too quickly and when the computer can't process all the data, it freezes. This is how I felt."[56]

During the remainder of his stay in Peru, Icke continued to engage in various spiritual activities by visiting numerous sacred sites, including the infamous Machu Picchu, to "channel" and interact with "earth energies" and such. It was while he was near the river of Machu Picchu that he channeled an entity he refers to as Rakorski. Although the fact isn't generally recognized today, Icke was heavily involved in channeling around the time of his "spiritual awakening," even to the extent of practicing full trance channeling. This is where a medium enters a trance state and loses awareness of the outside world while they allow an entity to "take over" their body and communicate through them. Icke describes the process: "Eventually, your mouth begins to speak, but they are not your words. They are the words, or rather thought energy, of the communicator."[57] One theory holds that the messages merely originate from the unconscious mind of the medium, a theory to which Stan Gooch subscribed.

Icke arrived back in England in early 1991, marking the beginning of what he calls his "turquoise period"—a reference to his fondness at the time for dressing in turquoise-colored clothes. He says he was "walking around in a daze"[58] and could "hardly tell you what planet I was on,"[59] so unstable was his mind from the experiences he'd had in Peru. Central to his life during this period was a psychic by the name of Deborah Shaw. Together they conducted channeling sessions, primarily by means of automatic writing. They soon became intimately involved. Icke's then wife, Linda, must have been an extremely tolerant woman, because she permitted Shaw to move into the family home, which was occupied not only by her and Icke but also their young children Kerry and Gareth. This polygamous arrangement was short lived. The situation became ugly when Shaw fell pregnant to Icke, giving birth to a daughter named Rebecca in December 1991.

Having been a well-known and respected TV personality, the events of Icke's life during his "turquoise period" received extensive media coverage, and he found himself the focus of "unimaginable national ridicule." No matter where he went in public, people would jeer and laugh at him. "I would stop at traffic lights to look across and see whole families laughing at me in the cars alongside . . ."[60] Some of the ridicule stemmed from bizarre comments made by Icke during a press conference held in early 1991. After referring to himself as "a son of the Godhead," he said the world would soon be ravaged by a series of devastating natural disasters (which, of course, never came).

Much of the information relayed during the conference came from channeled messages received by him and Shaw as they acted as a mediumistic team. According to Icke he was not in full control of his mind at the time he gave the conference; the "energies" he'd been channeling had "taken him over." He explained, "I was speaking the words, but all the time I could hear the voice of the brakes in the background saying: 'David, what the hell are you saying? This is absolute nonsense.'"[61]

Icke's experience as a target for ridicule reached its peak following an appearance he made on Terry Wogan's prime-time chat show, *Wogan,* on April 29, 1991. To the great amusement of Wogan and his audience, Icke, dressed in a terribly unfashionable pink-and-turquoise shell suit, repeated the same themes that he put forward during the press conference. At one point during the interview he poignantly remarked, "Survey the world, ladies and gentlemen. Is the force of love in control of this world, guiding the planet at this time? Of course not! The negativity, the thoughts I'm talking about that are very destructive, are pouring out of this planet every day." To this Wogan replied in a deadpan fashion: "Was it a great shock for you to discover this at the age of thirty-eight?"[62]

Icke, now in his sixties, believes the Wogan interview had a beneficial impact on his character by endowing him with the strength and courage to no longer care what other people think of him.

One of the world's foremost conspiracy researchers—with around twenty books under his belt and a website that attracts over a quarter million visitors per month—Icke regularly travels the world giving lectures to audiences of thousands. His books sell in the tens of thousands and have so far been translated into eleven languages. In 2011 his website, Davidicke.com, "entered the top 5,000 most visited in the United States and the top 6,500 in the world."[63] As a result of his growing popularity, the mainstream media despise him and have launched scathing attacks on his reputation, calling him a "fruitcake" and an "oddball" and his ideas "crazy" and "wacky." "Icke's brand of loopiness has proved more resilient and lucrative than anyone could have guessed," says an article that appeared in the *Daily Mail* in 2012.[64]

I was fortunate to attend one of his lectures while conducting research for this book; the lecture was held at Melbourne's Astor Theatre in late 2011. Since the event was to be an all-day affair (his lectures often run for up to ten hours), I had to arrive at the venue early, joining the throng of Icke fans lined up outside the doors. After entering the venue and taking our seats, a short clip accompanied by spiritual music appeared on the screen in front of us.

Moments later Icke stepped onto the stage, drawing great applause from the audience. A well-built man with gray hair, blue eyes, and a casual, confident manner, he reminded me almost of a charismatic schoolteacher. Here was someone, I felt, who had their feet firmly planted on the ground—not even a tsunami would be able to knock him over. Possessed of a keen sense of humor, Icke had no trouble making the audience laugh with comments like, "The whole world's bloody mad!"

For the next seven to eight hours the audience sat spellbound as Icke presented his alternative ideas on the nature of reality, science, history, politics, banking, medicine, and religion. We were told reality is an illusion, similar in nature to a hologram, which is decoded into existence by the brain. Furthermore, the world is secretly controlled by a race of sinister reptilian entities who operate primarily beyond the physical realm by controlling the minds of people in key positions of authority, such as Barack Obama, Hilary Clinton, and Henry Kissinger—members of what he calls the "global elite." He included those who constitute the global elite are genetically unique from the rest of humanity, belonging to special reptilian-hybrid bloodlines that can be traced all the way back to the ancient world, to the civilizations of Sumer and Babylonia. We were told global events have been and continue to be covertly manipulated to further the aims of the reptilian force in control of this planet, including, for example, the September 11 attacks. In addition, the members of the global elite, who are little more than puppets for these reptilian forces, take part in ghastly satanic rituals, involving blood-drinking and child sacrifice; that the global elite are fast trying to establish a

"New World Order"—a totalitarian one-world government, whereby the population will be micro-chipped and kept under constant surveillance, rendered docile with the use of mind-control technology and prevented from exercising the most basic freedoms.

During the twenty or so years he's been active in the field of conspiracy research, Icke has constructed his own unique "system" of ideas. Vast and cohesive, it brings together such subjects as Atlantis, ancient astronauts, UFOs, mind control, secret societies, government conspiracies, and just about everything else. At the very core of Icke's system is the notion that the world is covertly controlled by a race of reptilian entities. While some embrace this idea with great enthusiasm, feeling that it makes perfect sense, others, not surprisingly, find it completely laughable, becoming almost offended at the very mention of it. Even many so-called alternative-minded people, although willing to accept the existence of UFOs, for example, consider Icke's reptilian hypothesis simply too farfetched.

Icke's system continues to grow in vastness and complexity with the addition of new ideas. It was only very recently that he came to adopt the notion that the Moon is an artificial construct, and of reptilian origin, no less. In 2009, while working on the book *Human Race Get Off Your Knees,* he felt "the atmosphere in the room [change]" and the thought came to him that "the Moon is not what we think it is."[65] He said this thought originated not from his own mind but from the "presence" he's been in "contact" with since 1990. Icke said, "I feel a powerful intuitive 'knowing' about something and the 'five sense' information then follows quickly afterwards. So it was with the Moon [being an artificial construct]."[66]

Following the experience, Icke decided to contact his friend and spiritual mentor Vusamazulu Credo Mutwa, a highly respected Zulu shaman who lives in South Africa, to see what he had to say about the Moon from the perspective of Zulu lore. Born in 1921, Credo is the official storyteller and historian of the Zulu nation. He's written a number of celebrated books on the myths, customs, and spirituality

Credo Mutwa and David Icke.
Photo courtesy of David Icke Books Ltd.

of his people. Icke calls him "a walking library of ancient legends and accounts."[67] Icke said he was careful not to share with Credo his own thoughts on the Moon, but simply asked him the question: "What do Zulu shamans say about the Moon?"[68] Icke was told by Credo that the Moon is not only a hollow object but also the home of the *Chitauri,* which means "children of the serpent" or "children of the python."

Let me preface here that Credo first told Icke about the Chitauri in 1998, very soon after they became acquainted. At the time, Icke was in South Africa conducting a speaking tour and his book *The Biggest Secret* had just been released. It was with the publication of this book that Icke first introduced to the world his idea of the reptilians. Having heard about Icke and his work, Credo called him to arrange a meeting. During the meeting, Credo asked Icke, "How do you know about the Chitauri?" Of course, Icke knew nothing about the Chitauri. Rather, it just so happened that the reptilians he described in *The Biggest Secret* were remarkably similar to the Chitauri of Zulu lore, seemingly confirming his research.

According to Credo, the Chitauri are a race of scaly reptilian beings of extraterrestrial origin, who, possessed of extremely advanced technology, came to Earth in the distant past for purely selfish and exploitative reasons. Employing the strategy of divide and rule, they scattered humanity across the Earth and gave them different languages, preventing them from communicating with each other. They also interbred with humans, giving rise to a race of reptilian-human hybrids. Members of these "royal bloodlines" were appointed by the Chitauri to rule over humanity, assuming positions of kingship and such.

Icke was further informed by Credo that the Moon hasn't always accompanied the Earth. It was brought here "hundreds of generations ago" by Chitauri leaders Wowane and Mpanku. The two were brothers and had "scaly skin like a fish." Having stolen the egglike Moon from the "Great Fire Dragon," Wowane and Mpanku emptied out its yoke, rendering it hollow. Next, they "rolled the Moon across the sky to the Earth."[69] Due to the arrival of the Moon, a canopy of water vapor that encircled the Earth—protecting it from the harsh rays of the Sun and so keeping the climate moist and humid and the temperature constant—fell to the ground in the form of a huge deluge of rain, resulting in massive and cataclysmic flooding. The Earth owes its present climate to the absence of this canopy of water vapor. Previously there were no deserts, only lush, green forests.

From the perspective of Zulu lore, there is good reason to be afraid of the Moon. We are told it is from the Moon that the Chitauri manipulate humanity. We are also told—or rather warned—never to upset the Moon on account of its harsh and unforgiving nature. These tidbits of Zulu lore, as passed down from Credo Mutwa, are remarkably similar in theme to the spaceship moon theory. Obviously, by describing the Moon as a giant egg whose yoke was removed, the Zulus may as well be saying the Moon is an artificially hollowed-out planetoid.

According to Icke, "all over the ancient world you find the same recurring theme of the serpent gods."[70] As examples, Icke cites not only the Chitauri of Zulu lore but also the Nagas of Hindu and Buddhist mythology, the Anunnaki of Sumerian mythology, the Nommo of Dogon mythology, and many others.

Being reasonably well acquainted with Buddhism, I'm certainly familiar with Nagas. Visit any traditional Buddhist temple, and you're bound to see an abundance of Naga imagery. Many statues of Buddha show the "awakened one" shielded by and seated on the coiled tail of a many-headed Naga cobra. Often this cobra has seven heads in reference to a story about Buddha and the snake king Mucalinda. The story states that Buddha, while meditating under the Bodhi tree, found himself in the middle of a huge downpour, which continued for seven days; however, the Buddha did not get wet. For, during those seven days, he was sheltered from the rain by the charitable Mucalinda.

The term *Naga* comes from the Sanskrit, meaning "serpent." They are described as a strong and handsome race of half-human, half-serpentine beings of semi-divine status. They are said to possess the ability to shape-shift, assuming either human or serpentine form. Though not necessarily malevolent, they can pose a danger to humans, especially when threatened. Nagas, it would seem, are not to be interfered with; they are extremely wrathful and protective. Being semi-divine, they are, in some respects, superior to humans, possessing magical powers.

Nagas are associated with all things watery, such as lakes, rivers, seas, and wells, and are said to be responsible for controlling the weather, especially rain. If offended, they can withhold the rain, causing droughts. Appeasing them will encourage them to release rain. The Nagas dwell not on Earth, we are told, but in an underground kingdom of their own where they greedily guard gems and other precious objects. Just as the Chitauri interbred with humans thereby producing "royal bloodlines," so too did the Nagas. As the *Encyclopedia Britannica* clearly states: "[T]he dynasties of Manipur in northeastern Asia, the Pallavas in southern India, and the ruling family of Funan (ancient Indochina) traced their origin to the union of a human being and a nagi."[71]

There are some obvious parallels between Nagas and dragons, and it would seem that both names refer to the same creature or race of beings. Dragons are featured prominently in both European and Asian mythology, not to mention in modern books, TV shows, and movies. Few mythological creatures capture the imagination of the public so strongly. They are commonly depicted as huge reptilian beasts, with scaly skin, bat-wings, a barbed tail, and the ability to breathe fire. Their reptilian nature extends to the fact that they even lay eggs. Earlier I mentioned the Babylonian goddess Tiamat, who is described in the *Enuma Elish* in very dragonlike terms, possessing both a scaly body and wings.

In Christianized Europe, dragons were seen as largely malevolent creatures, representative of sin and paganism. In the biblical Book of Revelation, the devil himself appears in the guise of a dragon, more specifically, as a "great red dragon" with "seven heads and ten horns." The devil is further described as "that old serpent . . . which deceiveth the whole world."[72] And of course, who isn't familiar with the Old Testament story of Adam and Eve in which Eve is "seduced" by the evil serpent into eating from the "tree of knowledge of good and evil"?

Curiously, of the twelve animals in the Chinese zodiac, all of them are factual except one: the dragon. Whereas in European mythology the dragon is a deceptive and dangerous, even evil, creature, Chinese mythology holds the opposite view. In China, dragons are regarded as auspicious and beneficent. They and are a symbol of power, strength, and good luck. To be born under the year of the dragon is considered extremely fortunate. Unlike their European counterparts, Chinese dragons are generally depicted as wingless. They are said to be able to fly by magical means, and hence have no need for wings. Another difference between the two is that Chinese dragons are longer and more serpentine in form, giving them a certain undulating, ethereal quality lacking in European dragons.

From a Taoist perspective, the dragon represents yang: the masculine, active principle in nature. In many pictures of Chinese dragons there is often a fiery, luminous pearl, usually red in color, positioned under the chin of the dragon or "hovering" near its mouth as though about to be swallowed. Called the "dragon pearl," there are numerous interpretations as to the symbolic significance of this profoundly precious object. Most likely it symbolizes the Moon and the watery, life-giving properties thereof. After all, Chinese dragons, like Nagas, are thought to preside over all things watery—the rain, lakes, rivers, seas—while, as we've already discussed, the association between the Moon and water is very extensive indeed. Furthermore, there is a fascinating link between pearls and the Moon. To quote G. Elliott Smith, "The pearls found in the oysters were supposed to be little moons, drops of the moon-substance (or dew) which fell from the sky into the gaping oyster."[73]

In China the dragon stands for nobility and lordliness. There is a close affiliation, dating back to ancient times, between the dragon and the Chinese imperial family, who used the dragon as their emblem. At one time this consisted of a yellow or golden dragon with five claws. The emperor, who was known as "the dragon," wore robes adorned with dragons, and the throne on which he sat was called the "dragon throne." Commoners, on the other hand, were not allowed to wear such dragon-themed clothing lest they incur the death penalty. Many Chinese emperors believed themselves to be descended from the dragon, and some were said to bear dragon-shaped birthmarks, to have dragon tails, or to have a face similar to that of a dragon.

By far the most "dragonesque" of Chinese emperors was Gaozu, the founder and first emperor of the Han Dynasty, who reigned from 206 to 195 BC. His personal name was Liu Bang, and he was born of a peasant family in what is today the province of Jiangsu. There is a very strange story concerning the nature of Liu Bang's conception. Apparently, just before he was born, his mother was caught in a rainstorm and so decided to take shelter under a bridge. All of a sudden, along came dark storm clouds, followed by thunder and lightning. Worried, Liu Bang's father set off to fetch her. When he arrived at the bridge he found his wife fast asleep. She was not alone. Looming over her body was a great scaly dragon, red in color. Months later, she gave birth to a baby boy, Liu Bang.

Was Liu Bang literally the offspring of a human female and a male dragon? It has been said of Liu Bang that he bore some resemblance to a dragon, and, strangely enough, portraits show a man who did look remarkably dragonesque with a very high forehead and brow bone. Apparently, too, his left leg featured seventy-two anomalous dark spots. (Presumably these were only moles.) To add to the mystery surrounding Liu Bang, friends of his are said to have perceived the silhouette of a dragon over his body whenever he was drunk.

In many respects, Icke is a proponent of the "ancient astronauts" theory—supported by authors and researchers like Erich von Daniken, Robert Temple, and the late Zecharia Sitchin. The basic idea behind this theory is that Earth was visited in the ancient past by technologically advanced extraterrestrial beings from outside our solar system, perhaps from the Draco constellation or the Sirius star system. After arriving in spaceships, they imparted a great wealth of knowledge to humanity and were considered gods. We still worship these same gods today while remaining largely ignorant of their extraterrestrial nature. They left behind such monuments as the Giza pyramids and the Great Sphinx. Furthermore, it is because of them we have science, mathematics, medicine, and so on.

As an elaboration on this theory, it has been suggested that these beings were advanced geneticists. With the intention of creating a slave race to serve them, they genetically manipulated humanity. These experiments bore the current "breed" of human being—a savage, petty, and "spiritually blind" creature possessed of only a fraction of its former potential and glory. It was because of these manipulative and meddling "gods" that humanity "fell from grace."

Although Icke has included some of these themes in his own work, his perspective on such matters is unique. Icke is convinced our ancient extraterrestrial visitors were reptilian in form—more specifically, reptilian humanoids. What's more, he believes they're *still* in control of the planet, operating not in the open as they once did, but from the shadows and from within.

According to Icke, when the reptilians genetically manipulated humanity, they did so by introducing some of their own DNA so that we ended up becoming more like them. One particular "gift" we acquired from these reptilian geneticists, he says, is what's known as the reptilian brain—the R-complex.

The reptilian brain lies at the base of the skull, and it's one of the oldest and most primitive parts of the human brain. From an evolutionary point of view, the reptilian brain is a remnant of our reptilian ancestry as we mammals evolved from reptiles roughly two hundred million years ago. And indeed, there is little difference between the reptilian brain of a human being and the entire brain of a modern-day reptile. The reptilian brain is made up of the brain stem (which itself consists of the medulla oblongata, the midbrain, and the pons) and the cerebellum. It is essential to our very survival, controlling such vital processes as swallowing, breathing, respiration, heart rate, digestion, movement, and balance.

Instinctive, reactive, and forever scanning the environment for possible threats, the reptilian part of our brain is engaged whenever we find ourselves in a dangerous situation, even if the danger is only imaginary. The fight-or-flight response is a reptilian brain function. Also, to react with "blind rage" is to succumb to the programming of the reptilian brain, to which the words "rational thought" mean nothing at all. The reptilian brain is rigid, obsessive, compulsive, selfish, greedy, ritualistic, paranoid, territorial, and knows only fear. Those who "live" exclusively in the reptilian part of their brain care nothing for the feelings of others and hold no moral or spiritual values. Rather, the focus of their existence is to obtain wealth, power, sex, and status. Explains one author, "With only this brain, you cannot recognize or even acknowledge the existence of others; everything is either prey or a predator. That's why some parental reptiles don't interact with their offspring, and the offspring don't interact either, for fear of being eaten."[74]

Icke believes the reptilian brain in humans (in its present form) is not natural or normal but a product of reptilian genetic manipulation. "It acts like an enormous microchip and locks us into their control system," he says. "Its primitive, emotional, fear-based sense of reality provides the perfect vehicle for collective control and the conflict and insecurity so essential to divide and rule."[75] If Icke's hypothesis is correct, it would mean the human species is part reptilian; or, put another way, each and every one of us is a human-reptilian hybrid.

However, Icke uses the term "reptilian hybrid" to refer to those among us who have an especially large percentage of reptilian DNA. These people are more reptilian than the general population, and the "hybrid bloodlines" to which they belong date all the way back to the ancient world. This would include members of the Chinese imperial family, members of the British royal family (who belong to the House of Windsor, formerly known as the House of Saxe-Coburg-Gotha), as well as members of the Darwin-Wedgwood-Galton family. Then we have prominent people like George Bush, Dick Cheney, Barack Obama, and Hillary Clinton—all of them, from Icke's perspective, are reptilian hybrids; therefore, they are less human than the rest of us.

According to Icke, the reptilian bloodlines came into being as a result of reptilian genetic engineering, and their purpose on this Earth is a very important one. It is by operating through these bloodlines, he says, that the reptilians control humanity, as they've been doing for thousands of years. For you see, the reptilians are indigenous to the Fourth Density, which exists within a different vibrational frequency range from our world, which is the Third Density. Our Third Density and the reptilian Fourth Density, despite differing slightly in terms of vibrational frequency (density), occupy the same space and hence interpenetrate each other. They are like neighboring radio stations. The reptilians, then, are not extraterrestrials in the true meaning of the term. They are extra-dimensional beings; they originate from outside this dimension. Icke says the reptilians have the technological capability to manifest in our density, but only to a very limited extent. And this is where the reptilian bloodlines come in to the picture. Those who belong to these bloodlines (the reptilian hybrids) serve as "physical vehicles," or mediums, for sinister Fourth Dimensional reptilian entities. Explains Icke:

> Reptilians operating outside of visible light basically "wear" these genetic hybrid holographic computers [the bodies of reptilian hybrids] to infiltrate human society in precisely the same way as portrayed in the movie *Avatar*. The far greater genetic capability means there is a far greater vibrational and frequency compatibility. This allows these human-reptilian hybrids of the Illuminati families to be "possessed," and their mental and emotional processes (actions) to be controlled from another reality.[76]

How strange to think that people like Bill Clinton might actually be reptilian hybrids controlled by reptilian entities from another dimension! The weirdness doesn't end there. Icke further claims the reptilian hybrids interbreed with each other to preserve their special bloodlines and to prevent dilution of their reptilian DNA with the rest of the population. Thus, reptilian hybrids make sure to marry, and have offspring with, other reptilian hybrids.

Such interbreeding is very carefully conducted, says Icke, because the reptilian hybrids want to ensure that their bloodlines remain potent, though not overly potent. After all, reptilian hybrids are shape shifters, meaning they can change between human and reptilian form. Most of the time, of course—and most certainly while in public—they are careful to conceal the reptilian side of their being. This would be difficult to achieve if they had too much reptilian DNA, and they might find themselves assuming reptilian form when they don't intend to. To counter this possibility, says Icke, the reptilian hybrids consume the blood of people, hence the legend of the vampire.

Taking the vampire comparison even further, Icke says the reptilians operating on the Fourth Density "feed" on the negative emotional energy generated by humanity—our fear, hate, suffering, and so forth. In fact, they depend on this energy for their sustenance. We are their food source.

There is little difference between Icke's reptilians and what are known among occultists as "lower astral entities"—so called because they are said to inhabit the lowermost level of the astral realm, or the lower astral. It is believed that, in terms of density, the lower astral is very close to the physical realm. Lower astral entities are little more than demons. Being so near the physical realm, to which they are extremely attached, the energy of the living is available to them. This they soak up, like the ravenous vampires they are, while we, lacking the ability to perceive them (except under special circumstances), remain largely ignorant of their existence. Whereas humans both give and receive energy, these beings are capable of only the latter; they lack the capacity to generate their own energy. Also, no amount of energy they receive is enough. Always they remain unfulfilled.

To recap, Icke says the reptilians are Fourth Density beings that possess the bodies of reptilian hybrids as their primary means of gaining access to this world: the Third Density. Expressed another way, the hybrid bloodlines serve as a bridge between the human Third Density and the reptilian Fourth Density. It's also possible for the reptilians to visit our world using very advanced technology. This is achieved, says Icke, using "interdimensional craft" and "interdimensional 'gateways,'" as well as "through energetic manipulation performed in satanic ritual."[77]

These "gateways," or "portals," can be accessed within secret underground bases on Earth, wherein dwell colonies of reptilians who, unable to cope in our density for long, remain confined to areas "technologically manipulated to vibrate closer to their frequency range."[78] Facilitating access between bases is an extensive network of tunnels and caverns. Icke refers to the Moon as "the center of reptilian operations" on Earth and says their spacecraft are "constantly to-ing and fro-ing between the Moon and underground facilities and bases on Earth."[79] He elaborates:

> The Moon is not just a "physical" phenomenon; it is a technologically generated interdimensional portal that allows Fourth Density reptilians and other entities and energies to enter Third Density reality. They can then travel in spacecraft or be teleported, still cocooned from the Third Density vibrational field, to underground locations within the Earth that give them protection from Third Density consequences for Fourth Density entities.[80]

As well as believing the Moon to be a "technologically generated interdimensional portal," Icke subscribes to the spaceship moon theory, calling it a "gigantic spacecraft." The most significant of his lunar theories by far is that of the Moon Matrix.

Much of Icke's research concerns the nature of reality, and he believes the physical world—the world we experience as "out there" through our five senses—to be an illusion. We live in a "virtual-reality universe," he says, the nature of which is holographic. The universe is, at its most fundamental level, a waveform information construct. It consists of energy in a state of vibration. Our five senses pick up vibrational (waveform) information and decode this into electrical information, or electrical impulses, which are then sent to the brain. These electrical impulses are processed by the brain to construct the "world" we experience as solid and external. Thus, everything we taste, hear, touch, see, and smell is really just an experience in the brain. Icke sums this up by saying that the "'world' only exists as an illusion in our 'heads.'"[81] In a moment I'll explain what this has to do with the Moon.

There is a fascinating but little-known science-fiction film called *They Live* (1988), and it's one Icke frequently refers to in his work, particularly in relation to his Moon Matrix theory. Written and directed by John Carpenter, whose films include *The Thing* (1982) and *Starman* (1984), it fits all the criteria of a B-movie. The film was made on a modest budget, and the acting within it is very substandard though nonetheless charming in its own way. The main character in the film (played by former professional wrestler Roddy Piper) is a drifter without a name, whom others refer to as Nada. The story, set during a period of considerable economic decline in America, begins with Nada wandering the streets of Los Angeles in search of work.

Nada manages to land a job at a construction site, where he befriends fellow worker and battler John Armitage, who invites him to stay at a nearby shantytown. Here people live in makeshift houses and survive on food provided by a soup kitchen. Nada's attention is alerted to a series of suspicious activities in connection with a church across the road. Inside the church, Nada finds chemicals and scientific equipment and, more significantly, a hidden room filled with cardboard boxes. Shortly thereafter both the church and shantytown are brutally raided by police. Homes are bulldozed to the ground, and some of the residents are assaulted by police officers dressed in riot gear.

The following morning Nada enters the church again and secretly removes one of the cardboard boxes. He opens it to discover that it's filled with pairs of sunglasses. Placing one of the pairs in his pocket, he takes a casual stroll down the street on what appears to be a perfectly ordinary day. He decides to try on the sunglasses, only to realize, to his utter amazement, that they aren't normal sunglasses. With the sunglasses on, the world looks completely different. For one thing, what previously looked like billboard advertisements are found to be subliminal commands, such as "consume," "obey," "watch TV," "stay asleep," "conform," and "no independent thought." Looking at a banknote, he sees the words "this is your god." The sunglasses show the world as it truly is.

Especially disturbing is the fact that some of the people he sees—

who tend to be the wealthy and upper-class members of society—have alien-looking faces. Nada soon discovers that they are, indeed, aliens, and they've been living unseen among the human population all along, exploiting the planet and manipulating society. Even the president of the United States is an alien.

When the aliens notice he can see them, Nada becomes a wanted man. He eventually comes into contact with a small group of rebels who know about the aliens and are working to expose them—the same group who developed the special sunglasses. It turns out the aliens are keeping humanity in a state of hypnotic slumber by broadcasting a signal from a TV station that masks peoples' ability to perceive them and the world in its true state. The signal feeds humanity an entirely false reality. The sunglasses work by interfering with, or bypassing, this signal.

Nada gains entrance to a secret base located beneath the city—a major center of alien operations on Earth. To his great shock, the base is occupied not only by aliens but by humans who've "sold out" and are working in collaboration with the aliens. While here, he discovers that the aliens originate from Andromeda, and, furthermore, they possess amazingly advanced technology that allows them to travel, almost instantaneously, from one point in the universe to another. The Earth is one of many planets on which they've established an exploitative presence. The film ends with Nada shooting down the aliens' broadcasting antenna (which is disguised as a satellite dish). Suddenly, with the signal gone, people "wake up." At last they're able to see the aliens living among them.

Icke likens what is happening to humanity to the scenario presented in *They Live* and says the Moon is affecting our minds in a similar fashion to the broadcasting antenna in the movie. He describes the Moon as a "receiver-transmitter and broadcasting system," which has transmissions feeding the collective human mind with a false version of reality.[82] The Moon operates by hacking in to, or intercepting, the waveform information from which we decode reality, he says. From the Moon is transmitted a distorted and suppressed version of reality. This "fake reality broadcast" is what he calls the Moon Matrix. "The reptilians are broadcasting a false reality from the Moon that humans are decoding into what they think is the physical world," Icke explained. Therefore, the reality we're experiencing is not the reality we ought to be experiencing. "We are living in a dream-world within a dream-world—a Matrix within the virtual-reality universe. . . ."[83]

The reptilian brain plays a fundamental role in our being "locked into" the Moon Matrix, says Icke. Allow me to explain this using an analogy of my own. As the name implies, radio-controlled (RC) vehicles operate by means of radio signals. These signals, sent from a handheld transmitter, get picked up by a device located inside the vehicle, called a receiver. When the "go" button is pressed on the transmitter a signal is sent to the receiver instructing the vehicle to drive forward.

For the transmitter to be able to communicate with the receiver, it's essential that both devices operate on the exact same frequency. This is generally around 27 MHz. Some RC vehicles have a pair of removable crystals: one that plugs into the transmitter, the other into the receiver. These come in matching pairs, and they determine the frequency on which the radio system operates. If you had an RC vehicle with a 27.953 MHz crystal in the transmitter and a 29.822 MHz crystal in the receiver, its radio system would be unable to function; the crystals have to be identical in frequency.

The Moon Matrix operates on a certain frequency much like the radio system of an RC vehicle. Icke says that, as part of the reptilian genetic manipulation to which the human species was subjected, we were given the reptilian brain and for a very important reason: it is via the reptilian brain that we are tuned to the Moon Matrix. If we imagine, for example, the Moon Matrix exists on a frequency of 23 MHz, then the reptilian brain is the "23 MHz crystal" that enables us to pick up this specific broadcast.

With his Moon Matrix theory, Icke has taken the notion of a "lunar effect" on humanity to a whole new level—and beyond. In 2012 he expanded the theory to include the gas giant Saturn, the second largest planet in the solar system. He now calls it the "Saturn-Moon Matrix." Orthodox opinion would have us believe that Saturn's rings are composed chiefly of icy debris, from when a moon collided with the planet in the distant past. Icke, on the other hand, argues that its rings are a product of reptilian technology and are made up of crystal (of a type not known on Earth). Saturn, he says, is a "ginormous broadcasting system" that operates in conjunction with the Moon.[84]

In Icke's Saturn-Moon Matrix theory, the role of the Moon has been diminished in importance, with Saturn being "the master control center." No longer is the Moon described as responsible for hacking into the waveform information broadcast from which we decode reality. Rather, it is Saturn that occupies this role. As for the Moon, this "amplifies the Saturn broadcasts and beams them specifically at Earth."[85]

Could any of this actually be true? Are we really trapped in a false reality due to some influence we're receiving from the Moon? Gurdjieff certainly taught as much. It is his teachings regarding the Moon that we're going to take a look at now.

7

FOOD FOR THE MOON

The moon is man's big enemy. We serve the moon. . . .
We are like the moon's sheep, which it cleans, feeds and
shears, and keeps for its own purposes. But when it is
hungry it kills a lot of them. All organic life works for the
moon.

G. I. GURDJIEFF (1866?–1949)

Every era has its share of spiritual teachers, occultists, gurus, mystics, and messiahs, some of them more legitimate than others, all of them controversial and influential. If I had to compose a list of important spiritual figures of the past two hundred years, I'd be sure to include the following: Joseph Smith Jr. (1805–1844), Aleister Crowley (1875–1947), Madame Blavatsky (1831–1891), Rudolf Steiner (1861–1925), Osho (Bhagwan Shree Rajneesh) (1931–1990), Chogyam Trungpa (1939–1987), L. Ron Hubbard (1911–1986), and last but not least the aforementioned George Ivanovich Gurdjieff (1866?–1949).

These remarkable human beings are all deceased, but their teachings, philosophies, and religions live on, influencing thousands, even millions, in the deepest way possible. Although the teachings and ideas of Gurdjieff, particularly those that pertain to the Moon, constitute the focus of this chapter, it's important to place Gurdjieff in the proper context by first taking a brief look at some of the other names listed above.

It's common knowledge that Hubbard was both a science-fiction and fantasy author as well as the founder of the Church of Scientology. The organization, whose roots extend back to the 1950s, currently operates in more than 150 countries and claims to have millions of members worldwide. (Though a more accurate figure would be somewhere in the tens of thousands.)

A comparable organization is The Church of Jesus Christ of Latter-day Saints. Commonly called the Mormon Church, it has a global membership of millions (14.1 million and growing, according to the church itself). Mormonism began in America in 1830 with the publication of The Book of Mormon. According to its "author," Joseph Smith Jr., the text wasn't written by normal means but "translated" under divine guidance from a set of mysterious gold plates.

Then we have Blavatsky's Theosophical Society, which was established in New York City in 1875 and is by no means as popular as it once was. As a spiritual seeker in my early twenties, I attended several "study group" meetings at the Melbourne branch of the organization. I was rather underwhelmed by the experience and so didn't become a member. The meetings, held weekly, were attended by an average of five people, all of them elderly except me, and consisted of reading aloud from Blavatsky's esoteric tome The Secret Doctrine (1888).

Continuing with my zealous "spiritual search," I next became a member of the Rosicrucian Order (AMORC), as founded in New York City in 1915 by Harvey Spencer Lewis (1883–1939). Being a member of AMORC meant participating in their home-study course, by which educational booklets were sent to me in the mail on a frequent basis. They weren't cheap, either; to keep my membership active and continue receiving the booklets, I was required to pay hundreds of dollars a year.

The booklets—or monographs, as they're called—supposedly

contain profound spiritual knowledge dating back to the ancient Egyptian mystery schools. They are classified according to a student's level of attainment, or grade; one starts off as a "neophyte," then progresses to "temple," and so on. In addition to lessons, the booklets feature rituals, ceremonies, and exercises to perform. One preliminary exercise was designed to enable the student to perceive the aura surrounding their hand. Another instructed the student to fill a glass with water, charge it with their aura, then drink the contents immediately. It was claimed that the water would induce a sensation of well-being and revitalize the mind and body. (In my case the exercise was a failure; the water in the glass remained completely unremarkable.)

About three months into my home-study course, I was invited to attend an initiation ceremony of sorts at AMORC's Harmony Lodge in Melbourne. The lodge proved to be a rather unimpressive brick building located in a dingy-looking suburb. As for my fellow Rosicrucians (albeit ones of a higher grade than me), they seemed reasonably pleasant and insisted I eat plenty of cake. I found their excessive chattiness a little grating and sensed an indefinable phoniness about them.

At one point I was approached by a confident and well-spoken gentleman in his fifties. During the course of our conversation (not that I did much of the talking), he boasted that he'd recently been on a spectacular holiday to Peru with his wife. The holiday was expensive, he said, but paying for it wasn't a problem. He went on to explain that by following certain Rosicrucian exercises it's possible to attract wealth into one's life, in addition to other desirable things. He mentioned I too would be able to learn these exercises as soon as I'd reached the appropriate grade.

Prior to the ceremony, one of the lodge's senior members checked to make sure my membership was active. This being the case I was permitted to enter the temple, a small, solid room decorated with various Egyptian symbols and knick-knacks. Everything was arranged in a very elaborate and sophisticated manner, suggesting something "scientific" rather than religious. I can't remember the specific details of the ceremony, except that it involved special breathing exercises, the use of "sacred sounds," and a long period of visualization—all intended to induce an altered state of consciousness. I exited the temple feeling pleasantly relaxed.

Soon after this experience my feelings about AMORC grew increasingly suspicious, and I stopped paying my dues altogether. I am no longer a member of the organization. Being naive and idealistic at that age as well as eager to develop my "psychic potential," I was drawn to AMORC like a pauper to a lottery ticket.

The next esoteric group with which I became involved—for a longer period of time than any of the others—was the Melbourne branch of the Gurdjieff Society of Australia. This brings us, of course, to the main topic of this chapter: Gurdjieff's fascinating and bizarre ideas regarding the Moon. First, we ought to know something about the enigmatic Mr. Gurdjieff himself.

Gurdjieff was a Greek-Armenian mystic, philosopher, and guru who rose to prominence in the first half of the twentieth century. The origin of his unique system, called the "Fourth Way," is not fully known; when asked, Gurdjieff simply referred to it as "esoteric Christianity." It was while living in Moscow, Russia, during the early 1900s that Gurdjieff became a guru—a powerful spiritual teacher with a devoted group of followers.

He later moved to France, where, at a large country estate, he established an esoteric community of sorts called the Institute for the Harmonious Development of Man. During its heyday the institute attracted a considerable number of visitors from the West, among them the famous novelist Katherine Mansfield and the British intellectual Alfred Richard Orage. Gurdjieff taught his students sacred dances and other exercises and directed them to undertake exhausting physical tasks—all done as a means to awaken them, to bring them to a higher state of consciousness. But there was plenty of enjoyment to be had as well. It was not uncommon for Gurdjieff, an expert in the kitchen, to prepare sumptuous feasts for his students.

George Ivanovich Gurdjieff (1866?–1949)

Much like the system he brought into being, Gurdjieff was an austere, "no nonsense" individual for whom the words "tough love" couldn't be more fitting. Possessed of a strong, solid body, piercing eyes, a handlebar mustache, a shaven, egg-shaped head, and a feline manner of movement, his appearance was most unusual. According to one writer, "almost everyone who met Gurdjieff was struck by his powerful personality and commanding presence. His physical attributes, personal magnetism and immense knowledge together created an impression of great strength and mastery."[1] Gurdjieff could be cold and harsh, generous and compassionate, mischievous and eccentric, and often downright baffling and unpredictable. He had a knack for keeping his students on their toes, occasionally taking them on frightening car trips through the French countryside; Gurdjieff was a maniac at the wheel.

Gurdjieff was no doubt an extraordinary man, but he was certainly no saint. Nor was he some kind of "god" as many of his students believed. He drank and smoked, was fond of rude jokes, had a tendency to manipulate and dominate others, and slept with many of his attractive female students (the majority of them already married). "I myself have met a professor at an American University who told me he was one of Gurdjieff's natural sons, and by no means the only one," wrote Colin

Wilson in *The Occult*.[2] (In his autobiography, Wilson refers to Gurdjieff "as by far the greatest teacher of the twentieth century."[3])

On one occasion the English esoteric author and engineer John G. Bennett (1897–1974), who studied under Gurdjieff for many years, brought a large group of students to meet the great man. Prior to the occasion, he gave them some cautionary advice.

> I must warn you that Gurdjieff is far more of an enigma than you can imagine. I am certain that he is deeply good, and that he is working for the good of mankind. But his methods are often incomprehensible. For example, he uses disgusting language, especially to ladies who are likely to be squeamish about such things. He has the reputation of behaving shamelessly over money matters, and with women also. At his table we have to drink spirits, often to the point of drunkenness. People have said that he is a magician, and that he uses his powers for his own ends. . . . I do not believe that the scandalous tales told of Gurdjieff are true, but you must take into account that they may be true and act accordingly.[4]

So murky is Gurdjieff's past that even his exact date of birth is unknown. The *Encyclopaedia Britannica* gives an approximate date of 1877; other sources state January 13, 1866. What is certain, though, is his place of birth: the Russian-Armenian town of Alexandropol (now Gyumri). A garrison town, it helped to defend Russia's border with Turkey, making it a tense and difficult place to live. It was also a place of considerable cultural diversity, with both Eastern and Western characteristics.

Gurdjieff's mother was Armenian and his father, Ioannas, Greek. Along with Gurdjieff, the eldest, they had four other children: one son and three daughters. For many years Ioannas was a successful herdsman and the family was reasonably wealthy. But that all changed when Gurdjieff was seven. The family's herds of cattle were wiped out by a plague, bankruptcy resulted, and they were forced to sell their property. The family moved into a far humbler house, and Ioannas took up business as the owner of a lumberyard and carpenter's shop, which unfortunately never thrived.

Though unwise as a businessman, Ioannas was wise in other areas. He was a skilled and well-known ashokh, or bard, and thus belonged to an ancient oral tradition. As Gurdjieff makes clear in his semi-autobiography, *Meetings with Remarkable Men,* he had much respect for his father and considered him to be spiritually accomplished. (Little mention is given to his mother as the culture in which he was raised was highly patriarchal.)

Ioannas raised Gurdjieff to be very tough indeed, forcing him to get up early in the morning, splash cold water all over his body, and run around in the nude. "And if I tried to resist he would never yield, and although he was very kind and loved me, he would punish me without mercy."[5] Gurdjieff claims that on other occasions his father would place a frog, worm, mouse, or other creature into his bed and also make him handle non-poisonous snakes, so as to rid him of "fastidiousness, repulsion, squeamishness, fear, timidity and so on."[6]

Gurdjieff and his family moved from Alexandropol to the mountain town of Kars in 1878, a year after Alexandropol became part of the Russian empire. Here Gurdjieff attended a Russian municipal school and excelled in his studies. In his spare time he contributed to the family budget by helping his farther in the workshop. Around this time Gurdjieff made the acquaintance of a wise and intelligent priest by the name of Dean Borsh. Father Borsh, a lean, frail man in his seventies, was dean of the Kars Military Cathedral "and the highest spiritual authority for the whole of that region."[7] When he recognized Gurdjieff's giftedness and potential, he took it upon himself to tutor the boy. Gurdjieff, he thought, would make a fine priest and physician.

During his childhood, Gurdjieff developed an interest in the para-normal and the mysterious. This interest was augmented by the tragic death of his sister—an event that caused him to question the possibility of life after death. One evening, while in the company of a group of educated men, a discussion came up about spiritualism and table-turning. Each man had a different opinion as to how the phenomenon was caused. One of the men, an army engineer, attributed the phenomenon to "spirits." Others suggested "magnetism," "the law of attraction," and so on.

Table-turning is an early (and outdated) form of mediumship involving the use of a small, wooden table. Participants place their hands on the table, and the "spirit" with whom they're communicating "taps out" its answers. The number of taps correspond to the various letters of the alphabet: one tap equals "A," two taps equal "B," and so forth.

Gurdjieff and the men decided to try their own table-turning experiment. After twenty minutes the table began to move on its own, tapping out the ages of each person present. Gurdjieff was truly amazed by what he witnessed and quickly began to search for an answer to the mystery. Among those he consulted was Father Borsh, who dismissed it as "nonsense." Later Gurdjieff encountered other incidents of paranormal phenomena and was left feeling equally puzzled and astonished. He was surprised too that none of the intelligent and well-educated adults with whom he discussed such things were able to adequately explain them. Some, like Father Borsh, didn't even try.

One summer, while staying at his uncle's house in Alexandropol, Gurdjieff witnessed something truly extraordinary. Hearing a shriek in the distance, he ran to go help and came across a boy standing in a circle drawn on the ground. The boy, who belonged to a special religious sect called the Yazidi, was clearly upset and for some reason could not escape from the circle. "The child was indeed trying with all his might to leave this magic circle, but he struggled in vain," explains Gurdjieff.[8] A group of boys stood nearby and laughed; they were the ones who had drawn the circle. As soon as Gurdjieff erased part of the circle, the boy was no longer "imprisoned" by it and immediately ran off.

The Yazidis, who live in small, segregated communities throughout

Armenia and the Middle East, are falsely reputed to be "devil worshippers." Always eager to experiment, Gurdjieff ended up testing what he'd heard from others and witnessed himself—if a circle is drawn around a Yazidi he cannot escape from it. He confirmed this as true. He also discovered that although a Yazidi is unable to escape of his own volition (due to some strange force that keeps him inside the circle) it's possible for someone else to drag him out of the circle; however, this cannot be achieved by one man alone. At least two strong men are required. Stranger still, the Yazidi, once forcibly removed from the circle, will fall into a state of catalepsy; when brought back inside the circle, the effect is instantly reversed.

Keen to solve the Yazidi mystery, Gurdjieff sought the opinion of a respected physician by the name of Dr. Inanov and was told the phenomenon is due to "hysteria"—yet another empty explanation. He writes, "The more I realized how difficult it was to find a solution, the more I was gnawed by the worm of curiosity. For several days . . . I thought and thought of one thing only: 'What is true? What is written in books and taught by my teachers, or the facts I am always running up against?'"[9]

In 1883, at the age of only seventeen, Gurdjieff left home and moved to Tiflis, the capital of Russian Georgia. Had he fulfilled his father's wishes he would have studied at the Georgian Theological Seminary, which is located in Tiflis, and continued to become a priest. But Gurdjieff was too unconventionally minded, too adventurous, and probably too rebellious, to spend his days cooped up in a seminary. Instead he took up a job with the Transcaucasian Railway Company, working on and off as a stoker, while at the same time pursuing various spiritual aims. At one point he undertook a pilgrimage on foot to the Armenian holy city of Echmiadzin.

In Tiflis, Gurdjieff met and became close friends with two "remarkable" men: one an Armenian by the name of Sakris Pogossian, the other an Aisor named Abram Yelov. A seminary graduate, Pogossian had decided that the priesthood wasn't his calling and became a locksmith instead; eventually he transitioned to a marine engineer. Yelov was a scholar and bookseller and went on to become an expert in philology. All three men shared a passionate interest in spiritual matters and ancient esoteric knowledge, eventually forming a group called the Seekers of Truth. Members were united by a common aim: to recover fragments of such knowledge and try to piece them together.

The above details are related in Gurdjieff's *Meetings*—a book that, in all honesty, could hardly be called an autobiography. Though parts of it are obviously factual, much of it is allegorical. The difficulty—or impossibility—lies in trying to determine which is which. We have no way of knowing, for example, if Gurdjieff truly did belong to a group called the Seekers of Truth, or if Pogossian and Yelov were real people. The best approach, perhaps, is to take most of Gurdjieff's stories with a large grain of salt.

According to those who follow the Fourth Way, Gurdjieff never intended *Meetings* to be taken in an exclusively literal sense but used it as a vehicle to convey certain spiritual truths. Concerning *Meetings,* the author and philosopher Jacob Needleman wrote, "Although there is no reason to doubt the accuracy of his account, the fact remains that the principal aim of Gurdjieff's writings was not to provide historical information but to serve as a call to awakening and as a continuing source of guidance for the inner search that is the *raison d'être* of his teaching."[10]

In addition to *Meetings,* which was published in 1963, more than a decade after his death, Gurdjieff wrote two other main books: *Beelzebub's Tales to His Grandson* (1950) and the (never completed) *Life Is Real Only Then, When "I Am"* (1974). Taken together they form a trilogy called *All and Everything. Beelzebub's Tales*—the first book in the trilogy and by far the longest—was Gurdjieff's magnum opus. Contained within this weighty tome are some important spiritual insights regarding the Moon, but we'll address these later.

In 1886, during one of his many expeditions in search of age-old esoteric knowledge, Gurdjieff and Pogossian spent days digging among the ruins of the ancient city of Ani, once the capital of Armenia. At one point they came across an underground passage blocked at the end with fallen stones; behind the stones was a long-abandoned monastic cell. Inside the cell they discovered a pile of old parchments written in ancient Armenian. The parchments proved to be letters written by one monk to another monk, and one letter made reference to a certain Sarmoung Brotherhood, "a famous esoteric school which, according to tradition, was founded in Babylon as far back as 2500 BC." Adds Gurdjieff, "This school was said to have possessed great knowledge, containing the key to many secret mysteries."[11]

Having concluded that the school might still exist, "somewhere between Urmia and Kurdistan," Gurdjieff and Pogossian set out on an expedition to find it. During their journey they "had thousands of adventures" and encountered many dangers.[12] On one occasion they were almost torn to pieces by a pack of savage sheepdogs. Also, depending on the region through which they happened to be passing, Gurdjieff and Pogossian found it necessary to don certain disguises, dressing up as Aisors, Turks, Persians, and Transcapian Buddhists. These disguises helped to protect them from religious and ethnic persecution.

Just as Gurdjieff and Pogossian were close to where they thought the monastery would be located, Pogossian received a bite in the calf by a yellow phalanga, a type of tarantula that is extremely venomous. Gurdjieff managed to save Pogossian's life by cutting off a piece of his calf and thus removing most of the venom. But Gurdjieff accidentally removed too much flesh, leaving Pogossian with a huge wound and in desperate need of medical attention. As a result, Gurdjieff and Pogossian were forced to put their plans on hold and head in a different direction.

About two days later they were fortunate to make the acquaintance of an Armenian priest who offered to let them stay at his house so that Pogossian could be nursed back to health. They accepted his offer, staying almost a month. During this time the priest casually mentioned he had an ancient parchment with a map on it in his possession and that he was visited by a wealthy Russian prince once who'd wanted to buy it from him for a huge sum. Because he did not wish to sell the parchment, he let the prince copy it instead for about half the amount.

Filled with curiosity, Gurdjieff asked to see the parchment, to which the priest obliged. When he finally worked out what was drawn on the parchment, he was "seized with violent trembling, which was all the more violent because I was inwardly trying to restrain myself and not show my excitement."[13] Gurdjieff claims that the parchment was nothing less than a map of pre-sand Egypt—the Egypt of about 7,500 BC or earlier, when the region was green and lush rather than the desert it is today.

Gurdjieff desperately wanted a copy of the map, but unlike the Russian prince he didn't have money to burn. And so, with typical slyness, he waited until the priest was absent from the house, took the parchment from his room and, with Pogossian's help, spent all night copying it. Before leaving the priest's house, he returned the parchment to its rightful spot and concealed the copy in the lining of his clothes. Whatever Gurdjieff saw on the map (he didn't say specifically) must have been terribly important because instead of continuing his search for the Sarmoung monastery, he went straight to Egypt's Giza Plateau.

While there, Gurdjieff supposedly worked as a guide by taking tourists to see the Giza pyramids so as "to earn the money I needed to carry out what I had planned." As to what Gurdjieff "had planned," no one quite knows, since he never explained it adequately. All Gurdjieff tells us is that, using the map as an aid, he spent time searching for "an explanation of the Sphinx and of certain other monuments of antiquity."[14]

Some researchers and authors, such as William Patrick Patterson, have suggested Gurdjieff's map of pre-sand Egypt featured the Sphinx, which would further suggest the Sphinx was carved around 7,500 BC or earlier. Certainly this would explain why Gurdjieff experienced such excitement upon seeing the map, why he considered it an object of supreme value, and why he then journeyed directly to the Giza Plateau to investigate the monument.

Gurdjieff claims that while working as a guide in Egypt he

encountered another remarkable man, a wealthy Russian prince. He was, in fact, the exact same prince who'd tried to buy the map from the Armenian priest. Named Prince Yuri Lubovedsky, he also became a Seeker of Truth. Another member was found in Professor Skridlov, a Russian archaeologist. In *Meetings,* Gurdjieff tells us he and the prince went on various expeditions together, journeying to India, Tibet, other parts of Asia, and—most significant of all—to the ancient Egyptian city of Thebes. Here, among the temples and tombs, they were joined by Skridlov, who helped them with their excavations. Gurdjieff mentions the three of them spent three weeks together "in one of the tombs."[15] As to what their excavations involved, Gurdjieff prefers not to say.

Apparently, in 1898, Gurdjieff finally gained access to the secret and mysterious Sarmoung monastery, the appearance of which reminded him of a "fortress." To get there he and a friend named Soloviev (also a Seeker of Truth) were blindfolded and escorted by guides, journeying for twelve days on horseback through the mountains of Turkestan. While staying at this secret and enchanting place, which may or may not have existed, Gurdjieff was free to explore as he pleased. He makes particular mention of seeing priestesses perform sacred dances—the same dances, apparently, that he later taught to his students.

According to James Moore, Gurdjieff's main biographer, the events that occurred during Gurdjieff's "missing twenty years"—1887 to 1907—are completely unverifiable. His visit to the Sarmoung monastery is one such event, as are most of the events described in *Meetings.* Gurdjieff went to great lengths to conceal his past, and there's a simple reason why his exact date of birth will forever remain a mystery: before a trip to America in 1930, Gurdjieff destroyed all of his private papers and documents, including birth certificates and passports. Like a number of other famous occultists, such as Aleister Crowley,* it would appear that Gurdjieff worked as an intelligence agent.

It is believed that, beginning in 1901, Gurdjieff was employed as

---

*The two "magicians" did in fact meet on one occasion, with discordant consequences.

an espionage agent in the service of Tsar Nicholas II and was sent to live in Tibet. While there, he supposedly married a Tibetan woman, with whom he had two children, and spent time studying with the Red Hat Lamas. He also may have worked for Lama Aghwan Dordjieff, a Russian spy and close associate of Thubten Gyatso, the thirteenth Dalai Lama. If true, it would explain why Gurdjieff referred to himself as having been a "tax collector" or "fee collector" for the Dalai Lama. Perhaps he served this role through Dordjieff.

John Bennett must have done a thorough job of investigating his teacher's past, because he allegedly saw an extensive dossier on Gurdjieff's activities as an agent. If, as many believe, Bennett was an agent himself, it's no wonder he was able to access such information. That Gurdjieff had worked as a tsarist agent and participated in "The Great Game"—the conflict and competition between the British Empire and the Russian Empire for control of Central Asia—helps explain why the British government regarded him with suspicion, at one point denying him entry into the country.

It has been suggested that the reason Gurdjieff worked as an agent, if at all, was not for political or patriotic reasons, since Gurdjieff was known to abhor politics, but as a means to pursue his search for ancient esoteric knowledge. Such a job would have given him the opportunity to travel to remote and exotic locations, such as Tibet, to make contact with important people as well as to earn a presumably good living. By all accounts Gurdjieff was an intelligent, resourceful, and cunning man, possessed a wide range of skills and was capable of just about any job.

When it came to earning the money he needed to continue his spiritual quest, Gurdjieff did whatever possible, no matter how unscrupulous. On one occasion, he boasts, he made a large sum selling "American canaries." Only they weren't authentic canaries, but sparrows he'd caught in traps, clipped, and dyed yellow. On another occasion he ran a successful business of repairing and remodeling machinery, toys, and appliances, among other things. He called it his "Traveling Workshop." Gurdjieff isn't ashamed to admit he occasionally tricked wealthy customers out of money, much like a dishonest mechanic who pretends to fix items that do not need to be fixed. Another way he earned money with ease was by selling carpets, oil, and other goods.

Around 1907, Gurdjieff arrived in Tashkent (then under Russian rule), where he is said to have established himself as a "professor-instructor" in "supernatural sciences." It's fair to assume one of these "sciences" was hypnotism—a practice at which Gurdjieff was extremely skilled. A formative, experimental period in Gurdjieff's career as a spiritual teacher gave him the opportunity to study his Russian "guinea pigs," as he occasionally referred to his students. During this time, Gurdjieff apparently acted like a charlatan on purpose. According to Moore, it was also around this time when Gurdjieff began to "assemble" his system, piecing together the fragments of esoteric knowledge that he'd managed to uncover throughout his travels.

The serious phase of Gurdjieff's new career began in 1912, when he moved to Moscow and started recruiting students. Also in 1912, Gurdjieff read a book called *Tertium Organum* by an esoteric philosopher and journalist named P. D. Ouspensky, a resident of St. Petersburg. Ouspensky's book, which is still in print today, deals with topics such as theosophy, cosmic consciousness, mystical states, and the "fourth dimension." Gurdjieff was impressed with the book and recognized its author as an ideal recruit. After all, he didn't wish to accept just anyone as a student, only the bright and dedicated—the crème de la crème of spiritual seekers. And if they happened to be well connected, as Ouspensky was, then so much the better.

While lecturing in Moscow in 1915 about his "search for the miraculous" in India, Ouspensky was approached by two of Gurdjieff's students, who insisted he meet their teacher. They revealed the group to which they belonged was involved in various "occult investigations." Ouspensky was skeptical of such talk and had no desire to meet Gurdjieff. But, so persistent was one of the men, he eventually relented. Their meeting occurred in a small, noisy, backstreet cafe. Ouspensky writes, "I saw a man of oriental type, no longer young, with a black mustache and piercing eyes, who astonished me at first because he seemed to be disguised and completely out of keeping with the place and its atmosphere."[16]

The two men spoke at length about esoteric matters; Gurdjieff answered Ouspensky's questions in a "careful and precise" manner. Ouspensky was interested in and impressed by everything Gurdjieff said. He could tell Gurdjieff possessed knowledge of a very rare and intriguing kind. Later the same day, Gurdjieff took Ouspensky to a large, empty flat to meet some of his students. While Gurdjieff sat on the couch, drinking coffee and smoking, Ouspensky was made to listen to a long story called "Glimpses of Truth," read aloud by one of the students.

The story, which Ouspensky says "was evidently written by a man without any literary experience," touched on some of Gurdjieff's ideas but was, overall, quite vague.[17] Gurdjieff inquired as to whether the story could be published in a paper, so as "to acquaint the public with our ideas," but Ouspensky said it could not; the story was too long for a newspaper to publish.[18] Before leaving, says Ouspensky, "the thought flashed in my mind that I must *at once, without delay,* arrange to meet him [Gurdjieff] again, and that if I did not do so I might lose all connection with him" [his emphasis].[19]

Ouspensky's experience in the flat was immensely odd, and he couldn't help but wonder if Gurdjieff and his pupils had put on a kind of performance or if the intention had been to test him somehow. If a test it had been, Ouspensky must have passed it because he quickly became one of Gurdjieff's most enthusiastic pupils and remained that way until the 1920s, before becoming a Fourth Way teacher himself. "The break between the two men, teacher and pupil, each of whom had received much from the other, has never been satisfactorily explained," wrote the late John Pentland, once president of the Gurdjieff Foundation of New York.[20]

In addition to *Tertium Organum,* Ouspensky penned a number of other highly regarded books, among them *A New Model of the*

Universe (1931), a semi-autobiographical novel called *The Strange Life of Ivan Osokin* (1947), and the posthumously published *In Search of the Miraculous* (1949). The last, a classic, is a detailed account of the time Ouspensky spent as a student of Gurdjieff from about 1914 to 1924. Much of the book consists of talks given by Gurdjieff to his students— talks that, remarkably, Ouspensky managed to commit to memory and years later put down on paper. Ouspensky quotes Gurdjieff at length. But because these are only remembered quotes—not direct quotes—the words belong to Ouspensky.

In *In Search,* Ouspensky describes a conversation he had with Gurdjieff in Moscow not long after they first met. Ouspensky brought up the topic of mechanization in cities. The phenomenon, he said, was getting out of control, and people themselves were "turning into machines." Gurdjieff responded by saying people are machines already. Pointing down the street, he remarked, "Look, all those people you see are simply machines— nothing more."[21] On another occasion he stated, "Man is a machine. All his deeds, actions, words, thoughts, feelings, convictions, opinion, and habits are the result of external influences, external impressions. Out of himself a man cannot produce a single thought, a single action. Everything he says, does, thinks, feels—all this happens."[22]

The idea that people are machines is fundamental to Gurdjieff's system of the Fourth Way. According to Gurdjieff, the vast majority of people on Earth are simply lumps of flesh who, rather than being in control of their lives, are thrust this way and that by events occurring around them. A suitable metaphor would be a ship cast adrift on the ocean without a captain to steer it. Of course we think we're in control of our lives, but this is only an illusion. Gurdjieff further taught that the average person, being but a lump of flesh, ceases to be after death; meaning they have no existence beyond the physical body.

Gurdjieff's philosophy would be a bleak and mechanistic one were it not for the fact that he also believed man can become more than his physical body. Gurdjieff taught that man has the potential to develop three additional bodies: the "astral body," the "mental body," and the "causal body," each of which is composed of progressively finer matter. Generally speaking, we are not born with these bodies; they must be earned, as it were, through rigorous inner work. Ouspensky quotes Gurdjieff as saying, "Only the man who possesses four fully developed bodies can be called a 'man' in the full sense of the word. This man possesses many properties which ordinary man does not possess. *One of these properties is immortality*" [his emphasis].[23] To have all four bodies is to have a "soul."

Gurdjieff called his system the "Fourth Way" to distinguish it from the other three "ways": the way of the fakir, who works on himself by means of the body; the way of the monk, who works on himself by means of his emotions; and the way of the yogi, who works on himself by means of the mind. Each way, said Gurdjieff, has the same aim: to acquire immortality. But because each of the three ways focuses on one area only, each can lead to a person being highly developed in one area and not so developed in the other two.

A fakir, for example, may have considerable will over his body to the extent that he's able to stand in the same awkward position for days, yet he could be emotionally limited and possess little knowledge. He may not even know how he is able to perform such remarkable feats. The fourth way, on the other hand, is a far more balanced approach; it enables a person to work on all three sides of their being evenly and simultaneously. "The fourth way is sometimes called *the way of the sly man*. The sly man knows some secret which the fakir, monk, and yogi do not know. Perhaps he found it in some old books . . . perhaps he stole it from someone," Gurdjieff explains.[24]

It's perhaps no wonder that Gurdjieff's system, with its emphasis that "men are machines," emerged during WWI, a time of unprecedented slaughter, carnage, and destruction. With the Bolshevik revolution a year or two away, Ouspensky asked Gurdjieff the question: "Can war be stopped?" He was expecting to receive a response in the negative, but Gurdjieff replied, "Yes, it can. But the whole thing is: how? War," he continued, "*is the result of planetary influences*" [his emphasis].[25]

According to Gurdjieff, the planets, including the Moon, are living beings that grow and develop. And, as is the case with people, tension can result if two or more planets happen to pass within close proximity to each other. Because all our actions and movements are caused by planetary influences, this "planetary tension" manifests in the form of tension among people on Earth and, consequently, war. Everything we do, especially on a collective level, is governed from outside by forces beyond our control.

Gurdjieff used the term "organic life on Earth" to refer not only to humanity but plants and animals as well—all forms of life. Organic life on Earth, he said, is subject to influences from the planets, from the Moon, and from the Sun as well as from the stars. These various influences act simultaneously, but, depending on the moment, one influence tends to predominate. Gurdjieff made the point that we cannot avoid these influences; we are dependent upon them. What we can do, though, is make a choice between influences—a topic to which we'll return.

Of all the cosmic influences listed by Gurdjieff, he attached particular importance to the Moon. Concerning the topic of war and planetary influences, he told Ouspensky that the Moon "plays a big part in this."[26] On another occasion he said, "the influence of the Moon upon everything living manifests itself in all that happens on the Earth. The Moon is the chief, or rather, the nearest, the immediate, motive force of all that takes place in organic life on Earth."[27]

From another perspective, the author Charles Fort (1874–1932) once wrote: "The Earth is a farm. We are someone else's property."[28] Had Gurdjieff read these words, I'm sure he would have agreed with them. Gurdjieff was opposed to the popular view of Earth as humanity's servant. He did not think it possible for humanity to control or dominate nature. "Humanity, like the rest of organic life, exists on earth for the needs and purposes of the earth," Ouspensky quotes him as saying.[29] The Earth, then, is the boss, and we are its servants. Or, perhaps more appropriately, we are its *inmates,* because in many ways Gurdjieff considered the Earth as a kind of prison. "Humanity as a whole can never escape nature," he said.[30] Gurdjieff further held that everything we experience as a species—whether we happen to be thriving, dying in large numbers, evolving, degenerating, or struggling against nature—happens in conformity with nature's purposes. This would suggest, for example, that the current population of the Earth—huge and seemingly unsustainable though it is—is what nature requires at this moment in time.

And for those who like to believe the human species is currently evolving or will eventually evolve, Gurdjieff offers yet another bunch of sour grapes: "The evolution of large masses of humanity is opposed to nature's purposes."[31] He is careful to point out that this rule only applies to a significant percentage of the population; the evolution of a small percentage may be what nature requires—it all depends. Gurdjieff goes so far as to propose if a large number of people were to evolve, it would be detrimental to the Earth and other planets. This is why "there exist . . . special forces (of a planetary character) which oppose the evolution of large masses of humanity and keep it at the level it ought to be."[32]

Once again, Gurdjieff makes special mention of the Moon:

The evolution of humanity beyond a certain point, or, to speak more correctly, above a certain percentage, would be fatal for the *moon*. The moon at present *feeds* on organic life, on humanity. Humanity is part of organic life; this means that humanity is *food* for the moon. If all men were to become too intelligent they would not want to be eaten by the moon. [his emphasis][33]

Gurdjieff placed the Moon at the very bottom of a diagram he called the "ray of creation." This diagram shares a number of similarities with the Qabalistic Tree of Life and was used by Gurdjieff to represent the universe. Unlike the Tree of Life, it consists of not ten but seven "worlds," each representing a single note (or tone) in the great descending octave of the universe. (I am referring, of course, to the seven-tone musical scale, called the heptatonic.)

The ray of creation begins with the Absolute, which is synonymous with God. According to Gurdjieff, "everything in the universe can be weighed and measured," and so too can the Absolute.[34] Like everything else, it is made up of matter in a state of vibration. The rate at which matter is vibrating determines its density. The higher the rate of vibration, the least dense it will be; the lower, the denser. The Absolute is the least dense of all seven worlds, and each world is denser than the one preceding it. Nothing in the universe is immaterial; there are only varying degrees of materiality.

The Absolute exists in a state of perfect unity. It is a complete trinity, possessing one will and one consciousness. It operates under one law and is designated by the number one. Through a process of division, the Absolute gives rise to the next world in the chain, All Worlds. All Worlds is made up of matter that is slightly denser than the Absolute, and, because it is governed by three laws, it is designated by the number three. At this point in the chain a mechanical element makes its appearance; there is no longer unity but division and no longer one law but three. Anything separate from the Absolute is, by its very nature, mechanical; and the greater this degree of separation, the more mechanical it is.

Next in the chain is All Suns (number six); followed by Sun (number twelve); followed by All Planets (number twenty-four); followed by Earth (number forty-eight); and, lastly, Moon (ninety-six).

Absolute (1)
All Worlds (3)
All Suns (6)
Sun (12)
All Planets (24)
Earth (48)
Moon (96)

The further a world is separated from the will of the Absolute, the greater the number of laws under which it operates and, consequently, the more mechanical it is. As shown above, the world in which we live, Earth, is extremely far from the will of the Absolute and operates under forty-eight laws. The Moon is further still and operates under twice as many laws. Gurdjieff compared our place in the universe to northern Siberia, where life is bleak and difficult; there is little sunshine, and nothing comes easily. The forty-eight laws under which we're forced to struggle are like lead weights attached to our feet, restricting us from gaining freedom; that said, things could be a lot worse. Stated Gurdjieff: "If we lived on the moon . . . our life and activity would be still more mechanical and we should not have the possibilities of escape from mechanicalness that we now have."[35]

Gurdjieff described the ray of creation as a living, growing system like the branch of a tree. He spoke of the Moon as the tip of this branch—the part that is young and still developing. Always eager to challenge scientific orthodoxy, Gurdjieff's Moon is not a cold, dead satellite that was once internally active but, on the contrary, "an unborn planet, one that is, so to speak, being born."[36] The Moon, said Gurdjieff, is gradually warming up and will one day become like the Earth, eventually giving birth to a moon of its own. This new moon will be another link in the ray of creation. At the same time the Earth is also warming up, and may, in time, become like the Sun, the intelligence of which is "divine."

Without pointing out the obvious flaws in Gurdjieff's theory, he further declared the Moon, because it is growing and developing, requires a great deal of sustenance. This idea brings us back to what Gurdjieff stated earlier: the Moon feeds on organic life on Earth. To quote from *In Search:* "Everything living on the Earth, people, animals, plants, is food for the Moon. The Moon is a huge living being feeding upon all that lives and grows on the Earth."[37]

Gurdjieff revealed two distinct ways by which organic life on Earth provides sustenance for the Moon. One relates to the reception and transmission of planetary influences. The other relates to death. Whenever something dies—be it a plant, animal, or person—a certain amount of energy is released as a result. This energy is what kept the living thing "animated." Rather than simply being wasted, it is absorbed by the Moon, thus contributing to the growth of the ray of creation. Gurdjieff makes the point that the universe is entirely economical; energy, having served its purpose on one plane, goes to another. And so it is that the "soul" energy of Earth's organic life is consumed by the hungry, infant Moon.

Continuing with the theme of death, Gurdjieff described the Moon as a place akin to purgatory or hell—a place where "souls" remain trapped under ninety-six laws for "immeasurably long planetary cycles."[38] Gurdjieff explained, "the Moon is 'at the extremity,' at the end of the world; it is the 'outer darkness' of the Christian doctrine 'where there will be weeping and gnashing of teeth.'"[39] Those with familiarity of the New Testament may recognize the following words from Matthew 25:30 (King James Version). The verse in its entirety reads: "And cast ye the unprofitable servant into outer darkness: there shall be weeping and gnashing of teeth."[40]

Gurdjieff believed, as stated earlier, wars and other events involving large groups of people are caused by planetary influences. The fact that we're subjected to planetary influences is due to Earth's lowly position in the ray of creation (specifically, between All Planets and the Moon). Because energy flows from the Absolute to the Moon, and the Earth is an intrinsic part of this system, it is our job, as organic life on Earth, to pass this energy along down the line. The ray of creation could be likened to a cosmic electrical circuit within which we are a tiny but necessary component. We are, to quote from *In Search,* a "transmitting station of forces" between the planets and the Earth.[41] By transmitting and receiving energy, we keep the "circuit" functioning.

We are also "Earth's organ of perception"; and "like a sensitive film which covers the whole of the Earth's globe."[42] Gurdjieff even said different types of organic life belong to the three major categories of vegetable, animal, and human and work with specific types of planetary influences. So that, for example, "a field merely covered with grass takes in planetary influences of a definite kind and transmits them to the Earth," while the "population of Europe" works with another kind, and so on.[43]

It would appear, from what we've discussed so far, that organic life on Earth has three main functions: to take in planetary influences for the benefit of the Earth, to act as the planet's "organ of perception," and to provide nourishment for the Moon (primarily by "completing" the ray of creation). Of these three functions, the first two involve the reception of energy, whereas the last involves transmission. Gurdjieff stated that we not only transmit energy to the Moon but to the Sun and the planets as well.

Remarked Gurdjieff: "Everything that happens on earth creates radiations [energies] of this kind. And many things often *happen* just because certain kinds of radiation are required from a certain place on the Earth's surface" [his emphasis].[44] This seems to imply, rather disturbingly, that unpleasant and destructive events such as wars are made to happen on Earth so that our celestial neighbors don't go hungry. Equally disturbing is Gurdjieff's insistence that the Moon governs our every action, and we are entirely under its spell and domination. From *In Search:*

All movements, actions, and manifestations of people, animals, and plants depend upon the Moon and are controlled by the Moon. The sensitive film of organic life which covers the earthly globe is entirely dependent upon the influence of the huge electromagnet that is sucking out its vitality. Man, like every other living being, cannot, in the ordinary conditions of life, tear himself free from the Moon. All his movements and consequently all his actions are controlled by the Moon. If he kills another man, the Moon does it; if he sacrifices himself for others, the Moon does that also. All evil deeds, all crimes, all self-sacrificing actions, all heroic exploits, as well as all the actions of ordinary everyday life, are controlled by the Moon.[45]

Here Gurdjieff is saying the Moon controls our movements and, in turn, our actions. Oddly, no mention is made of the Moon controlling our thoughts or emotions. So why would Gurdjieff liken the Moon's influence to that of a huge electromagnet? An electromagnet is a type of magnet powered by means of an electric current, which is distinct from a permanent magnet as it does not require an electric current. Perhaps by using the term "electromagnet" rather than simply "magnet" Gurdjieff was trying to convey the idea that in order to control us the Moon needs energy to do so—the very energy it steals from us by "sucking out our vitality." Was Gurdjieff implying that the Moon generates no energy of its own? That it is entirely vampiric?

Gurdjieff spoke of our relationship with the Moon as one of mutual dependency. Without its influence—its powerful magnetic pull—organic life on Earth would perish. It would "crumble to nothing." If the Moon's influence were to suddenly cease, our ability to move would presumably cease too. Like puppets without a puppet master, our strings would slacken and we'd slump to the ground and die. Gurdjieff and his students compared the Moon and organic life on Earth to an old-fashioned clock and the weight that keeps it in motion. This type of clock is driven by the gravitational pull of a weight (rather than a spring or electric motor). Take the weight away and the mechanism of the clock will stop moving. The Moon is the weight, and organic life on Earth is the mechanism of the clock.

So we know, according to Gurdjieff, the Moon's influence is mechanical in nature and is restricted to the mechanical realm; and, further, the only reason the Moon can control us is because we ourselves are mechanical beings—"machines." To quote from *In Search:* "If we develop in ourselves consciousness and will, and subject our mechanical life and all our mechanical manifestations to them, we shall escape from the power of the moon."[46] To cease being a machine is to be liberated from the Moon; is to no longer be useful to the Moon; is to no longer be food for the Moon; is to no longer contribute to the growth of the ray of creation. No wonder Gurdjieff said, "the way of development of hidden possibilities is a way *against nature, against God*" [his emphasis].[47]

Mentioned earlier was Gurdjieff's idea that although we're dependent on influences from the Moon, the Sun, and the planets—but most especially the Moon—we can make a choice between influences. To be able to make such a choice, though, we must first become less mechanical and more conscious. In one of his lectures, Ouspensky stated, "So development means passing from one kind of influence to another kind of influence. At present, we are more particularly under the influence of the Moon. We can come under the influence of planets, sun, and other influences, if we develop."[48] Ouspensky adds it's possible to come under influences higher even than the Sun. Another means of transcending the influence of the Moon is by creating a second Moon within ourselves. Ouspensky referred to this second Moon as "a permanent center of gravity in our physical life."[49] Someone who possesses such a Moon, he said, is no longer dependent on the physical, external Moon.

Gurdjieff spoke of the Moon as being primarily responsible for humanity's un-evolved, or "fallen," condition. In *In Search,* Gurdjieff alludes to certain "forces"—at one point identifying these forces as "planetary" in nature—which both oppose the evolution of humanity and keep humanity in a state of hypnotic sleep. Careful reading of the book reveals one of the "forces" he refers to is, if not the Moon itself, closely linked to the Moon. On account of these forces, "man is hypnotized and this hypnotic state is continually maintained and strengthened in him."[50]

Gurdjieff taught that we spend almost every second of our so-called waking life in a state of hypnotic sleep. And by this he meant we are literally asleep. Gurdjieff said that hypnotic sleep—distinct from "normal," natural sleep—is unnatural, abnormal, and of no benefit to our well-being. The most obvious symptom of our being asleep, he added, is that we fail to "remember ourselves." Most of us are only familiar with two states of consciousness: sleep and the normal waking state.

But according to Gurdjieff there are—in addition to these—two higher states of consciousness: self-remembering (also called self-consciousness) and objective consciousness.

Ordinarily, of course, our attention is entirely consumed by whatever we happen to be observing at the time, whether it's something within ourselves or something outside of ourselves. If I see a shiny red Ferrari cruise down the street, for example, it immediately captures my attention and for a moment I forget about everything else. A second later, something else captures my attention, and so on. This is a state of passive, one-way attention. Self-remembering, on the other hand, is a state of active, divided attention. When we remember ourselves, our attention is not only directed toward the observed phenomenon—the red Ferrari—but also toward ourselves. Our attention goes both ways.

We assume we possess self-consciousness and always remember ourselves, but in fact we do not. States of self-consciousness (if we're lucky to experience them at all) are rare and fleeting experiences in our lives brought about by accident. They can occur, for example, when we head off on a holiday and are overcome by a feeling of excitement and optimism. (Drugs too can bring them about.) The only way we can achieve such a state on any kind of consistent basis is by sheer effort alone. As for objective consciousness, this is known as enlightenment. But to describe it in words is impossible.

The life of an average human being is dominated by sleep—the sleep that occurs in bed at night and the sleep we mistake for wakefulness. During the sleep we mistake for wakefulness, we go about our days failing to remember ourselves, unable to control our thoughts, emotions, or attention, all the while perceiving the world not how it truly is but how we *imagine* it to be, and all the while imagining ourselves to be awake. Gurdjieff once remarked, "Man's possibilities are very great. You cannot conceive even a shadow of what man is possible of attaining. But nothing can be attained in sleep. In the consciousness of a sleeping man his illusions, his 'dreams' are mixed with reality. He lives in a subjective world and he can never escape from it."[51]

So why would certain forces want to keep humanity asleep and hyp-notized? Rather than answering this question directly, Gurdjieff chose to do so in the form of an Eastern tale. The story concerns a rich, mean magician who owned many sheep, which he used for their skin and flesh. To avoid their fate, the sheep kept escaping, wandering off into the forest and getting into all sorts of mischief. So low and miserly was the magician that instead of hiring shepherds to take care of his sheep or building a fence around their pasture he decided to hypnotize them.

By hypnotizing his sheep, the magician made them believe the following: he was a kind master who loved them dearly; they weren't going to be harmed and had no reason to be concerned; and, lastly, they weren't actually sheep. He made some of them believe they were lions, eagles, men, or magicians. With his sheep under a hypnotic spell, the magician had no problems with them. They ceased trying to escape and in fact remained calm and passive, waiting around happily until the day their flesh and skins were required. "This tale is a very good illustration of man's position," said Gurdjieff.[52]

Gurdjieff associated the Moon with the force known as Kundalini. Most of us would know of this term in relation to tantric yoga. Kundalini is said to be a form of cosmic sexual energy, pictured as a coiled ser-pent lying at the base of the spine, which can be awakened and, once awakened, lead to bliss and enlightenment; however, Gurdjieff argued that this definition of the term is entirely incorrect. He said Kundalini is far from something positive and beneficial; it actually helps to keep humanity in a state of hypnotic sleep. According to Gurdjieff's defini-tion of the term, Kundalini is "the power of imagination." It is "a force put into men to keep them in their present state. . . . A sheep which considers itself a lion or magician lives under the power of Kundalini."[53]

The way Gurdjieff describes Kundalini—as a force that has been "put into men"—doesn't seem to make sense. Would it not be more accurate to say this force "acts upon" men? Is this force meant to be internal or external? The answer is quite surprising.

In a talk given in New York in 1924, Gurdjieff boldly stated, "the

Moon is man's big enemy. We serve the Moon." He then went on to mention something called "Kundabuffer." He explains, "Kundabuffer is the moon's representative on earth. We are like the moon's sheep, which it cleans, feeds and shears, and keeps for its own purposes. But when it is hungry it kills a lot of them. All organic life works for the Moon."[54]

So what might you ask is Kundabuffer? The term, coined by Gurdjieff himself, is a combination of the words "Kundalini" and "buffer." We already know what is meant by Kundalini. As for *buffer*, this is something "designed to lessen or absorb the shock of an impact." They are used, for example, between railway carriages so that no shock or damage occurs when the carriages make contact.

Gurdjieff taught that buffers of a sort also exist within human beings. We create them ourselves, and their purpose is to keep us calm and asleep by preventing us from feeling our inner contradictions. "Buffers help a man not to feel his conscience," Gurdjieff once remarked.[55] By conscience Gurdjieff meant a state in which a person feels at once everything about themselves, no matter how unpleasant or contradictory. Buffers must be destroyed if a person wishes to awaken and grow in conscience, said Gurdjieff.

To find out what Gurdjieff meant by Kundabuffer it is necessary to consult *Beelzebub's Tales*—easily the strangest, most deliberately obscure book ever written. Gurdjieff began writing the book shortly after his mysterious car accident in 1924, which left him battered and bruised from head to foot and in a comalike state for up to six days. (Gurdjieff, as I mentioned earlier, was by no means a timid driver.) Considering the seriousness of his injuries, it remains a mystery as to how Gurdjieff managed to pull himself from the wreckage and position himself on the nearby ground with his head resting on a seat cushion taken from the car and a blanket draped over his body.

On the morning prior to the accident Gurdjieff made sure his Citroen was thoroughly inspected by a mechanic, "especially the steering wheel." What makes this detail significant is that the accident was caused in part (more accurately, worsened) by the steering wheel detaching. Had Gurdjieff's assistant, Olga de Hartmann, accompanied him on the drive as per the original plan, it's likely she would have been injured as well. But Gurdjieff instructed her to take the train. All of this seems to indicate Gurdjieff somehow had foreknowledge of the accident.*

There is reason to believe Gurdjieff may have been altered by the accident, that it rendered him more eccentric than he was beforehand—perhaps even mildly insane. If true, it would explain why *Beelzebub's Tales* was described by one reviewer as "the nonsensical ramblings of a deranged guru." Others consider the book to be comparable in profundity to the bible. At over one thousand pages in length, and replete with invented words that defy pronunciation, such as "Heptaparaparshinokh" and "Almznoshinoo," completing the book is no easy task, as I know from experience. The first sentence of chapter 1, "The Arousing of Thought," reads:

Among other convictions formed in my common presence during my responsible, peculiarly composed life, there is one such also—an indubitable conviction—that always and everywhere on the earth, among people of every degree of development of understanding and of every form of manifestation of the factors which engender in their individuality all kinds of ideals, there is acquired the tendency, when beginning anything new, unfailingly to pronounce aloud or, if not aloud, at least mentally, that definite utterance understandable to every even quite illiterate person, which in different epochs has been formulated variously and in our day is formulated in the following words: "In the name of the Father and of the Son and in the name of the Holy Ghost, Amen."[56]

---

*The possibility that Gurdjieff possessed psychic abilities is intriguing and difficult to ignore. Many of his students were convinced that he did, and with good reason. Ouspensky, by no means a credulous man, relates an incident in *In Search* where, while in a heightened state of consciousness, he and Gurdjieff conversed on an apparently telepathic level.

Beelzebub's Tales is an allegorical work loosely structured around a science-fiction story. The story's protagonist is Beelzebub—the devil. Unlike the traditional Judeo-Christian devil, Gurdjieff's Beelzebub is a redeemed being who has, as it were, paid for his sins and attained a level of high spiritual development. Some interpret the book as an autobiography of sorts, suggesting Gurdjieff is Beelzebub himself.

At the beginning of the book we find Beelzebub, accompanied by his grandson Hussein and his personal assistant Ahoon, aboard the spaceship *Karnak*. The ship is on a course from Beelzebub's home planet, Karatas, to the planet Revozvradendr, located in another solar system. During the long journey, Beelzebub uses the time to teach young Hussein about the troublesome "three-brained beings" (humans) who dwell on the planet Earth and whom he studied and lived among.

The book consists of long, dry talks by Beelzebub on topics of benefit to Hussein's education, with a focus on the various troubles of humanity. Some of the topics include the history of the Earth and the "fall" of humanity as well as hypnosis, objective art, war, sexual hygiene, and polygamy. I even recall mention of a famous German scientist who devised a way to magically create "delicious chicken soup". . . or something along those lines. (Parts of the book are so bizarre and seemingly ridiculous that the reader is left completely perplexed.)

Gurdjieff wrote *Beelzebub's Tales* as an attempt to preserve his esoteric knowledge in a form that would remain undistorted over time—hence the unique, coded style of his writing. Much of it was dictated by Gurdjieff, with some of his students assisting as scribes. As to why the book is so difficult to read and understand, apparently Gurdjieff did this on purpose in the hope it would challenge and awaken the reader. The book is long and complicated to ensure that those who could be bothered reading it have a sincere interest in Gurdjieff's ideas. "I bury the bone so deep that the dogs have to scratch for it," Gurdjieff once said about the book.[57]

An entire chapter of *Beelzebub's Tales* (albeit a very brief one) is devoted to the genesis of the Moon. Amazingly, the events described by Gurdjieff are extremely similar to those posited by the Giant Impact hypothesis—a theory that didn't emerge until the 1970s, two decades after Gurdjieff's death. Gurdjieff's explanation for the origin of the Moon is almost identical to today's most widely accepted theory of lunar origin, which is miraculous, to say the least. While the Giant Impact hypothesis has a Mars-size body responsible for the giant impact in question, Gurdjieff blames the comet "Kondoor." Rather, he blames "certain sacred Individuals concerned with the matters of World-creation and World-maintenance," who, due to a calculation error, let the paths of the two bodies cross.[58] The creation of the Moon, then, was a huge accident. It was never meant to happen.

As related in *Beelzebub's Tales,* the collision between Earth and Kondoor was so huge in magnitude that two large fragments broke off the Earth. The larger fragment, Loonderperzo, became what we know as the Moon. The smaller fragment, Anulios, also exists to this day, yet we fail to notice it. This is due in part to the small size of Anulios, the remoteness of its orbit, and the fact that contemporary humanity is accustomed to perceiving "only unreality." Apparently the inhabitants of the former civilization of Atlantis knew of the existence of Anulios and referred to it by this name.

At the time of the collision, the Earth was still a young, unpopulated planet that hadn't yet developed an atmosphere. Due to the formation of the Moon and Anulios, it was decided by various archangels and other sacred beings that the only way to maintain these two fragments—and prevent them from leaving their positions and causing "irreparable calamities" in the future—was to establish life on Earth. Thus, the human species came into being. But little did humans know that the purpose of their existence was to transmit "sacred vibrations" (also called "askokin") to the Moon and Anulios as a means of nourishment (to produce "food for the Moon").

Before long, members of the "Most High Commission" (a committee of sacred beings) began to have concerns about the long-term maintenance of the Moon and Anulios. They were worried humans might evolve to the point where they attained "objective reason" and thereby ascertain the miserable purpose of their existence. Should this happen, the commission concluded, the human race would lose its drive to continue living. Not only that, but it would destroy itself, and the Moon and Anulios would cease receiving the sustenance they required.

The commission soon came up with a solution to the problem: they would alter humankind through the implantation of a "special organ" called Kundabuffer. Presumably by means of genetic alteration, this organ was made to grow within human beings. The organ was located at the base of the spinal column. (Or, put another way, at the root of the tail, which humans then possessed.) The properties of this organ were such that it caused humans to perceive reality in a "topsy-turvy" fashion. It also made them more attached to sensations of "pleasure and enjoyment."

Put simply, the organ Kundabuffer turned humans into pleasure-seeking "animals" with a limited and distorted perception of reality. While the specific term "hypnotic sleep" does not appear to have been used by Gurdjieff in relation to the organ Kundabuffer, it can only be surmised that the organ did precisely that: placed humanity under a hypnotic spell akin to a form of sleep. The organ, having fulfilled its intended purpose, was eventually removed from humanity. But because its properties became so crystallized—so fundamental to our nature— they remained. Through a process of heredity, the traits were passed from generation to generation. These undesirable properties are still within us today.

With this fable Gurdjieff has provided a compelling explanation for the "fall of man." The reason why, in Gurdjieff's words, "'men' are not men" was due to a series of unfortunate events in the past, most of them simply accidents on a very big scale. If the comet Kondoor had never struck Earth, resulting in the genesis of the Moon and Anulios, the human race would be in a far better position than it is at present. Our evolution would never have been hijacked.

But if organic life was created on Earth to provide nourishment for the Moon, don't we owe our very existence to the fact that Kondoor struck Earth and gave birth to the Moon? In other words, isn't it contradictory to state that the human race would be better off had these events never occurred? Gurdjieff implies in *Beelzebub's Tales* that the establishment of organic life on Earth (human beings included) was part of the plan all along. But because of the unexpected and accidental genesis of the Moon, it became necessary to put this plan into action sooner. In which case, the birth of the Moon had a disruptive effect on what was supposed to have been a steady and orderly process.

Why would Gurdjieff describe the organ Kundabuffer as "the Moon's representative on Earth"? The answer to this question is simple when you consider the intended function of this organ: to render humans blind to the fact that the purpose of their existence is to provide food for the Moon. As the Moon's representative on Earth, the Kundabuffer *acts on behalf* of the Moon. Gurdjieff couldn't have chosen a more accurate word to define the relationship between the two.

We are told in *Beelzebub's Tales* that the vast majority of planets in our solar system are populated by "three-brained" (humanlike) beings of various physical forms. The beings who inhabit Mars are described as possessing long broad trunks, "enormous protruding and shining eyes," small, clawed feet, and large wings. As a result of their impressive eyes, they have the ability to "see freely everywhere."[59] They can also fly long distances, even beyond the atmosphere of their planet. As for the beings that dwell on Saturn, these are described as birdlike, with an appearance similar to that of a falcon.

Apparently even the Moon is not devoid of intelligent life. The inhabitants of the Moon, we are told, have "very frail" bodies and look like large ants. But despite their frailty, they have a "very strong spirit" and "an extraordinary perseverance and capacity for work." Forever busy, they managed to "tunnel" the entire Moon within a short period of time. On account of the Moon's abnormal climate of extreme hot and cold, these beings dwell both within and on the surface of their world. But there are also periods of fine weather on the Moon, during which the "whole planet is in blossom," and the Moon beings have more food than is needed.[60]

Gurdjieff, being the witty teacher that he was, probably wrote these words with his tongue in his cheek. But might they still contain a grain of truth? Perhaps his Moon beings, with their "very frail" bodies, are not of flesh and blood. Could they, instead, be nonphysical entities existing in a "realm" beyond ordinary human perception? Do these entities feast on the energy of the living and the souls of the dead? Does this account for the Moon's insatiable appetite? Another question worth asking: From where did Gurdjieff acquire the idea of a hollow, tunneled Moon populated by antlike entities? Either Gurdjieff was very familiar with H. G. Wells' *The First Men in the Moon,* which was published long before *Beelzebub's Tales,* or he came up with the idea independently.

It's difficult to know what to make of certain of Gurdjieff's lunar ideas. When he stated the Earth has two moons, was he speaking factually, metaphorically, or on another level entirely? The most obvious interpretation is that Gurdjieff was referring both to the physical, external Moon and to the smaller, metaphorical moon we ought to develop within ourselves. According to Richard Myers, in an article titled *Gurdjieff, the Moon and Organic Life,* "Gurdjieff maintained that all of his ideas could be taken in seven different ways, one of which is factual."[61]

At least one of Gurdjieff's lunar ideas has an obvious factual element, having been confirmed scientifically: the Moon draws energy from the Earth. As was discussed in an earlier chapter, the gravitational attraction of the Moon affects the Earth's ocean water in such a way it takes on the shape of an ellipsoid. The two "ends" of this ellipsoid correspond to tidal bulges. As the Moon orbits the Earth, one of these tidal bulges remains closest to the Moon, not quite facing it but slightly ahead of it, while the other faces away. The pull of the Moon on this nearest tidal bulge slows down the rotation of the Earth. But since gravity works both ways, this tidal bulge also pulls on the Moon, bringing it ahead in its orbit and causing it to speed up.

This means that the rotation of the Earth is gradually slowing down while the rotation of the Moon is gradually speeding up, consequently becoming more distant. Energy, if you like, is slowly being transferred from the Earth to the Moon. It would not be inaccurate to say, then, that the Moon does indeed draw energy from the Earth. Less likely, though, is Gurdjieff's idea of the Moon using this energy for its own growth and development. According to the orthodox scientific view, the Moon and the Earth are around the same age (4.6 billion years). It is no celestial spring chicken. Nor does it appear to be warming up or forming an atmosphere. Any activity on the Moon is thought to have ceased long ago.

From a scientific point of view it is difficult to say if the Moon is dependent on organic life on Earth, yet the opposite cannot be disputed. Gurdjieff was correct to some extent when he said organic life on Earth would "crumble to nothing" were it not for the influence of the Moon. As we already know, the gravitational pull of the Moon has a stabilizing, or anchoring, effect on the Earth, keeping the obliquity of the planet at 23.5 degrees. Because of this, the climate here is relatively stable, and life thrives to the extent that it does. An Earth without a Moon would be an extremely hostile place for many forms of life.

How interesting, then, Gurdjieff should compare the Moon to a giant electromagnet as well as to a weight on an old-fashioned clock in relation to its influence on Earth's organic life. Both instruments are supportive in nature—and so too is the Moon; it supports organic life. The Moon is like a pair of crutches providing support to a man with an injured leg. The crutches prevent the man from falling over while also enabling him to walk, but it is he who must still do the walking.

And what if the man was required to use the crutches for a long period of time? It's possible he'd become so accustomed to them that an unhealthy dependency of sorts would develop. He might even find he can no longer walk normally without them. What Gurdjieff said about the Moon is very similar: the fact we are dependent on it—like a pair of crutches—is the very thing that makes us weak and inhibits our evolution. The only way to evolve, then, is to learn how to walk on our own legs. In other words, by reducing our dependency on the Moon and eventually transcending it altogether. Certainly human beings would have evolved very differently if the Moon had never come into being. I wonder, would we be stronger as a species in terms of both mind and body?

What are we to make of Gurdjieff's idea that the Earth is orbited by not one but two natural satellites? A small amount of scientific evidence suggests something similar to "Anulios" may in fact exist, or did in the past. According to an article that appeared on the SPACE.com website in 2011, a pair of researchers from the University of California (Santa Cruz), Martin Jutzi and Erik Asphaug, proposed a new theory to account for the dramatic geological differences between the near and far sides of the Moon.

Their theory adds an additional chapter to the Giant Impact hypothesis by suggesting that the collision in question not only resulted in the formation of the Moon but also in a number of smaller moonlike bodies, one of which survived for tens to hundreds of millions of years. This hypothetical mini-moon had a diameter of around 750 miles and was suspended between the Moon and the Earth in a position of balanced gravitational attraction. As a result of becoming unstable, it collided with the Moon at a very slow speed, producing the highlands that dominate the far side. "According to our simulations, a large 'moon-to-Moon' size ratio and a subsonic impact velocity lead to an accretion pile rather than a crater," concluded Jutzi and Asphaug.[62]

There is an asteroid named 3753 Cruithne, discovered in 1986, that shares Earth's orbit around the Sun and is incorrectly referred to by some as "Earth's second Moon." Only 3.1 miles in diameter, it moves around our planet in a horseshoe configuration every 770 years. The object is in fact a quasi-satellite, of which there are about five in our solar system. Such objects do not orbit the Earth directly. Cruithne could be Gurdjieff's Anulios, though I very much doubt it.

Gurdjieff's idea of the Earth having two moons—both of which originated during the same huge collision—has a great deal of merit to it. But to explore it in any depth is beyond the scope of this book. I should add too that the idea has been around since at least 1870, when it appeared in Jules Verne's famous novel *Round the Moon,* and did not originate with Gurdjieff. Perhaps "Anulios" will be discovered one day. Or perhaps it will never be discovered—not because it doesn't exist, but because, as Gurdjieff stated, we've lost the ability to perceive it.

Gurdjieff's "food for the Moon" is one of his more ambiguous and unusual ideas and for that reason has attracted considerable speculation and ridicule. Many consider the idea absurd. Even many Fourth Way students and authorities, when writing about Gurdjieff and his system, appear not to take the idea seriously and often avoid all mention of it. For example, in John Shirley's *Gurdjieff,* a 320-page biography, the idea is barely acknowledged. I say barely, because the idea is discussed very briefly in the penultimate appendix in the book.

> Some people are stopped dead in their reading of Gurdjieffian "Work books" when they come upon the idea . . . about humanity being "food for the moon". . . . My personal—I emphasis *personal*—interpretation of this teaching is that the moon in this teaching represents the inert, more-or-less lifeless part of the cosmos. It stands for all of the lower cosmos. The Moon is the first stop in a kind of "energy sink" where our life energies will go if we don't work on ourselves.[63]

In addition to being a musician of some note, Shirley is a successful American novelist and screenwriter specializing in the genres of science-fiction/fantasy and horror. He is also a student of the Fourth Way. It comes as no surprise that many of his novels, such as *Demons* and *The Other End,* explore Gurdjieffian themes. I got in touch with Shirley in 2011 to ask for his thoughts on Gurdjieff's "food for the Moon," as well as to ask why, in his biography on Gurdjieff, he mentions the idea only briefly—and, it would seem, only out of necessity.

I sent an e-mail asking this question on April 19 and received a reply from Shirley the next day:

> Regarding "food for the moon," no I'm not trying to soft soap or gloss over Gurdjieff's scariest ideas. . . . I really do think that if there's anything to the moon story, in his ideas, it's what I described in my appendix. I am not a Gurdjieffian literalist (I mean, his biggest book, *Beelzebub's Tales,* is a novel after all), and I feel that some of his ideas are more clearly supportable than others. Some are, to me, along the lines of allegory. . . . He [Gurdjieff] said believe what you can verify—and I agree with that. I believe in the scientific method. . . . I do not believe in the supernatural (as we've always understood the term), but only in an unknown level of the natural.[64]

In a later e-mail (April 22), Shirley added:

> I don't know if Gurdjieff thought of the moon as "stealing" energy from the Earth; it was more like harvesting, and part of a big "food chain" of sorts, a cosmological ecology, and just more of the natural order. *If* Gurdjieff was right about it, it's sort of the energy that arises with any degree of consciousness, anything that isn't "bottled" (my term) in a soul created by spiritual work. . . . He [Gurdjieff] didn't teach that stuff ["food for the Moon"], as I mentioned before, after the 1920s.[65]

It was generous of Shirley to have answered my questions. Few authors with whom I've communicated are as gracious as he. Still, I couldn't help but detect in his response—and in his Gurdjieff biography—an unwillingness to address Gurdjieff's lunar ideas in a deeper, more direct manner. Something seemed to be holding him back, but what? And why did he find it necessary to insist Gurdjieff ceased teaching "food for the Moon" after the 1920s? Was this meant to imply that the idea was one of Gurdjieff's least valuable?

Although the specific phrase "food for the Moon" doesn't appear in *Beelzebub's Tales* (published in 1950), the idea is still very much present in the book. So it's not as though Gurdjieff stopped teaching the idea; rather, he stopped teaching the idea in such an explicit form. This is why, in *Beelzebub's Tales,* we find such coded sentences as: "the . . . Moon is just a part of this Earth and the latter must now constantly maintain the Moon's existence."[66] It's obvious that Gurdjieff considered the Moon to have profound spiritual importance, and what he taught regarding the Moon is a pivotal aspect of his system. Why, then, is the subject not discussed more openly by those who follow the Fourth Way?

I have reason to believe the lunar component of Gurdjieff's system is so esoteric that only senior Fourth Way members understand its true meaning. I further believe this knowledge is purposely withheld from the general public for fear that it might be distorted and misinterpreted. As with many occult organizations, the Fourth Way is semi-secret; its members share some, but not all, of the knowledge passed down by Gurdjieff. If you wanted to gain "complete access" to Gurdjieff's system then you would need to belong to a genuine Fourth Way group (connected to the Gurdjieff Foundation). Having once belonged to such a group, I know for a fact that much of their knowledge—especially as regards the practical side of the system—is not made available to nonmembers.

In May of 2011, I wrote to William Patrick Patterson asking if he'd be willing to answer some of my questions in relation to Gurdjieff's "food for the Moon" idea. Patterson is the founder and director of the Gurdjieff Studies Program, which is based in California, and has written numerous books on Gurdjieff. He was a student of John Pentland, who was a student of Gurdjieff. Now a Fourth Way teacher himself, Patterson is very much an authority on the system.

In a letter I received from Patterson (dated July 4, 2011), he cryptically wrote:

In these eclectic piglet times, rife with chakra-talk and ego-tized chest thumping and shakti-antics, whatever experiences beyond the mundane so quickly psychologized and leveled to whatever personalized identity is found comforting and self-calming . . . well, y [sic] friend, one has to be careful in opening esoteric doors for general consumption. . . . There is reluctance then to speak about Mr. Gurdjieff's placing of the two Moons, the laws that govern, and how one must make a Moon to have a Soul; all this being given in great though somewhat hidden detail in his magnum opus [Beelzebub's Tales].[67]

Patterson's reluctance to discuss these ideas came as a surprise to me, because his books display an openness about Gurdjieff's system that is rare among Fourth Way authorities. His letter confirms my opinion that the lunar component of Gurdjieff's system is highly esoteric and cannot be disclosed in full to the general public. Patterson was kind enough to point me in the right direction, by revealing that Beelzebub's Tales is a thorough (perhaps even complete) source of information on the topic. This is valuable to know, though one significant problem remains: the book is so abstruse that one could spend an entire lifetime trying to make sense of it. Gurdjieff "buried the bone" very deep indeed—some might say too deep.

Gurdjieff seems to have hinted at some kind of lunar intelligence, a sinister, manipulative, and exploitative "force" that works to keep humanity in a state of sleep. We will now attempt to ascertain the nature of this force.

8

LIFE ON THE MOON

The astronomical accomplishments of Sir William Herschel, to whom the reader has already been introduced, are well recognized by scientists and historians. Born in Hanover, Germany, in 1738, Herschel immigrated to England at the age of nineteen. Originally a composer of music, he shifted his interest to astronomy in his mid-thirties. He constructed his own reflecting telescopes from mirrors he had ground and polished himself. His discovery of Uranus in 1781 brought him instant fame. A year later he was appointed as an astronomer to King George III. Along with Uranus, he discovered two of its major moons: Titania and Oberon. He was particularly intrigued by Saturn; in fact, the discovery of Mimas was his. He also discovered Enceladus, another of Saturn's moons.

Herschel's discoveries weren't limited to astronomy but also included physics. In an experiment he conducted in 1800, light was passed through a prism, and a thermometer was placed just beyond the red end of the visible spectrum. To Herschel's astonishment, the thermometer showed a higher temperature reading in this position than when placed within the visible portion of the spectrum. From this he hypothesized the existence of what we now call infrared radiation (heat). He interpreted it to be a form of light outside the visible spectrum. Were it not for his open-minded nature, it's unlikely Herschel would have discovered infrared rays, which are, of course, invisible to the human eye. Scientists tend to be quite dismissive of phenomena that cannot be seen directly.

Indeed, critics accused Herschel of being excessively open-minded and of engaging too much in wild speculation, especially with regard to the question of life on other planets. In Herschel's opinion, it seemed only natural that intelligent life should exist on every planet in the solar system, even on the Sun itself. "We need not hesitate to admit that the Sun is richly stored with inhabitants," he once wrote.[1] He believed life thrived on the Moon and, by means of careful telescopic observation, detected what he thought were signs of such life. This included trees, forests, pastures, canals, roads, and villages—details of which he recorded in his journals but never published.

Herschel fathered only one child, the equally brilliant and distinguished Sir John Herschel. Born in 1792, John Herschel followed in his father's footsteps. Though first and foremost an astronomer, he made some important contributions to the fields of chemistry, mathematics, and physics. In 1842 he invented the cyanotype photographic printing process.

Beginning in 1834, John Herschel, accompanied by his family, spent four years living near Cape Town, South Africa. The purpose of the expedition was to take advantage of the clear southern skies for astronomical purposes, and indeed it proved very fruitful. He recorded the locations of thousands of stars and observed galaxies only visible in the Southern Hemisphere.

On August 25, 1835, while John Herschel's expedition was still in progress, an exciting article appeared in the *New York Sun* carrying the headline "Great Astronomical Discoveries Lately Made by Sir John Herschel." Part of it read: "we have just learnt [*sic*] from an eminent publisher in this city that Sir John Herschel . . . has made some astronomical discoveries of the most wonderful description, by means of an immense telescope of an entirely new principle."[2] The article claimed to have gained its information from the *Edinburgh Journal of Science,* a journal that had been rendered defunct some years before. The lunar discoveries attributed to John Herschel were extremely "wonderful" indeed, and they created a huge sensation among the public. They included "conglomerated masses of polygon crystals," "brown quadrupeds, having all the external characteristics of the bison," "pygmy zebra," and—most amazingly of all—"bipedal winged creatures called man-bats."[3]

The Great Moon Hoax, as it later became known, consisted of six articles in total, and it helped boost sales of the *New York Sun.* The stories were reprinted in newspapers around the world, fooling a great many credulous readers. "Not one person in ten discredited it," commented author and critic Edgar Allan Poe (1809–1849), "and (strangest point of all!) the doubters were chiefly those who doubted without being able to say why—the ignorant, those uninformed in astronomy, people who would not believe because the thing was so novel, so entirely 'out of the usual way.'"[4]

A certain Richard A. Locke, who worked as editor of the *New York Sun* at the time the stories were published, was identified as the perpetrator of the hoax, although it's been suggested that others were involved as well. When news of the hoax reached John Herschel by letter while still on expedition in South Africa, his initial reaction was amusement. Later he began to complain about the incident and said, "I have been pestered from all quarters with that ridiculous hoax about the Moon—in English, French, Italian, and German!"[5]

Although we can be fairly sure the Moon isn't occupied by "man-bats," something very strange exists there; it is stranger, I would suggest, than anything the human mind could possibly imagine. During March 1787, William Herschel reported seeing three "bright spots" on the dark side of the Moon. About a month later another strange incident caught his attention, consisting of what he perceived to be three "volcanoes" on the dark side of the Moon, one of them much brighter than the others.

The "volcanoes" were still there when he checked the following night, although the brightest of the three was "even brighter and at least three miles in diameter."[6] Whatever Herschel saw through his telescope, it's unlikely to have been lunar "volcanoes," because the Moon is supposed to have been volcanically dead for at least a billion years. What, then, did he see?

Events like the ones reported by Herschel are known as transient lunar phenomena (TLP), and they remain a mystery even today. Reports of such phenomena go back about a thousand years, becoming increasingly common with the invention and widespread use of the telescope. As I stated earlier, TLP are short-lived lights, colors, or changes in appearance on the surface of the Moon.

Prior to the Moon landings NASA went to the trouble of assigning a group of experts, including Patrick Moore, to compile an extensive list of TLP events. The result was Technical Report R-277 titled *Chronological Catalogue of Reported Lunar Events,* published in 1968. The document lists more than 570 lunar anomalies that took place from 1540 to 1967. "The purpose of this catalogue is to provide a listing of historical and modern records that may be useful in investigations of possible activity on the moon," stated the authors.[7]

Some of the incidents described in the document are highly anomalous indeed and completely defy conventional explanation. For example, on July 4, 1881, several observers reported that "two pyramidal luminous protuberances appeared on the Moon's limb. . . . These points were a little darker than the rest of the Moon's face. They slowly faded away . . ." In October 1878 a "faint bright shimmer like thin white cloud" was seen in the crater Plato. The incident was reported by H. Klein. An observer by the name of James C. Bartlett reported a total of six incidents of unusual activity within the crater Aristarchus for the months of June, July, and August of 1950 with such descriptions as "blue glare," "bright blue glare," "strong bluish glare," and "intense blue-violet glare." On June 21, 1964, Harris, Cross, and Helland observed, for a period of over two hours, a "moving dark area" in the Ross crater.

Other incidents include sightings of "mists," "fogs," "bright spots," "blackish spots," and "flashing" and "blinking" lights.[8]

Sure enough, reports of TLP are given little adequate attention by scientists and astronomers. After all, the Moon is supposed to be a dead world, and such phenomena strongly suggest otherwise, challenging scientific orthodoxy. Then there's the fact that some of these reports are very bizarre in nature, featuring elements found in UFO sightings and the like. For example, what is a conventional scientist to make of a "black cloud surrounded by violet color"?[9] Such an observation was made by a group of observers in Canada while studying the Mare Tranquillitatis region of the Moon on September 11, 1967. They might explain the sighting as an "optical illusion." Fortunately there is a growing body of evidence to suggest that the Moon is a richer and more alive environment than previously thought, which is forcing scientists to look at the Moon with greater open-mindedness—and take TLP seriously.

As we know, the Moon has been identified as an electrostatically active environment, which can possibly create dust storms. This prompts the question: Might some TLP be explainable in terms of electrostatic activity, and, in particular, dust storms? The answer is yes, according to Gary Olhoeft, professor of geophysics at the Colorado School of Mines in Golden. "It may be that [TLP] are caused by sunlight reflecting off rising plumes of electrostatically lofted lunar dust."[10] Another theory offered to explain TLP is out-gassing, whereby gas is released from underground cavities, either slowly or in a sudden, explosive manner. There have been other theories suggested as well, which don't require mention here. Clearly, if we're willing to accept the Moon's lively and active nature, a modest portion of TLP is not at all difficult to explain. The mystery, to some extent, vanishes.

I use the words "to some extent" because clearly not all TLP can be attributed to natural causes. There is the very strange case of Linné crater, for example. Located in the northwestern region of Mare Serenitatis, the crater is listed on authoritative lunar maps of the early nineteenth century as having a diameter of 6.2 miles and a depth of 0.2 mile, making it completely unremarkable as far as craters go. It remained that way until, in 1866, the respected German astronomer and mapmaker Julius Schmidt (1825–1884) announced with great astonishment that the crater had disappeared; where it had once been was now a brilliant white patch. The observation was confirmed by other astronomers.

Later a tiny crater pit was observed to appear in the middle of this feature, which is how Linné looks today when viewed through powerful Earth-based telescopes. High-resolution photographs of the crater were obtained during the Apollo missions, revealing its diameter to be 1.5 miles and its depth 0.4 mile. The Linné crater mystery has never been solved, at least not to the satisfaction of all. Of course features on the Moon can undergo significant changes in form depending on the way sunlight strikes them, and this explanation had been used in the case of Linné to account for its apparent disappearance. But this doesn't explain why the crater was perceived larger in size prior to 1866 compared to the size it is today. It's as if Linné was demolished—hence its temporary disappearance—and then replaced by a smaller version of the original.

Worth mentioning in connection with the topic of TLP is the case of the mysterious "bridge" on the Moon. John Joseph O'Neill (1889–1953) was science editor of the *New York Herald Tribune* and an acquaintance of the Serbian-American inventor and electrical engineer Nikola Tesla (1856–1943), on whom he wrote a biography. He was also an amateur astronomer. While observing the Moon using a four-inch refracting telescope on July 29, 1953, O'Neill spotted what he thought was an immense land bridge spanning two mountaintops in the western region of Mare Crisium.

O'Neill reported his find by letter to the aforementioned Hugh Percy Wilkins in the hope of having it confirmed. Wilkins was skeptical but nonetheless curious. His skepticism vanished when, on the evening of August 26 while using a much more powerful instrument than O'Neill's, he saw the apparent structure. He described it as "a bridge with sunlight streaming under it, and the shadow of the arch cast on the surface of the plain."[11] Sadly, O'Neill passed away that year, before he had the chance to receive Wilkins's letter of confirmation. Further confirmation came from Patrick Moore, a longtime friend of Percy. He too saw the "bridge," which, so far as he could tell, had "popped up," almost overnight.[12]

During a BBC radio interview that aired on December 23, 1953, Wilkins announced the "bridge" is "real," adding:

Its span is about twenty miles from one side to the other, and it's probably at least five thousand feet or so from the surface beneath. . . . It looks artificial. It's almost incredible that such a thing could have been formed in the first instance, or if it was formed, could have lasted during the ages in which the Moon has been in existence. . . . It looks almost like an engineering job. . . . Yes, it's most extraordinary.[13] (See plate 9.)

The following year on June 17 Wilkins paid a visit to Mount Wilson Observatory, where, using its one-hundred-inch reflecting telescope, he saw the "bridge" again. Sure enough, the "bridge" became the subject of a large controversy, because other astronomers had denied its existence. They labeled it as an optical illusion formed by an unusual arrangement of light and shadow, not a bridge. Even in the face of great ridicule, Wilkins stuck to his original conclusion: the "bridge" was real. The situation became so ugly that by 1955 Wilkins was forced to resign as director of the Lunar Section of the British Astronomical Association (BAA). He passed away roughly five years later, on January 23, 1960.

Apparently the matter of the lunar "bridge" was finally put to rest in the 1960s after NASA's Lunar Orbiter program brought back photographs of the lunar surface showing no such structure. Only those who think of NASA as an honest and open organization would find this explanation satisfying. There is a joke among conspiracy researchers that NASA stands for "Never A Straight Answer," which nicely sums up the situation.

There is an interesting footnote to this story—the irony of which wouldn't have been lost on Wilkins. In 2010, NASA announced the discovery of two natural bridges on the Moon's far side: one of them 66 feet long and 23 feet wide and the other about half the size of the first. The bridges are located adjacent to each other and were discovered in images taken by NASA's Lunar Reconnaissance Orbiter. They are thought to be pieces of lava tube roof, from when the crust around them caved in. Funny, isn't it, that Wilkins was deemed crazy for claiming to see a bridge on the Moon? He never said it was artificial, only that it *looked* artificial, and now such a structure has been found.

The year 1947 ushered in the modern era of the UFO with Kenneth Arnold's famous sighting on June 24 near Mount Rainier, Washington.

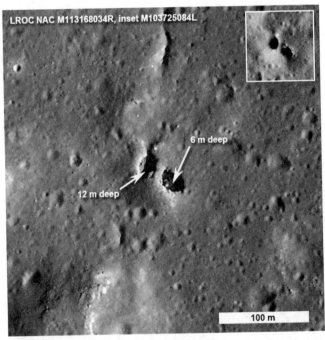

A "bridge" on the far side of the Moon discovered by NASA's Lunar Reconnaissance Orbiter (LRO) and thought to be a collapsed lava tube. Photo courtesy of NASA.

Arnold observed nine mysterious crescent-shaped objects that moved, he said, "like saucers skipping on water."[14] He estimated their speed at well over one thousand miles per hour—a speed unheard of by aircraft at the time. The event was followed in July by the equally significant Roswell UFO Incident, which involved the supposed crash and recovery of a "flying saucer" on a ranch near Roswell, New Mexico. I will refrain from providing a summary of the UFO phenomenon, under the assumption that most people reading this already have a good understanding of the topic and have accepted the reality of UFOs. Worth quoting at this point, I feel, are the first two sentences of Jim Marrs's definitive *Alien Agenda:* "The controversy over the existence of UFOs is over. UFOs are real."[15]

With the emergence of UFOs into public consciousness and with the Moon still unvisited by humans—and hence still mysterious—it seems only natural that the two would become linked. And indeed, such a connection is present in the UFO literature of the 1950s. A good example is Harold T. Wilkins's *Flying Saucers on the Attack* (1954), in which he states, "the Moon . . . long has been a stopover for what we call flying saucers." (Not to be confused with the astronomer Hugh Percy Wilkins.) Envisaging man's first trip to the Moon and the nature of the Moon itself, Wilkins seems to borrow from the fertile imagination of H. P. Lovecraft, describing it as a world containing "massive portals" leading to "great sub-lunar tunnels," inhabited by "beings of other unknown worlds in space."[16]

Donald E. Keyhoe, an American Marine Corps naval aviator and one of the leading UFO researchers of the 1950s, draws a connection between UFOs and the Moon in his book *The Flying Saucer Conspiracy* (1955). The book includes a brief discussion on TLP, which Keyhoe interprets as strong evidence of an alien presence on the Moon. Perhaps, he suggests, the aliens who occupy the Moon aren't there on a permanent basis but use it as a "space base for travel to other planets," coming and going as they please. "The intermittent use of the moon as a space base would explain the strange lights of the past two centuries, as well as the mysterious radial cracks or lines which might be caused by intense heat from 'blastoffs.'"[17] Keyhoe, like many of the great pioneers of UFO research, is no longer among the living; he passed away in 1988.

Another such pioneer is Morris K. Jessup (1900–1959), whose efforts in this field remain sadly unrecognized. An academically accomplished man, Jessup had a master of science degree in astronomy, and started, though never completed, his doctorate in astrophysics. He is known to have worked at the Lamont-Hussey Observatory in Bloemfontein, South Africa, and to have taken part in various archaeological expeditions in South America. For reasons unknown, Jessup renounced his career in astronomy, thereafter earning his living as an automobile-parts salesman as well as a photographer among other things. His first book, *The Case for the UFO*, was published in 1955, and his fourth and last, *The Expanding Case for the UFO*, in 1957. The latter is of relevance here, because in it Jessup suggests that the Moon is "a base for UFO activity."[18]

The book contains a long and exhaustive chapter on TLP, and many of the incidents described have obvious UFO connotations. One section of the chapter is dedicated to cases of bright lights on the Moon, which is one of the rarer forms of TLP. Jessup is careful to distinguish between such lights and the aforementioned "bright spots." The former is a "star-like, glittering point or gleam," sometimes great in intensity, while the latter is an "area that is bright as compared to the surrounding surface but does not have the appearance of a lamp, arc light, or blast furnace." According to Jessup, "No single indication of UFO activity on the Moon is more intriguing than the unexplained, intermittent lights." In many cases, he says, the behavior of these lights suggests some form of "intelligent control." "Sometimes they fluctuate in a manner unlike the steady glare of reflected sunlight. Sometimes they appear suddenly, shine a few minutes or hours, and as suddenly disappear."[19] In view of Jessup's observation that the lights display intelligent behavior, it's difficult to account for them in terms of "sunlight reflecting off rising plumes of electrostatically lofted lunar dust," to requote geophysicist Gary Olhoeft. No mundane explanation fits.

So far as I know, Jessup was the first author and researcher to study (and write about), in a serious, scientific fashion, the connection between UFO activity and TLP. Because he was an astronomer himself, his conclusions on the matter carry weight—although outdated by half a century. For a book written so long ago (indeed, prior to the Moon landings) we're bound to find, in *The Expanded Case for the UFO*, plenty of ideas with regard to the Moon that today seem ludicrous or silly. For example, speculating on the nature of the "Selenian Moon dwellers," Jessup suggests they might be giants. "Some life forms on the Moon might have grown to giant stature, because of the low lunar gravity which would permit bone structures to carry much larger bodies. There are many Bible references which hint (e.g., Genesis 6) that some such race of giants arrived from space."[20]

Jessup died at the age of fifty-nine in Dade County, Florida, from an apparent suicide caused by carbon monoxide poisoning; he was found dead in his station wagon, with a pipe running from the exhaust to the rear window of the vehicle. Prior to this, events in Jessup's life had taken a turn for the worse. His books were selling poorly, and he'd recently separated from his wife. According to friends, his behavior seemed gloomy and unstable; however, there are some who suspect foul play in the case of Jessup's death. According to one theory, he was murdered as a result of discovering too much in regard to the alleged Philadelphia Experiment, which is said to have involved the teleportation of a U.S. Navy ship and its crew in October of 1943. But that's a topic for a whole other book.

The year 1969 marked the first Moon landing—an event that, for many, completely dispelled any lingering hope that the Moon might actually be inhabited. The idea that the Moon was a dead world, completely devoid of even the simplest form of life, became accepted as fact. No atmosphere, no flora, no fauna, no moisture, no sound, no weather, and hardly any color was observed on the Moon. The image of the Moon presented to the public was of a pile of rocks and dust that had been sitting around, undisturbed, for billions of years. If NASA had wanted to make the Moon sound uninspiring, they couldn't have done a better job of it. Astronaut James Lovell's famous description of the lunar surface as "looks like plaster of Paris, or sort of a grayish beach sand" makes a trip to the lavatory sound more interesting and appealing than a trip to the Moon.[21] But just as there have always been those who don't question the status quo or what they're told by the media and government, so there have always been those who do ask questions.

In the wake of the Apollo program, members of the latter camp began to question whether NASA had been entirely open and honest with the public concerning discoveries made on the Moon. Had evidence of advanced life on the Moon been found and withheld, so as not to alarm the public? Had NASA pulled the plug on the Apollo program, canceling missions eighteen through twenty, not because of budgetary constraints, as NASA had proclaimed, but because they'd been warned off the Moon by whoever, or whatever, dwelled there? It was questions such as these that occupied the minds of many during the 1970s. One result was the publication of books like Don Wilson's *Our Mysterious Spaceship Moon* (1975). Wilson, referring to the cancelation of the remaining Apollo missions, asks: "Did the astronaut encounters with UFOs worry NASA officials that another trip might bring startling encounters or revelations that might blow their cover of secrecy?"[22]

Similar in content to *Our Mysterious Spaceship Moon* is George H. Leonard's provocatively titled *Somebody Else Is on the Moon* (1976). In it, Leonard claimed to have found conclusive evidence that the Moon is "occupied by an intelligent race or races which probably moved in from outside the solar system."[23] He arrived at this conclusion, he said, after carefully examining hundreds of official NASA photographs of the Moon (not classified photographs, but ones available to the public). He said:

The Moon is firmly in the possession of these occupants. . . .
Evidence of their presence is everywhere: on the surface, on the near
side and the hidden side, in the craters, on the maria, and in the
highlands. They are changing its face. *Suspicion or recognition of*

that triggered the U.S. and Soviet Moon programs—which may not really be so much a race as a desperate "cooperation" [emphasis in original].[24]

Included in the book are thirty-five black-and-white lunar photographs, featuring what Leonard believes are things such as "manufactured objects," "control wheels," "immense mechanical rigs," "strange geometric ground markings and symbols," "vehicle tracks," "towers" "pipes," and "conduits" on the lunar surface. Although I'd like to be impressed by these photographs, I'm not; what are, according to Leonard, artificial objects look quite natural to me. Overall, the photos do little to prove the existence of an alien presence on the Moon.

At the beginning of the book Leonard becomes acquainted with a former NASA scientist living in California. He refers to this person by the pseudonym of Dr. Samuel Wittcomb and describes him as "an engineer who had gone on to get his PhD in physics and an astronomy freak to boot."[25] During their first meeting, which takes place at Wittcomb's house, Wittcomb informs Leonard that there's "an intelligent race on the Moon," which is definitely nonhuman and "probably not from within the solar system," and those in the "inner circles of government" are aware of this fact but have kept it secret from the public.[26]

Wittcomb refuses to tell Leonard everything he knows about the Moon because it would be "too easy." Rather, like some kind of wise old Zen master, he encourages Leonard to look for the (photographic) evidence himself. From then on Wittcomb assumes the role of mentor to Leonard by providing him with the occasional "lead" and cryptic piece of information. One gets the impression that Wittcomb is entirely fictitious and that a significant percentage of the book was fabricated.

One of the "hints" provided by Wittcomb is that much of the activity seen on the lunar surface (in photographs) involves mining of the Moon's plentiful resources. In fact, this theme is strongly emphasized in the book, while many of Leonard's "artificial" objects are described as serving a mining-related function. One such object, or "rig," is what Leonard calls an "X-drone." These have the appearance, he says, of "two cross earthworms . . . vary in size from under a mile to three miles in any direction," and are quite abundant on the Moon. Furthermore, "whenever there is a lot of work to be done in certain parts of the Moon, the chances are excellent you will find these big X's slaving away; ripping and slicing crater rims . . . [and] lifting hundreds of tons of weight at one time."[27]

Among the many objects discussed by Leonard, nothing is quite as funny as his "mammoth yurts," which abound, he says, on the floor of Tycho. (A yurt is a dome-shaped tent used by nomads in Central Asia in countries like Mongolia.) Comments Leonard, in an almost tongue-in-cheek fashion, "The temptation is to call this a dwelling place. How neat! We have seen evidence of the Moon occupants at work, and now we see Their abodes! But I tend to think most of Them live underground. . . ."[28]

Some mystery surrounds Leonard's background and, in particular, his motivation for writing *Somebody Else Is on the Moon.* According to a short biographical entry in the *Encyclopaedia of Science Fiction,* he was born in Malden, Massachusetts, in 1921, died in 2004, and wrote a total of five books during his lifetime. As it turns out, *Somebody Else Is on the Moon* was his only work of nonfiction. The remaining four all belong to the genre of sci-fi adventure: *Sexmax* (1969), *Beyond Control* (1975), *Alien**(1977), and *Alien Quest* (1981). He must have been embarrassed by *Sexmax,* because it was published under the macho pseudonym of Hughes Cooper. The cover illustration of this bizarre and erotic novel features two naked men and three naked women; the women are seemingly suspended above the lunar surface.

There is no indication that any of Leonard's books were a notable success, or that he "made it" as an author. Also, given that he was forty-eight when his first novel was published, it could be assumed that writing was his secondary occupation and not the primary means by which

---

*Not related to the Ridley Scott movie of the same title.

he earned his living. (However, we cannot exclude the possibility that he earned money by writing freelance stories and articles for magazines.) Supporting this assumption is the fact that Leonard is referred to by the space journalist and CSICOP member James E. Oberg, in his book *UFOs and Outer Space Mysteries* (1982), as a "retired government health worker" and "amateur astronomer."[29] (Strangely, he makes no mention of Leonard's fictional work.) Another curious detail is that Leonard may have been educated at Harvard—a claim he makes, in passing, in *Somebody Else Is on the Moon.*

The purpose of my discussing Leonard's book is not to ridicule it; numerous others have done so already. And besides, the book is an extremely easy target—suspiciously easy, you might say. I've noticed that whenever skeptics wish to discredit the idea of intelligent life or ruins on the Moon, they often refer to Leonard's book as an example of how ignorant people delude themselves into believing they can see artificial objects in lunar photographs (or photographs of Mars), when, clearly, such objects have no basis in reality. Leonard's book is considered one of the ultimate works of Moon-related pseudoscience ever published. Oberg, who writes articles for the *New Scientist* and *The Sceptical Enquirer,* with titles such as "The Failure of the 'Science' of Ufology," provides a lengthy criticism of Leonard's book in *UFOs and Outer Space Mysteries.* At one point he states, "In mid-1981 Leonard wrote to me and asked me not to publicly criticize his book anymore, implying it has been an honest mistake."[30] An "honest mistake"? Or was the book an intentional mistake?

It's possible that Leonard was a clever con-artist, who, using his skills as a novelist, wrote the book purely for financial gain, knowing full well that he was exploiting the gullibility of his readers. Then again, perhaps he was employed to write the book by an intelligence agency, as part of a secret plot to misinform the public—meaning that the book is disinformation. It's long been my belief that a small percentage of the books available on alternative topics (UFOs, ancient mysteries, and so on) are disinformation, meant to "muddy the waters" and deliberately lead the public down the wrong path. These books are a clever combination of truth and falsehood. Certain books, especially if they're popular, have the potential to discredit an entire field of research.

Intentionally or not, Leonard's book did much harm to the theory of an alien presence on the Moon. We can be sure the "debunkers" won't forget about it for a long time to come. However, it would be wrong to write off the book completely. Scattered within its pages are some compelling ideas and hypotheses with regard to the supposed inhabitants of the Moon. The following strikes me as an intelligent insight: "Although the occupants of the Moon are self-sufficient in most respects, it is probable that They are, to some extent, parasitic on Earth."[31]

For a man who presents himself as an eccentric, fringe author of the *Chariots of the Gods?* variety, Leonard was well connected. In *Somebody Else Is on the Moon,* he hints at an acquaintance between him and Farouk El-Baz. In one instance in the book he describes talking with El-Baz about lunar rays. In another instance he describes meeting with El-Baz "in a third-floor loft above the National Air and Space Museum's gift shop" for a private, hour-long discussion. The purpose of this get-together was to share with El-Baz the evidence he'd gathered suggesting the Moon is inhabited. He explains, "Some of the objects and Moon features I discussed with him he saw the same way. Others he did not. . . . I was pleased that he saw some of the . . . objects the same way, although I realized that if I had pressed him for his interpretations, some would probably have differed from mine."[32]

That El-Baz met with Leonard to discuss such far-out matters and, what's more, agreed with some of Leonard's conclusions, runs counter to common sense. Why would a highly respected NASA geologist give someone like Leonard the time of day, let alone meet with him privately to discuss "mammoth yurts" on the Moon? Some conspiracy researchers interpret El-Baz's interest in Leonard's research as an indication that Leonard found genuine evidence of an alien presence on the Moon. Speculates Mike Bara in *Ancient Aliens on the Moon,* "I've often thought that he [El-Baz] did this [met with Leonard] to perhaps not only see what he [El-Baz] and his team had missed in terms of Ancient Alien evidence, but also to see where they [El-Baz and his team] had slipped up in *hiding* that evidence" [emphasis in original].[33] Bara could be right, though I very much doubt it.

In July of 2013, I got in touch with Farouk El-Baz via e-mail to inquire about his association with Leonard. It turns out that El-Baz's recollection of their meeting at the National Air and Space Museum differs markedly from Leonard's. In his friendly message to me he explained, "I do remember talking with George Leonard while I was at the Smithsonian's National Air and Space Museum (1973–82). I do not recall if that was a phone call or a meeting. I told him that all features on the Moon can easily be interpreted as natural landforms. I also said that there were a few things that we did not fully understand, which is normal in science."[34]

Another book featuring official NASA photographs of alleged artificial objects on the Moon is Fred Steckling's *We Discovered Alien Bases on the Moon* (1981). The book has been compared to *Somebody Else Is on the Moon,* with its blurry, poor-quality photographs showing what are more likely to be photographic defects and perfectly natural objects than anything of an artificial nature. Hoagland and Bara, commenting on the book in *Dark Mission,* describe it as "Steckling's non-stop comedy of errors, misidentified photographic defects and general mish mash of mal-illustrated misinformation."[35] The authors reveal that Steckling, who passed away in 1991, once worked for the CIA.

Not long after emigrating from Germany to the United States in 1960, Steckling became a close associate of the famous "contactee" George Adamski. Steckling's devotion to Adamski and his "teachings" persisted after Adamski's death in 1965, as he served for a period as director of the G.A.F. International/Adamski Foundation, based in California. Adamski's sensational claims of meeting handsome and benevolent "space brothers" from Venus, and of being taken aboard their spacecraft on jaunts to other planets in the solar system, including the Moon, are not taken seriously by any ufologist with the slightest ounce of credibility. Adamski tried to back up his claims by providing photographs of the supposed craft, none of which look remotely genuine. There are indications that Adamski was used as a pawn by the U.S. government as part of a secret effort to reduce the credibility of UFO research. It's been noted by Jacques Vallee that "Adamski's major supporter abroad was a former intelligence officer with the British Army. . . . And according to a man who hosted Adamski during his tour of Australia, he was traveling with a passport bearing special privileges."[36]

There is good reason to be suspicious of Steckling and his farcical *We Found Alien Bases on the Moon,* especially in light of his association with Adamski and his involvement at one point with the CIA. As with Leonard's *Somebody Else Is on the Moon,* it would be a mistake to write off the book entirely. Respected alternative researcher Joseph P. Farrell, author of books such as *The Giza Death Star* (2001), defended the book in an article discussing the possibility of a secret base on the Moon, stating, "Regardless of one's biases toward the late Mr. Steckling . . . one cannot come away from his book without at least a glimmer of a suspicion that the idea might have some merit, for it is chock full of NASA unmanned and Apollo photos, and rare and hard-to-find information."[37] Humorously, Hoagland found the book helpful, in a strange kind of way, during his search for evidence of artificial structures on the surface of the Moon.

In 1992, while taking a close look at the photos in the book, Hoagland found nothing of value, except in the case of one particular image. Labeled by Steckling as "frame AS 10-32-4810," the image was taken during the *Apollo 10* mission in the vicinity of Ukert crater (positioned between Mare Vaporum in the north and Sinus Medii in the south). The image, as it appears in Steckling's book, features an arrow pointing to the supposed "entrance to an underground alien base."[38] This is but one of many such arrows—each arrow pointing at something amazing that has no existence outside Steckling's imagination.

While studying the image, Hoagland's attention was drawn to a compelling formation on the southeastern side of Ukert—a "curious, rectilinear arrangement of small hills, dark grooves and suspiciously parallel lines." For reasons unknown, the presence of this formation seems to have escaped Steckling's attention, because it isn't marked by an arrow. Hoagland was left wondering why Steckling had "failed to call the attention of his readers to this highly fascinating, highly anomalous, highly organized formation" yet had included "all those silly arrows pointing at absolutely nothing."[39]

There is a very geometric quality to the formation in question, the most striking aspect of which consists of what looks like a huge square. It is not easy to ignore the extremely triangular shape of Ukert crater itself. Overall, this is one very anomalous region of the Moon. The manner in which Hoagland discovered this formation—hidden in one of the many images featured in Steckling's book—is not without a touch of mystery. To add to the mystery, Hoagland says he ordered the frame (AS 10-32-4810) from NASA and received not the frame he expected to receive, but one that had been taken during the same sequence of photographs, which showed the formation from a slightly different angle. In other words, what was supposed to be AS 10-32-4810, according to Steckling, was another frame altogether. Fortunately, because Hoagland had a rare copy of the official *Apollo 10* photographic catalogue, he was able to ascertain the real identity of the image featured in Steckling's book (AS 10-32-4819). Did Steckling try to conceal the identity of the image? Or did he simply mislabel it by mistake? We'll probably never know.

Of the numerous lunar anomalies detected by Hoagland in NASA photographs, the vast majority are located in the Ukert/Sinus Medii region of the Moon. Sinus Medii (or Central Bay) is a small lunar mare, approximately 120 miles wide. Located at the intersection of the Moon's equator and prime meridian, it is the point closest to Earth. Someone standing on the Moon at this point would see the Earth positioned directly overhead, without any movement of the Earth through the heavens. Whereas we, on Earth, see the Moon move through the heavens, someone on the lunar near side would see no such movement take place with regard to the Earth. For them the Earth would remain relatively stationary, although it would be seen to "wax and wane," just as the Moon does as viewed from Earth.

Located on the southwest perimeter of Sinus Medii, near a large crater called Flammarion, is a spirelike object with an estimated height of more than 1.5 miles, named by Hoagland "the shard." The object appears in an image taken by *Lunar Orbiter 3* in mid-February 1967 (specifically, frame LO-III-84M). Hoagland and Bara describe it as a "highly anomalous, defiantly upright, 'bowling-pin' shaped structure—with an irregularly-pointed apex, a swollen middle 'node' and a narrowing 'foot.'"[40] It's hard to imagine that such an amazing object, towering so high above the lunar surface, could be anything other than a photographic defect. Yet, this explanation is contradicted by the fact that the object casts a shadow on the ground consistent with the angle of the Sun at the time the image was taken; only physical objects cast shadows.

Independent geologist and paleontologist Bruce Cornet, Ph.D., in a detailed study of the shard and other anomalies found in NASA lunar photographs, concluded it to be "an obvious structure," having an "overall irregular spindly shape containing a regular geometric pattern. . . ." He argues that the object, "if natural, has got to be a wonder of the Universe. No known natural process can explain such a structure."[41]

Frame LO-III-84M features not one but two anomalous structures. Located above and to the left (south) of the shard is what appears to be, at first glance, nothing more than a strange smudge. Hoagland has dubbed this the "the tower/cube." A keen eye is needed to appreciate the object, which indeed has the appearance of a tower extending high above the lunar surface, its topmost portion consisting of a large cube. The tower itself is quite faint, while the cube is easy to discern. "The tower represents an enigma of the highest magnitude," explains Cornet, "because it rises more than five miles above the surface of the Moon, and has been photographed from five different angles and two different altitudes. In all four photographs the same structure is visible and can be viewed from two different sides." Cornet, commenting on the cube, says it "appears to be composed of regular cubes joined together to form a very large cube with an estimated width of one mile!"[42]

Hoagland believes neither the shard nor the tower/cube is an independent, stand-alone structure. Rather, he says, they are "remaining fragments of a once *far-larger,* also clearly *artificial* structure, apparently once composed of glass (a lot of it) . . . and, attached to some kind of darker, vertical, overarching structural framework."[43] This former, "far-larger" structure, he suggests, was nothing less than a giant glass dome that enclosed the whole of Sinus Medii.

According to Hoagland's "ancient lunar domes hypothesis," there once existed numerous such domes on the Moon, constructed millions (possibly billions) of years ago by technologically advanced extraterrestrial beings. These structures now lie in ruins due to long-term damage sustained by incessant micrometeorite bombardment. As Hoagland understands it, the extraterrestrials who built the domes are former inhabitants of the Moon. Thus there is no extraterrestrial presence on the Moon today. All that remains are the ruins of their civilization—consisting, for the most part, of a great deal of glass (or, if not glass specifically, some kind of transparent medium with glasslike properties).

Because glass is an extremely fragile material, the idea of using it to construct a giant dome seems absolutely ludicrous. More exactly, it is strong in terms of compression but weak in terms of tension. As Hoagland and Bara point out in *Dark Mission,* glass produced in an airless, moisture-free environment such as the Moon would be immeasurably stronger than the normal terrestrial glass with which we're familiar. Glass produced under earthly conditions is fragile because of a phenomenon called hydrolytic weakening, whereby the silicates that make up the glass are prevented from bonding to their full potential due to the presence of moisture in the air. Explain Hoagland and Bara, "The hard-cold vacuum enhances the strength of lunar glass to the point that it is approximately *twice as strong as steel* under the same stress conditions" [emphasis in original]. The authors go on to state that scientists themselves "have suggested that lunar glass is the ideal substance from which to construct a domed lunar base."[44]

Hoagland has managed to acquire, during his many years of persistent research, a large collection of NASA photographs featuring what look like faint and reflective architectural forms suspended above the lunar surface—more evidence in support of his "ancient lunar domes hypothesis." Many of these photos were difficult to find, some of them not even available to the public—or rather, only available to the public in a "sanitized" format, so that anything unusual in the photo had been carefully edited out. For obvious reasons, it was original and untouched versions of NASA photos that most interested Hoagland and proved to feature the most remarkable anomalies.

In 1995 Hoagland was given an *Apollo 16* photographic print by a certain individual connected to NASA. This individual (whom Hoagland doesn't name) claimed to have stolen the photo from the administrator's office at NASA headquarters back in 1972, a few days after *Apollo 16* had returned to Earth. He'd held on to the precious photo for twenty-three years before passing it on to Hoagland. The photo (frame AS16-121-19438), looking down on Mare Crisium, shows what looks like a tall tower rising from the surface, with an estimated height of several miles. Dubbed by Hoagland the "Crisium Spire," its presence in the photo is impossible to miss. Also present in the photo, both on the surface and hanging in the sky, are more of Hoagland's "glass dome ruins."

During a visit to the National Space Science Data Center (NSSDC) in Greenbelt, Maryland (NASA's permanent archive for space science mission data), Hoagland managed to track down an official negative of frame AS16-121-19438. Wanting to compare it with the one he had, he ordered a large print of the photo (which would have been produced from the official negative). When the photo arrived, he discovered some major discrepancies between the two versions. Missing from the NSSDC version, he says, was all of the "marvellous, 'glittering detail' hanging in the sky over Crisium." This had been "replaced by a dead, dead, 'black'—which had obviously been airbrushed in on the copy negative, to represent what someone thought space above the lunar surface *ought* to look like."[45] Also missing from the image was the "Crisium Spire."

Once a photographic negative has been altered, such as with the use of an airbrush, the image in its original form is lost forever, and any prints produced from that negative will contain those alterations. Certainly NASA would have airbrushed photographic prints, so as to clean them up before being released to the public and press. The airbrushing of negatives, on the other hand, sounds extreme and completely unnecessary.

So has NASA really altered photographic negatives pertaining to the Moon as part of a deliberate effort to prevent the public from discovering what really exists there? Is the true nature of the Moon being covered up? Allegations made by former NASA employee Ken Johnston argue strongly in the affirmative. Johnston, as mentioned, worked as manager of the Data and Photo Control Department at NASA's Lunar Receiving Laboratory (LRL) during the Apollo program. (The LRL was part of the Manned Spacecraft Center [MSC] and today is known as the Lyndon B. Johnson Space Center [JSC].)

One of Johnston's responsibilities at LRL was to catalog and archive all of the Apollo photographs. The LRL ended up developing, he says, not one but four complete sets of Apollo photographs (taken during orbit and with the use of handheld cameras), amounting to tens of thousands of photographic negatives and prints. On one occasion, claims Johnston, while passing through a classified building at JSC (a building he rarely entered), he witnessed artists airbrushing photographic negatives. Significantly, the focus of this airbrushing was in the sky of those negatives. (This would appear to explain the mystery surrounding frame AS16-121-19438.)

On another occasion, it was Johnston's job to organize a private screening, for the benefit of astronomer Thornton Page, Ph.D., and his associates, of footage captured during the *Apollo 14* mission. *Apollo 14* and its crew had just returned to Earth and the footage recently processed in the NASA photo lab. A sequence projector, operated by Johnston, was used to play the film. One segment of the footage, Johnston says, showed the *Apollo 14* command module "coming around the backside of the Moon . . . approaching a large crater." Half of the large crater was in shadow, and in that shadowed portion, "down inside the rim," was a "cluster of about five or six lights." This was followed by a "column or plume or out-gassing or something, coming up above the rim of the crater."[46] Page showed great interest in the unusual phenomena, instructing Johnston to go back and forth over the footage several times. After the viewing, Johnston placed the footage back in storage at the photo lab, as was customary.

The following day Johnston procured the footage for another viewing; the audience, on this occasion, was rank-and-file engineers and scientists at the MSC. Among those present were several of his friends. Johnston, once again serving the role of projector operator, was eager to show his friends the amazing phenomena captured on film on the far side of the Moon, telling them, "wait until you see this view." Johnston explains what happened next: "And, as we were approaching the same crater . . . and we went past the crater—*there was nothing there!*" Johnston says he stopped the projector to examine the film but could find "no evidence of anything being cut out."[47] The rest of the story is best told in Johnston's own words:

> That afternoon, I ran into Dr. Page over at the Lunar Receiving Laboratory and asked him what had happened to the "the lights and the out-gassing or stream we saw," and he kind of grinned and gave me a little twinkle and a chuckle and said: "There were no lights. There is nothing there." And he walked away. And we were so busy . . . I didn't get the chance to question him again.[48]

It's worth noting that Page was a member of the CIA-created "Robertson Panel," a scientific committee that first met formally on January 14, 1953, in Washington, with the intention of finding a way to diffuse the huge buildup of public interest in UFOs. The Robertson Panel recommended, among other things, "that the national security agencies take immediate steps to strip the Unidentified Flying Objects of the special status they have been given and the aura of mystery they have unfortunately acquired."[49] One can't help but wonder if the "cluster of about five or six lights" filmed on the far side of the Moon by *Apollo 14* were UFOs, and if Page, having recognized them as UFOs, arranged for the objects to be promptly erased from the footage by means of airbrushing.

As we've seen, there are two main camps when it comes to the subject of intelligent life on the Moon. One camp supports the notion that the Moon is currently inhabited by extraterrestrial beings and views TLP in terms of UFO activity. The other camp, to which Hoagland belongs, supports the notion that the Moon was inhabited by extraterrestrial beings in the very ancient past and that traces of this former inhabitancy are there to be found on the lunar surface in the form of amazing architecture and so on. Common to both camps is the belief in a vast and highly organized conspiracy, directed from within the government, to keep the general public in the dark with regard to this (either former or present) lunar extraterrestrial presence. Certainly Johnston's allegations point toward such a conspiracy, and we'll examine more evidence for this in a moment.

Before we move on to the idea of a current alien presence on the Moon, we need to look more closely at the idea of a former alien presence on the Moon. The latter goes hand in hand with the awareness of alien artifacts—an awareness that, surprisingly, has been receiving some mainstream attention of late. Professors Paul Davies and Robert Wagner of Arizona State University suggest, in a 2011 paper, that alien artifacts might exist on the Moon. They even recommend we search for evidence of such artifacts among the huge database of images taken by NASA's LRO (which, at the time of writing this, is still in orbit around the Moon).

"Although there is only a tiny probability that alien technology would have left traces on the moon in the form of an artifact or surface modification of lunar features, this location has the virtue of being close, and of preserving traces for an immense duration," explain Davies and Wagner.[50] The scientists list four possible types of alien artifacts that might exist on the Moon: a message, instruments, trash, and large-scale changes to the lunar surface. An ideal place to look for "alien trash," they suggest, would be inside one of the Moon's lava tubes. They explain further:

> Lava tubes have been proposed as an ideal location to establish a human base, as they would provide protection from radiation and meteorites; perhaps aliens would come to the same conclusion. Furthermore, the same factors that make lava tubes attractive as a habitat imply that any artefacts left behind would endure almost indefinitely, undamaged and unburied. The downside is that there is no way to really investigate this possibility from orbit, so any confirmation or refutation will require a new robotic or human mission to the surface.[51]

It's encouraging to know that at least two mainstream scientists take seriously the idea of alien artifacts on the Moon, even though they seem to be entirely ignorant of the amazing evidence we've discussed in this chapter—evidence that strongly suggests that alien artifacts have already been discovered on the Moon. I'm quite sure Davies and Wagner would never accept such evidence anyway, because, by doing so, they'd be forced to enter the realm of conspiracism. Now, by suggesting we search through LRO images for evidence of alien artifacts, Davies and Wagner are presupposing those images are entirely "untouched." But of course, if the images have been secretly tampered with by NASA to remove evidence of alien artifacts, then there'd be absolutely no value in such a project. What's the point of going hunting for Easter eggs in the garden if someone's already come along and swiped them all?

What constitutes an alien artifact? And if we came across an alien artifact, would we recognize it for what it is? Such questions arise in the case of the Blair Cuspids (a cuspid is a canine tooth), an unusual arrangement of seven spirelike objects of varying heights, located on the western edge of Mare Tranquillitatis, not far from the crater Aeriadaeus B. The objects were photographed by *Lunar Orbiter 2* on November 20, 1966, at an altitude of approximately twenty-nine miles above the lunar surface. Two days later, on November 22, NASA released an image of the objects (frame LOH-61H3). In this image the objects are shown casting long shadows, giving an impression of great height, while there seems to be a definite order as to their arrangement. As far as I know, the *Washington Post* was the only newspaper to cover the story. An article published November 23, 1966, announced "six statuesque and mysterious shadows on the Moon were photographed . . . by Lunar Orbiter 2." The article continues: "Ranging from one about twenty feet long to another as long as seventy-five feet, the six shadows were hailed by scientists as one of the most unusual features of the Moon ever photographed."[52] (I wish to emphasize the accurate number of objects is seven—not six.)

After seeing the NASA photo of the "Moon spires," William Blair, an anthropologist at Boeing, began to scrutinize the image. In his previous job as a specialist in physical anthropology and archaeology Blair had used aerial survey maps coupled with his knowledge of geometry—skills he used to identify potential archaeological sites. It's an obvious fact that artificial structures tend to be built according to geometric principles, whereas natural structures are generally devoid of geometric significance. With this principle in mind, Blair went over the image with a compass and protractor and was amazed to discover a number of interesting geometric relationships with regard to the arrangement of the "spires." The findings included—to quote from an article that appeared in *Boeing News* in 1967—"a basic x, y and z right angle coordinate system, six isosceles triangles and two axes consisting of three points each."[53] An isosceles triangle is one that has two equal sides (and two equal angles). This is rarer to find in nature than a scalene triangle, which has three sides of different length. Also discovered by Blair was a feature with the appearance of a rectangular-shaped depression, located directly west of the largest "spire."

Asked during an interview whether he thought the "spires" were the work of nonhuman intelligence, Blair responded, "Whoa! Do you want them to put me away? But I will say this: if such a complex of structures were photographed on Earth, the archaeologist's first order of business would be to inspect and excavate test trenches and thus validate whether the prospective site has archaeological significance."[54]

The Blair Cuspids, as they became known, eventually caught the attention of Russian space engineer and Egyptology enthusiast Alexander Abramov. Abramov conducted his own analysis of the NASA image, concluding, like Blair, the objects were arranged not randomly but in a geometric pattern. "The distribution of these lunar objects," he claimed, "is similar to the plan of the Egyptian pyramids constructed by Pharaohs Cheops, Chephren and Menkaura at Gizeh, near Cairo. The centers of the spires of this lunar 'abaka' are arranged in precisely the same way as the apices of the three great pyramids."[55]

Many scientists were quick to oppose Blair's hypothesis, among them Richard W. Shorthill, Ph.D., of the Boeing Scientific Research Laboratories. Shorthill argued the objects were large but perfectly normal rocks, their long shadows a result of the low angle of the Sun at the time the photo was taken. He did not recognize anything special about the way the objects were arranged, stating "there are many of these rocks on the Moon's surface. Pick some at random and you eventually will find a group that seems to conform to some kind of pattern." Countered Blair, "If this same axiom were applied to the origin of such surface features on earth, more than half of the present known Aztec and Mayan architecture would still be under tree and bush studded depressions—the result of natural geophysical processes. The science of archaeology would have never been developed, and most of the present knowledge of man's physical evolution would still be a mystery."[56]

The mystery of the Blair Cuspids remains unsolved. Image scientist

Mark Carlotto, Ph.D., an expert on the controversial Cydonia region of Mars, conducted an extensive "3D analysis" of the Blair Cuspids and surrounding terrain. He published his results in a 2002 issue of the journal *New Frontiers in Science*. Carlotto's study argues in favor of the theory that the structures are artificial in origin. He found that the tallest of the "spires" is around fifty feet (which is close to NASA's original estimate of seventy-five feet) and, further, that the objects "are conical or pyramidal in shape," and "do not appear to be typical rocks or outcroppings." He also detected "perpendicular alignments between five of the seven objects."[57] Clearly the only way to determine once and for all if the Blair Cuspids are artificial is by photographing the region up close and in high resolution. As of yet, no such photographs have been released by NASA.

Intrigued as I am by the Blair Cuspids—and, indeed, by the very possibility of artificial structures on the surface of the Moon—I suggest the lunar interior is where we ought to focus our attention should we wish to find evidence of an alien presence on the Moon. This, I'm fairly sure, is where the "real action" is to be found. Even Davies and Wagner stated that lunar lava tubes would be good places to search for alien artifacts, given they'd make ideal habitats for beings living on the Moon. Nowhere in the paper do the scientists suggest the Moon's lava tubes might still be inhabited by aliens, living a quiet and covert existence.

After all, if there are aliens living on the Moon—and if, as Leonard so rightly pointed out, these aliens are probably "parasitic on Earth"—why should they make themselves known to us? The fact is they wouldn't. They'd do everything in their power to prevent being discovered, as is the nature of a parasite. How come, when you have a bloodsucking leech attached to your ankle, you don't notice it's there straight away, but much later when you go to remove your shoes and socks? This is because, when leeches bite, they release an anesthetic to numb the wound (in addition to an anti-clotting enzyme). What makes the leech such a successful parasite is its ability to remain unnoticeable to its host.

Let us imagine that the Moon is currently inhabited by a subsurface-dwelling alien presence, which, being an exploiter of humanity and the Earth, does not wish to be discovered. Parasites are, by nature, elusive. Common sense would suggest the most effective way to observe and understand this alien presence would be to visit the Moon, its natural habitat; there, one might be fortunate enough to catch a fleeting glimpse of alien activity (maybe in the form of a "cluster" of lights positioned inside a lunar crater).

The majority of UFO enthusiasts are familiar with Steven Greer's Disclosure Project, which is officially described as a "non-profit research project working to fully disclose the facts about UFOs, extraterrestrial intelligence, and classified advanced energy and propulsion systems."[58] As part of the Disclosure Project, a large press conference was held at the National Press Club in Washington on May 9, 2001. During the event, more than twenty former government workers, many of them military and security officials, came forward to speak, each claiming to have witnessed undeniable evidence of UFO or alien activity—evidence that, they alleged, is suppressed by the government. All of the witnesses said they'd be willing to testify under oath before Congress.

One of the witnesses present was former NASA employee Donna Hare. Specifically, Hare said she worked as a photo technician with NASA contractor Philco Ford at the Johnson Space Center in Houston. Throughout this fifteen-year period, which encompassed the Apollo program and ended during the early years of the Space Shuttle program, Hare held a special clearance that allowed her access to areas that were normally restricted. One time during the 1970s, according to Hare, she entered a restricted area of the photo lab and saw a male employee putting together a mosaic of photos—a collection of small photos joined together to create a larger image. So far as Hare could tell, they were satellite photos of Earth. The employee directed Hare's attention to one particular photo, which featured what looked like a UFO with "a round white dot," with "very crisp, very sharp lines on it." On the ground beneath the "dot" was a shadow, indicating that it had to be a physical object rather than a photographic defect. Without saying so explicitly, the employee informed Hare the object was exactly what it looked like—a UFO. Hare continues the story: "So I said, 'What are you going to do with this information?' And he said, 'Well, we always have to airbrush them out before we sell them to the public.' And I was just amazed that they had a protocol in place for getting rid of UFO pictures on these things."[59]

According to Hare, she ended up meeting a number of employees at NASA who told her they'd heard or seen things of a UFO nature, and who were careful not to talk about the subject too openly because of the secrecy surrounding it. This included a security guard who, at one time, had been "forced to burn a lot of UFO pictures." Another man, a close acquaintance of hers (who later "disappeared off the face of the Earth"), had spent time in quarantine with the astronauts. He learned that "just about all" of the astronauts had "seen things when they went to the Moon." He was even informed by one of the astronauts that "craft were on the Moon at the time of the landing."[60]

If the testimony of Karl Wolfe can be trusted, there most likely is an alien base on the far side of the Moon. A Disclosure Project witness like Hare, Wolfe spent roughly four years as an employee of the U.S. Air Force, beginning in 1964. Positioned at Langley Air Force Base in Virginia, Wolfe worked as a precision electronics photographic repairman with a "top secret crypto clearance." In 1965, he says, while assigned to a project connected to the Lunar Orbiter program, he was required to visit a NASA facility at Langley to carry out repairs. While there, in the photo lab, Wolfe struck up a conversation with a fellow Airman Second Class. About thirty minutes into the conversation, Wolfe was surprised to hear the young airman say, "By the way, we've discovered a base on the backside of the Moon."[61] The rest of the story is best told in Wolf's own words:

I said, "Whose? What do you mean? Whose?" He said, "Yes, we've discovered a base on the backside of the Moon." And at that point

I became frightened and I was a little terrified, thinking to myself that if anybody walks in the room now, I know we're in jeopardy, we're in trouble, because he shouldn't be giving me this information. I was fascinated by it, but I also knew that he was overstepping a boundary that he shouldn't. Then he pulled out one of these mosaics and showed this base on the Moon, which had geometric shapes. There were towers, there were spherical buildings, there were very tall towers and things that looked somewhat like radar dishes but they were large structures. . . . Some of the structures are half a mile in size. So they're huge structures. . . . Some of the buildings were, as I said, very tall, thin structures. I don't know how tall they were but they must be very tall. They were angular shots with shadows. There were spherical and domed buildings that were very large. They stood out very clearly; they were large objects. It's interesting because I tried to relate them in my own mind to structures here on Earth, and they don't compare to anything that you see here in scale and structure. . . .[62]

According to Wolfe, he had mixed feelings about looking at the picture of the lunar base. As well as being amazed and eager to know more (indeed, who wouldn't be?), he was deeply afraid of what the consequences might be were someone to enter the room and catch him in the act. As he put it, "I felt that my life was in jeopardy."[63] He knew that by looking at the picture he was breaching security and that his fellow airman, by showing him the photo, was also breaching security. Thus, he resisted the urge to examine the picture more fully.

A clip of Wolfe describing the alleged incident, during his appearance at the Disclosure Project's National Press Club meeting in 2001, is available to watch on the Internet. He seems to come across as genuine, as anyone who has seen the clip will agree. While it would be exciting to think that Wolfe is telling the truth—that there really is an alien base on the far side of the Moon—we have no way to determine the veracity of his story. And the same goes for NASA "whistleblowers"

Johnston and Hare. The area of UFOs and government conspiracies is littered with professionals in disinformation, making it difficult to judge what's true and what's false.

It's a little known fact that, during the 1950s, the U.S. Army explored the idea of constructing a military/scientific base on the Moon. They even went so far as to publish, on June 8, 1959, a lengthy and detailed report on the matter titled *Project Horizon: A US Army Study for the Establishment of a Lunar Military Outpost*. As stated in the report:

> There is a requirement for a manned military outpost on the moon. The lunar outpost is required to develop and protect potential United States interests on the moon; to develop techniques in moon-based surveillance of the earth and space, in communications relay, and in operations on the surface of the moon; to serve as a base for exploration of the moon, for further exploration into space and for military operations on the moon if required; and to support scientific investigations on the moon.[64]

Responsible for Project Horizon was a group within the Army Ballistic Missile Agency (ABMA), which was headed by aeronautical engineer Heinz-Herman Koelle. ABMA developed the Redstone, Jupiter-C, and Pershing missiles, none of which would have been possible without the scientific expertise of Werner von Braun, who served as technical director (and later chief) of the program. With an estimated cost of six billion dollars, it was hoped that the army's "manned military outpost" would be operational by the end of 1966. Construction was to begin in April of 1965, with the arrival of the first manned landing of two men. The plan was to make the base big enough to support a crew of twelve. As for the location of the base, several suggestions were made, including Sinus Asteuum and Sinus Medii. It may or may not be coincidental that 1965 was the same year that Wolfe supposedly saw the photo of the base on the far side of the Moon.

Project Horizon was planned in detail but never implemented—officially speaking, at least. Given that the 1969 Moon landing was itself a struggle for the United States, technologically and financially, how could they have possibly established a base on the Moon by 1966? The assumption is that the military was overly optimistic and overly ambitious about the project—indeed, about the overall practicality of space travel. Even with the technology we have today, establishing a lunar base would be no easy task (though it would be entirely achievable). One theory holds that Project Horizon, although officially canceled, continued in secrecy; meaning the U.S. military does have a base on the Moon. Connected to this is the theory of a secret space program, utilizing technology far in advance of what is available publicly. "Over the years I have encountered no shortage of quiet, serious-minded people who tell me of their knowledge that there is such a covert program," stated Richard Dolan, a respected authority on UFOs. "Are there bases on the far side of the Moon? I do not know for sure, but I cannot rule it out."[65]

If there are "bases on the far side of the Moon," it's difficult to imagine how these bases could be human in origin. In Wolfe's account, the alleged base is described as having a distinctly nonhuman appearance. As he said, "I tried to relate them in my own mind to structures here on Earth, and they don't compare to anything that you see here in scale and structure. . . ." Again and again, we encounter the theme of an alien presence on the Moon. There is some evidence, as I explained before, that the astronauts saw an alien base on the far side of the Moon during the Apollo missions. However, is there any evidence to back up the claim that the astronauts witnessed UFO activity on the Moon?

Otto Binder (1911–1974), in a 1972 article he wrote for *Saga* magazine, offered the following undocumented anecdote with regard to the *Apollo 11* mission: "Certain sources with their own VHF receiving facilities that bypassed NASA broadcast outlets claim there was a portion of earth-moon dialogue that was quickly cut off by the NASA monitoring staff." This portion of dialogue supposedly occurred while Aldrin and Armstrong were on the lunar surface conducting activities some distance from the LEM and begins with Armstrong exclaiming: "What was it? What the hell was it? That's what I want to know." A moment later one of the astronauts reports to mission control: "These babies are huge, sir . . . enormous. . . . Oh, God, you wouldn't believe it! I'm telling you there are other spacecraft out there . . . lined up on the far side of the crater edge . . . they're on the Moon watching us."[66]

Binder's story has been taken as proof that not only were the *Apollo 11* crew confronted by UFOs, but that NASA censored transmissions from *Apollo 11* (and the other Apollo missions) by using, when necessary, radio channels unknown to the public. At the same time, a very large question mark surrounds the authenticity of the story, while Binder himself admitted it had never been confirmed "by NASA or any authorities." Without naming those "certain sources" of his, he added, "We cannot vouch for its authenticity but, if true, one can surmise that mission control went into a dither and then into a huddle, after which they sternly told the moonwalkers to 'forget' what they saw and carry on casually and calmly as if nothing had happened."[67]

Binder was an author of science-fiction, nonfiction, and comic books as well as the creator of the DC character Supergirl. A fan of UFOs and the unexplained, he wrote a number of books on such matters, one of them called *Mankind: Child of the Stars* (1974). The book, written in collaboration with Max H. Flindt, features a foreword by Däniken and argues in favor of the theory that "mankind on Earth may have had super-intelligent ancestors from outer space."[68] It's been erroneously stated by some that Binder was a "former NASA employee"—perhaps as an attempt to add weight to his *Apollo 11* UFO anecdote. In actuality, Binder's only connection to NASA is that he wrote, under contract to the agency in 1966, the "Mercury, Gemini, and Apollo programs in chart form for educational purposes."[69] Also, prior to this, he was awarded by NASA an honorary master's degree in astronautical science.

Offering some support to Binder's story are the claims of Maurice Chatelain, who states in his book *Our Cosmic Ancestors* (1987) that Armstrong and Aldrin were indeed confronted by UFOs on the Moon. Just prior to Armstrong's historic moonwalk, he says, "two UFOs hovered overhead," several photographs of which were obtained by Aldrin.[70] If Chatelain is correct, encountering UFOs was a normal part of the astronauts' journeys into space. "It seems that all Apollo and Gemini flights were followed, both at a distance and sometimes also quite closely, by space vehicles of extraterrestrial origin. . . . Every time it occurred, the astronauts informed Mission Control, who then ordered absolute silence."[71] It would be easy to dismiss Chatelain as yet another writer of sensational hogwash, with no reliable knowledge of the subject in question, were it not for the fact that he did work for NASA.

Chatelain, who was born in France, immigrated to America in 1955 after seven years of living in Casablanca, Morocco. At the time of his departure from Morocco, the country had been granted independence from France, with the result being "social and economic chaos." For a Frenchman such as himself, to remain in the country would have been unwise, even dangerous. Chatelain, accompanied by his wife and three sons, settled in San Diego, California, where he soon found work as an electronics engineer with the aerospace company Convair Astronautics. Later, while working for Ryan Electronics, Chatelain was involved in developing new radar and telecommunication systems, receiving numerous patents for systems that he designed himself. When, in 1961, North American Aviation became the chief contractor for the Apollo Command/Service Module (CSM), Chatelain—having established himself as a radar and telecommunications specialist—was put in charge of designing and building the Apollo Communication and Data Processing System.

Had NASA wanted to hide certain transmissions from Apollo from the public—ones concerning UFO activity, for example—it could quite easily have done so, as Chatelain's description of the technology indicates: "When Apollo arrived within proximity of the Moon, the communications systems previously used could not reach that far; so all communications went through one single, very powerful, transmitter with a directional antenna in the S band, between 2,106 and 2,287 MHz, with a great number of channels, each transmitting several signals at the same time through multiplexing. For instance, there were seven channels to feed medical information about the physical condition of the astronauts, nine to retransmit the stored telemetry data from the passage behind the Moon that could not be beamed directly."[72] With all those radio channels available to NASA, concealing certain transmissions wouldn't have been a problem.

Although Chatelain doesn't say explicitly that secret radio transmissions were exchanged between the astronauts and mission control, he does refer to the use of the codeword "Santa Clause" by the astronauts "to indicate the presence of flying saucers next to space capsules."[73] According to Chatelain, James Lovell, while aboard *Apollo 8* in 1968, wasn't the first of the astronauts to use the codeword; the first was Walter Schirra aboard *Mercury 8* in 1962. Chatelain makes no mention of the *Apollo 8* crew having seen an alien base on the far side of the Moon. Rather, he says that what they saw—and, indeed, what Lovell's "Santa Clause" statement referred to—were UFOs within close proximity to the capsule.

Chatelain uses general rather than specific terms when describing a number of encounters between UFOs and manned NASA spacecraft— stories that he presumably picked up from others during his years as a NASA employee. Says Chatelain, "It is very difficult to obtain any specific information from NASA, which still exercises a very strict control over any disclosure of these [UFO] events."[74]

According to rumors heard by Chatelain, the *Apollo 13* oxygen tank explosion was caused not by accident but intentionally, by a UFO that had been following the spacecraft. Apparently *Apollo 13* had been carrying a nuclear device on board, and the plan was to drop this device on the Moon as part of a seismic experiment to measure "the infrastructure of the Moon." (Yet more evidence of NASA's obsession with trying to shed light on the lunar interior.) By sabotaging *Apollo 13*, the UFO prevented NASA's planned detonation, one that "could possibly have destroyed or endangered some Moon base established by extraterrestrials."[75]

With all this talk of alien bases on the Moon and astronauts encountering UFOs on or near the Moon—add to this those startling reports of TLP—it's hard to believe that our nearest celestial neighbor is uninhabited. While it's true that some of the stories we've examined are impossible to verify and may even be partially or entirely false, the evidence, taken as a whole, strongly suggests that there is a nonhuman presence on the Moon. I base this hypothesis on the principle that "where's there's smoke there's fire," and in this case the smoke is overwhelming. It's up to us to confront the smoke, even though it makes our eyes sting, and locate the elusive fire.

Unfortunately most UFO researchers and enthusiasts subscribe to the simplistic theory that these objects represent extraterrestrial vehicles piloted by flesh-and-blood humanoid beings. Perhaps this explanation applies to some UFOs and not others. Certainly it doesn't apply to all UFOs, because the phenomenon is simply too strange and mysterious for that to be the case. I'm not denying that there's a physical component to the UFO phenomenon; there is, and it's worthy of serious scientific investigation. I'm saying that the phenomenon encompasses not just the physical but also the nonphysical and the spiritual. The famous "abductee" Whitley Strieber, who wrote a series of excellent books detailing his encounters with the "visitors," as he calls them, hit the nail on the head when he wrote, "The visitors are physically real. They also function on a non-physical level and this may be their primary reality."[76]

One is inclined to think of the alien presence on the Moon as consisting of small, humanoid beings living beneath its surface who occasionally visit Earth in their amazing spacecraft, perhaps when in need of resources. Such notions are familiar and comfortable and require little depth of thought. But we need to open our minds to the possibility that what we're dealing with might be life of an entirely different nature, a form of life that functions primarily on a nonphysical (and therefore invisible) level; one that makes up for its physical limitations by excelling in terms of awareness and mental ability. Whatever form of life inhabits the Moon, it appears to be mentally superior to humanity, to possess powers of mind that we do not. This is suggested in the following quote from Chatelain: "It was said that during their flights, our astronauts frequently felt as if some external forces were trying to take over their minds. They experienced strange sensations and visions."[77]

9

THE LUNAR SPHERE

*The demons and succubae of occult lore are said to exist
on the next dimension "above" ours—commonly affiliated
with the lunar sphere—and draw off subtle energy from
the material world.*

AEOLUS KEPHAS, *HOMO SERPIENS*

Ingo Swann, born on September 14, 1933, in Telluride, Colorado, is an
artist, author, and scientist with a certain unique ability (see plate 10).*
Although it would seem all humans have a psychic faculty, or "sixth
sense," in most of us this faculty is weak, unreliable, and inconsistent.
Swann, by comparison, is one of those rare individuals who was born
psychically gifted, and who, through training and dedication, has managed to harness this gift for practical purposes. Known as the father of
remote viewing (RV), Swann played a crucial and pioneering role in the
development of RV methodology.

Remote viewing, for those unfamiliar with the term, is a system

---

*Ingo Swann passed away on January 31, 2013, at the age of seventy-nine. This
chapter was written prior to his death.

of clairvoyance that was created and utilized by the U.S. government for intelligence gathering purposes. It is, according to the original Department of Defence (DOD) definition, "the learned ability to transcend space and time, to view persons, places, or things remote in space or time; to gather and report information on the same."[1] The U.S. government's classified RV program, known in its final years as STAR GATE, was discontinued and became known to the public in 1995.

As a child born to hardworking Swedish immigrants, Swann spent much of his time outdoors, which helped to foster his free-spirited nature. He learned to read at an early age; by age seven, he'd already read the entire *Encyclopaedia Britannica*. At age two he had an out-of-body experience while undergoing an operation at the hospital to have his tonsils removed. After being administered ether (as a general anaesthetic), he found himself floating about his body and was able to observe every detail of the operation. He saw the doctor place the removed tonsils into a small bottle, and then he saw a nurse move the bottle behind some rolls of tissue on a counter. When the operation was over and he'd regained consciousness, Swann demanded that the doctor give him back his tonsils. But the doctor said he'd thrown them away. "No you didn't," replied Swann, pointing to the rolls of tissue. "You put them behind those over there. Give them to me."[2]

After studying art and biology at Westminster College, Swann joined the U.S. Army, then after that in New York where he worked for the United Nations. Swann spent eleven years with the UN, from 1958 to 1969, and found it to be a "wonderfully broadening experience."[3] He resigned after becoming disillusioned with the organization. This was followed by a career as an artist, out of which Swann made very little money. "No one wanted my paintings, and I refused to paint what everyone wanted," he says.[4]

Oddly enough, Swann's interest and involvement in parapsychological research was encouraged, in part, by the acquisition of a pet chinchilla in 1970. (A chinchilla is a South American species of rodent.) Swann named him Mercenary and allowed him to roam free in his apartment during the day. At night he kept Mercenary locked in his cage. Mercenary had a habit of hiding at night to avoid being placed in his cage.

On one particular night, Swann sat watching television with Mercenary perched on his knee. Swann was surprised to notice that, as soon as the thought passed through his mind that he ought to lock Mercenary in his cage, Mercenary suddenly bolted from the room, as though having read his mind. Curious to see if Mercenary really could read his mind, Swann began to lock him in his cage at unexpected times. "The effect was superb," says Swann. "After what appears to have been a brief learning period, he refused to react but would merely sit back on his hind legs, his eyes blazing, his tail switching back and forth, while he mentally probed my mind to find out if I meant it or if this was just another testing sequence." This left Swann in little doubt that Mercenary "could perceive and apprehend my thoughts."[5]

In 1971 Swann became a close associate of the polygraph (lie detector) expert Cleve Backster, who is famous for his controversial research into "primary perception." In 1966 Backster conducted a novel experiment whereby he hooked up a polygraph to the leaf of a philodendron in order to measure how quickly water traveled from the plant's roots to its leaves. (He used only the galvanic skin response section of the polygraph, which measures electrical conductance.) This didn't seem to work. Instead, Backster observed something very odd: whenever he harmed or threatened to harm the plant, even if only mentally, the plant exhibited a response on the polygraph similar to that of a human under stress. Backster was able to prove—to his satisfaction at least—that plants are highly sentient, to the extent of possessing ESP. Swann spent a year working in Backster's laboratory.

Also in 1971, Swann was invited to take part in a series of experiments for the American Society for Psychical Research (ASPR). The experiments were conducted by parapsychologists Karlis Osis (1917–1997), research director of the ASPR, and Gertrude Schmeidler (1912–2009), a professor of psychology at City University of New York. In one type of experiment Swann was to attempt to undergo an out-of-body experience (OBE) in order to view an object (the target) that had been placed inside a lighted box. An OBE, in the proper meaning of the term, is supposed to involve a sensation of floating outside of one's body, such as what Swann experienced as a child. In occult terms it involves the separation of the astral body from the physical body. Because the box sat on a platform suspended from the ceiling, its contents were entirely hidden—except, of course, when viewed from the perspective of someone "hovering" above or near the box.

Without necessarily undergoing an OBE, Swann found he was occasionally able to perceive the target hidden inside the box. Either he would describe what the target looked like or he would sketch the target. "At first I was not very good at this kind of 'perceiving,'" says Swann, "but as the months wore on I got better at it."[6]

On one occasion Swann reported to the testers that he could perceive only darkness, shouting, "the goddamn light is out over the target!"[7] A quick inspection proved that Swann was correct—the light had indeed stopped working. On another occasion, during a rest between experimental sessions, Swann's mind began to drift, and he found himself "transported" outside, "looking" at people walk down the street. One individual stood out: a woman wearing a ridiculous-looking orange raincoat. Eager to confirm the impression, Swann, accompanied by assistant Janet Mitchell, raced outside. They arrived just in the nick of time to see a figure dressed in an orange raincoat turn the corner of Central Park West.

In a different kind of experiment, Swann was to attempt to "see" the weather conditions in distant cities. After Swann had reported and recorded his impressions, Mitchell would telephone the weather station for the particular city in question to check if Swann had scored a "hit" or "miss." Explained Swann, "The cities we were to play around with 'seeing' were selected by third parties. So unless I had somehow memorised all of the weather conditions in a large number of cities, it is reasonable to conclude that a 'distant seeing' had taken place if the response was fed back as correct."[8]

322   The Lunar Sphere

On December 8, 1971, the target selected was Tucson, Arizona. Swann recalled:

> When I first heard mention of Tucson, Arizona, a picture of a hot desert flashed through my mind. But then I had the sense of moving, a sense that lasted but a fraction of a second. Some part of my head or brain or perception blacked out—and there I was—there. Zip, bang, pop—and there I was . . . something I would refer to in years ahead as "immediate transfer of perceptions."[9]

Swann gave his impressions as follows: "I am over a wet highway, it is cold. There is a strong wind blowing and it is raining heavily. This is a real rainstorm!"[10] By phoning the Phoenix weather bureau, it was discovered that Swann had been correct. Tucson has a desert climate, yet it happened to be stormy that day. Initially, Swann suggested that experiments of this kind be referred to as "remote sensing." This was soon altered to "remote viewing," when it became apparent that Swann "didn't just sense the sites, but experienced mental-image pictures of them in a visualizing kind of way."[11]

Following his impressive work with the ASPR, Swann was invited to the prestigious Stanford Research Institute (today known as SRI International), based in Menlo Park, California, to participate in further parapsychological experiments. The man behind this research was a physicist named Harold Puthoff, who, at the young age of thirty-three, was granted a patent for an adjustable infrared laser he had invented. Puthoff, a former scientologist and current CEO of EarthTech International (a company that carries out research with advanced space propulsion concepts and zero-point energy), is recognized as something of a genius.

One especially notable experiment at SRI occurred on June 6, 1972, and involved a heavily shielded superconducting magnetometer housed in the physics department at Stanford University. Swann's task was to try to influence the device from a distance by using only his mind. To the astonishment of everyone present, as soon as Swann began to focus on the device, by attempting to remote view its interior, there occurred a significant disturbance in the magnetometer's field output as displayed on a chart recorder. This lasted for approximately thirty seconds. Asked to stop the magnetometer's field output altogether, Swann managed to do so for a period of approximately forty-five seconds. Puthoff was impressed by this apparent demonstration of psychokinesis; he was particularly blown away by Swann's astonishingly accurate RV sketches of the interior of the magnetometer.

About two weeks after the magnetometer experiment, Puthoff was approached by members of the Central Intelligence Agency (CIA) who had heard about his research and wanted to fund it. The result was a remote viewing project by the name of SCANATE (SCANning by coordinATE), which began at SRI on May 29, 1973, and ended late the following year. For this, Puthoff was assisted by physicist Russell Targ, a pioneer in the development of laser technology. SCANATE employed the psychic talents of not just Swann but of a former police commissioner named Patrick H. Price, who proved to be an extremely skilled remote viewer. At the time, with the Cold War at its height, it was known that the Soviets were busy conducting research into parapsychology (what they called "psychotronics"), whereas the Americans had done very little research in this area. The CIA, concerned about this apparent "psychic gap" between the two nations, was eager to find out more about psi— hence their involvement in Puthoff's research.

By the latter half of the 1970s, the remote viewing research at SRI was receiving its funding from the Department of Defense (DOD). The remote viewing protocols developed at SRI were put into practice by the U.S. Army beginning in 1977 with the establishment of a project named GONDOLA WISH. A small group of psychically gifted individuals, some of them soldiers, others civilians, were recruited and trained to become "psychic spies." The project, which was placed under control of the U.S. Army's Intelligence and Security Command (INSCOM), became known as GRILL FLAME in mid-1978 and operated out of

Fort Meade, Maryland. So successful was the project that it continued until 1995, at which point it was known as STAR GATE.

That remote viewing works is beyond dispute. To quote Major General Edmund Thompson, who approved the establishment of GONDOLA WISH while employed as the Army's Assistant Chief of Staff for Intelligence, "I never like to get into debates with sceptics, because if you didn't believe that remote viewing was real, you hadn't done your homework. . . . We weren't so much interested in explaining it as in determining whether there was any practical use to it."[12]

During April 1973, when remote viewing research at SRI was still in its infancy, Swann suggested to Puthoff and Targ that they allow him to remote view the planet Jupiter. After much coaxing, they agreed. Swann was excited by the experiment. For, up until that time, the targets he had been assigned to were very mundane. Certainly none of those targets had pertained to anything "off-world." The experiment took place on April 27, 1973, roughly eight months before *Pioneer 10* (launched March 3, 1972) was due to fly by Jupiter to send back data on the planet—the very first space probe to do so. "Wouldn't it be interesting," wrote Swann, "if a psychic probe of Jupiter could be compared to the eventual feedback from *Pioneer 10*?"[13] Swann's "psychic probe" of Jupiter resulted in an extremely detailed report, copies of which were sent to scientists, two of them astrophysicists at NASA's Jet Propulsion Laboratory (JPL).

"There's a planet with stripes," begins Swann's report. "I hope it's Jupiter. I think it must have an extremely large hydrogen mantle. . . . Very high in the atmosphere there are crystals, they glitter, maybe the stripes are like bands of crystals, maybe like the rings of Saturn, though not as far out as that, very close to the atmosphere." Swann's report goes on to list details such as "tremendous winds . . . very close to the surface of Jupiter," the color of the planet being "orangish or rose colored" and "greenish-yellow," the presence of a "thermal inversion," as well as "an enormous mountain range."[14]

Due to the limited amount of data obtained by *Pioneer 10,* it remained difficult to tell whether there was any truth to Swann's descriptions. It wasn't until 1979, when *Voyager 1* and *Voyager 2* both made close approaches to the planet, that Swann was found to be correct in a number of respects. The space probes confirmed, for example, that Jupiter does indeed have a system of rings. These are very faint—so faint that they cannot be viewed from Earth—and are thought to consist primarily of dust (not crystal). To find rings around Jupiter came as a huge surprise to scientists at the time; the discovery was entirely counter to what was expected.

Swann also talked about entering Jupiter's atmosphere and seeing "cloud layers," which he also described as "crystal layers." We now know that Jupiter's white clouds do consist of crystals—specifically, crystals of frozen ammonia. Swann was also correct to say Jupiter has a hydrogen mantle; the planet undergoes very strong winds; and one of its colors is orange. Overall, it's impossible not to be impressed by the results of Swann's psychic "visit" to Jupiter. Remote viewing, when conducted by someone as psychically talented as Swann, has the potential to yield extremely valuable information.

Swann is the author of numerous books on psi, one of them a bestselling novel called *Star Fire* (1978), which explores the topic of psychic warfare. By far his most gripping and mysterious book is a brief, semi-autobiographical work called *Penetration: The Question of Extraterrestrial and Human Telepathy* (1998), of which few copies were published. Sadly, unable to find a publisher for the book, Swann had no other choice but to publish it himself. He said more than twenty publishers turned down the project, despite the fact that he was earlier assured by his literary agent that "its successful publication was a sure and easy thing." Explains Swann in the preface to *Penetration:* "This blanket rejection on such a large scale remains . . . mysterious. . . . One possible explanation might be that as outrageous as the tale and telepathic considerations are, something in them moves too close to Someone's comfort."[15] In the first sixty pages of *Penetration,* Swann relates a very unusual experience, whereby, during the mid-1970s, he was employed by a profoundly secret intelligence "group" to remote view the far side of the Moon. What Swann discovered as a result of this assignment forever changed his view of Earth's satellite.

According to Swann, in early 1975, roughly two years after his stunning remote viewing experiment concerning Jupiter and still engaged in research at SRI, he was contacted by a mysterious individual under the alias of Mr. Axelrod. During their first conversation, which took place over the phone at 3 a.m., Swann was instructed to meet Mr. Axelrod's associates at the National Museum of Natural History, in Washington, D.C., and to tell no one where he was going. This he agreed to do, despite feeling somewhat uneasy about the situation. Once at the museum, Swann was met by a pair of young, handsome, well-built men who were, as far as he could tell, twins. One of the twins handed him a card that read, "Please do not speak or ask any questions. This is for our safety as well as yours."[16] To make sure they had the right man, the twins subjected Swann to a minor bodily inspection. After this, Swann was escorted by the twins to a car waiting outside.

Inside the car, Swann was asked—once again by card—to remain silent. He was then frisked and a hood placed over his head. The car journey was followed by a ride in a helicopter, which was followed by a downward ride in an elevator. The distance the elevator traveled was fairly far. By the time it reached its destination, Swann knew he must have been deep underground inside a secret facility of some sort. Here the hood was finally removed, and he found himself face-to-face with the enigmatic Mr. Axelrod, whom he describes as a "jolly-guy type, smiling, with kindly eyes."[17]

Swann was told by Mr. Axelrod, "I can answer no questions as to where you are or what we represent. . . ."[18] It was revealed that Swann had been brought to the facility to undertake a remote viewing assignment pertaining to the far side of the Moon at a rate of $1,000 per day (this, remember, was 1975). Normally Swann would be asked to sign a nondisclosure agreement. But because the group to which Axelrod belonged was so covert that it existed "without leaving a paper trail," there could be no such written agreement. Rather, the agreement was a verbal one, in that Swann was asked not to talk about the assignment for a period of ten years. Although Swann had the option to turn down the assignment, he was, he admitted, "hot on the trail of $1000-days."[19]

Before going ahead with the remote viewing, Mr. Axelrod wanted to know if Swann had heard of George Leonard. (At the time, Leonard's book *Somebody Else Is on the Moon* had not been published. It was published the following year, in 1976.) "I don't recall any George Leonard," responded Swann. "There's a Leonard at SRI, but I can't remember names very well. . . ."[20] To make absolutely sure that Swann was unfamiliar with Leonard, Mr. Axelrod showed him a number of photographs with faces on them—one of them, of course, was Leonard. Swann was unable to recognize the face. Only later, after reading *Somebody Else Is on the Moon,* did Swann discover who Leonard was. The connection between Leonard and Swann will become apparent in a moment.

Swann stayed the night at the "astonishing underground facility," where he was provided his own room and allowed access to a small gym and swimming pool. The first of his remote viewing sessions occurred early the following morning. After achieving a "psychic touchdown" on the Moon, Swann was surprised to see, in his mind's eye, features of an artificial nature, including tractor tread marks and rows of green lights. The lights were high up on "towers of some kind," and reminded Swann of "lights at football arenas."[21] Investigating these "light towers" more closely, Swann said they appeared to be "constructed of some very narrow struts of some kind, thin like pencils."[22] He estimated the height of each light tower to be more than one hundred feet. Swann also reported seeing a huge tower positioned on the edge of a crater, comparable in height to the United Nations Secretariat Building. (This contains thirty-nine floors, and stands 505 feet tall.) Only once the session had ended did it become apparent to Swann that he'd just seen artificial structures on the far side of the Moon—structures so huge and complex that they could not have been made by human beings.

According to Swann, an extraterrestrial presence on the far side of the Moon was confirmed by Mr. Axelrod, who knew far more about the matter than he was willing to disclose. From what Swann was able to gather, NASA and others within the U.S. government, although fully aware of the situation, were not sure how to deal with it. "No one has known what to do, and many mistakes have been made," as Mr. Axelrod put it.[23] Swann was curious to know why he'd been hired to psychically spy on the Moon's inhabitants, when, presumably, a simple satellite would have been far superior in terms of gathering intelligence. The reason for this, he discovered, was disturbing to say the least: the Moon's inhabitants had proved to be hostile toward humanity and had made it abundantly clear that they wanted us to keep our distance. Psychic spying, then, was the only option available.

During his next remote viewing session, Swann perceived additional signs of an alien presence on the Moon.

> I found towers, machinery, lights of different colors, strange-looking buildings. . . . I found bridges whose function I couldn't figure out. One of them just arched out—and never landed anywhere. . . . There were a lot of domes of various sizes, round things, things like small saucers with windows. These were stored next to crater sides, sometimes in caves, sometimes in what looked like airfield hangars. . . . I found long tube-like things, machinery tractor-like things going up and down hills, straight roads extending some miles, obelisks which had no apparent function. There were large platforms on domes, large cross-like structures . . . Holes being dug into crater walls and floors, obviously having to do with some kind of mining or earth-moving operations.[24]

Swann also perceived what looked like dwelling places, one of which he managed to "enter." This he found to be reasonably dark, the only source of illumination being a "lime-green fog or mist." It was also occupied. The occupants consisted of a group of male beings who looked identical to humans, and who appeared to be hard at work, "digging into a hillside or cliff." Swann knew for certain that the beings were male, because "they were all butt-ass naked."[25] Much to his astonishment, he noticed that some of the beings were aware of his presence, nonphysical though his presence was. Two of them even "pointed in my 'direction,'" he says. The experience left Swann convinced that the inhabitants of the Moon possess a very advanced form of telepathy—what he refers to as "telepathy plus." This Mr. Axelrod was able to confirm, stating that they have "capabilities we here are trying to understand."[26] His remote viewing work at an end, Swann was free to return home.

Swann says he received in the mail, sometime during the summer of 1976, a copy of Leonard's *Somebody Else Is on the Moon*. The envelope in which the book had been packaged featured no postmark or return address. Swann felt certain the book had been sent by Mr. Axelrod— and, what's more, that it was intended as feedback for the remote viewing data he'd obtained earlier. Swann, who had sketched some of the objects he'd seen during his remote viewing sessions of the Moon, says that "many of Leonard's sketches resembled some of mine." He adds, "Apparently, at the time of my ultra-secret visit, Mr. Axelrod had already known that [Leonard's] book was coming out, and of course he had been interested in whether I knew the author or not."[27] Obviously, had Swann known Leonard, his remote viewing data would have been of no value, since Leonard may have told him about his research. While it was sensible of Mr. Axelrod to investigate this possibility, there is something very strange about the whole matter. To me it suggests that Leonard and Swann moved in the same circles, and Leonard may have been an employee at SRI, perhaps even, as discussed in the previous chapter, a member of the intelligence community. I also find it rather odd that Mr. Axelrod was aware of the publication of Leonard's book well before it was due to be released. If Swann's claims have any basis in truth, then, clearly, Leonard's lunar research ought to be taken more seriously.

So weird and comical are some of the events described in *Penetration* that they strain the reader's credulity. In the later months of

1976, according to Swann, while shopping at a very large supermarket in Hollywood, California, he found himself in the presence of an extremely "ravishing woman," with large, barely covered breasts. Swann, wanting to get a closer look at the woman's impressive features, casually approached her. At that moment, for no apparent reason, he felt "an electrifying wave of Goosebumps throughout my whole body." Swann then "knew" the woman was an extraterrestrial. To add to the bizarreness of the incident, Swann also noticed the twins, whom he first encountered during his initial meeting with Mr. Axelrod, were standing nearby, keeping a very close eye on the woman. "The twins' presence, coupled with my psychic alert, confirmed that the woman was an ET," writes Swann.[28]

A few days after the incident and back in New York, Swann was instructed by an anonymous female caller—clearly one of Mr. Axelrod's associates—to go to Grand Central Station that night. Once there, Swann encountered one of the twins, who, disguised as a typical-looking vagrant, led him to a nearby telephone booth. Mr. Axelrod was on the other end of the line. Swann was informed by Mr. Axelrod of the reason for his having to use a public telephone: because it "scrambles our conversation." Swann was then questioned about the incident in the supermarket. He was forced to explain to Mr. Axelrod that his visit to the supermarket had been entirely innocent and that he didn't know the strange woman he'd seen there. By the end of the conversation, Swann had managed to alleviate Mr. Axelrod's suspicions. Swann was then warned by Mr. Axelrod that the woman "is very dangerous. If you ever see her again, especially if she approaches you, make every effort to put distance between you and her."[29]

Swann suggests there are, living undetected among us, a large number of extraterrestrial beings—beings who, like the woman he saw in the supermarket, look entirely human but with perfect and extremely attractive bodies. He further suggests these beings hail from the Moon, and some of them are biological robots. He even questions whether the twins themselves were biological robots—not from Earth. When he first met the men, explained Swann, he was quite sure they were twins. (In fact, he refers to them as twins throughout the entire book.) Only later did he realize they had different-shaped noses and even different accents. It soon dawned on Swann that he'd mistaken the men for twins because they moved virtually in unison, "as if of one mind."[30]

Of course, this sounds like the stuff of science-fiction. One is reminded, for example, of the humanoid Cylons from the popular 2004 TV series *Battlestar Galactica*. The Cylons are highly advanced biological robots who look identical to humans, enabling them to infiltrate human society. Being far stronger than humans, they are potentially very dangerous. Fictional matters aside, it would be wrong of us to dismiss Swann's story as nonsense simply because it sounds unbelievable. We must not forget that there are "more things in heaven and earth" than are dreamed of in our philosophy.

Swann describes in *Penetration* an event so spectacular and mind-boggling that it makes his encounter with the sexy "extraterrestrial" in the supermarket sound dull in comparison. That event occurred in July 1977, he said, and it turned out to be the last time he ever saw Mr. Axelrod. Swann also says he was escorted by Mr. Axelrod and the twins: first by jeep, then by Lear Jet, then by foot to a remote, mountainous location (presumably in Alaska). Here, according to Swann, the four of them witnessed a triangular-shaped flying saucer materialize over a lake. Once the object had "grown" to the size of about ninety feet across, says Swann, it proceeded to suck up water from the lake below it. Swann received the impression that the craft was some kind of supply ship from the Moon, its purpose for visiting Earth to stock up on resources (water, of course, being one of those resources).

Swann says that, for reasons which remain mysterious, the events of the "Axelrod affair" quickly began to fade from his memory. Having become aware of this inexplicable amnesia, he decided, in 1991, to recall as much as he could about those events and write it all down before it was too late. The end result, of course, was *Penetration*. Speaking of amnesia, mind control, and other matters of this nature, Swann

332	The Lunar Sphere

suggests "Earthsiders as a whole seem to be caught up in some kind of strange but broadly shared amnesia induced, perhaps, in some kind of wholesale way by means totally unrecognizable by human intellects. I seem to recall . . . a science fiction story of social-wide amnesia having to do with hypnotic commands to FORGET, FORGET what you have seen, and ATTACK AND DESTROY those who insist they have seen it."[31] Swann uses the term "Earthsiders" to refer to human beings and the term "Spacesiders" to refer to the inhabitants of the Moon.

It is Swann's view that because Spacesiders are more technologically advanced than Earthsiders, they must be more advanced in terms of consciousness as well; it is to say that technological advancement and mental advancement go hand in hand. And this brings us back to Swann's "telepathy plus," which, he says, "beyond being a channel for information exchange might also achieve something along the lines of mind bending and fried brains."[32] He speculates that the "extraterrestrial" he saw in the supermarket whom Mr. Axelrod referred to as "very dangerous" was dangerous because she possessed this faculty. In the final chapters of *Penetration,* Swann cautiously advances the notion that we—Earthsiders—are being collectively influenced by Spaceside telepathy plus, which operates on a subtle, subliminal level. The essence of Swann's idea is that humanity is secretly controlled by an extraterrestrial presence on the Moon, possessing powers of mind and technology far in advance of our own.

According to David Ovason's *The Zelator* (1999), the Moon's inhabitants aren't extraterrestrials, nor do they have anything to do with biological robots. Rather, they are something far stranger and considerably more terrifying. Ovason, who lives in England, is an expert on astrology and the work of Nostradamus. He wrote *The Zelator* in collaboration with the late artist and esoteric initiate Mark Hedsel, whom he first met in 1955. The book is essentially Hedsel's "spiritual journal." Hedsel, who passed away in 1997, followed what's known as the "Way of the Fool," which incorporates elements of Hermeticism, Qabbalism, and alchemy. The objective of the "Way of the Fool" is to cultivate the "Ego," which is described as "the sacred self" and "the seat of control over the Will."[33] (The Ego roughly corresponds to what Gurdjieff called the soul.)

Scattered throughout *The Zelator*—a rambling, cryptic work if ever there was one—are some valuable fragments of esoteric knowledge related to Moon. It is the job of the reader to piece together these fragments as best they can. In *The Zelator* a distinction is made between those who have adopted some kind of spiritual path with the intention of awakening, and those who, because they couldn't care less about awakening, have adopted no such path. Belonging to the latter group are the vast majority of humanity—the "Sleepers." The Sleepers, we are told, "are the human slaves of the Moon goddess, Selene."[34] They "are content with the realm of appearances, and only want to be left alone, to sleep."[35] Selene is associated with sleep, and the connection relates to the myth of Endymion, the shepherd. Selene "seduced the sleeping Endymion," as explained in *The Zelator,* "and kept him in that dream-like condition of the Sleepers, which everyone on the Path seeks to awaken from."[36]

In *The Zelator,* Hedsel discusses how, at one point during his life, he moved to New York to study under a certain esoteric teacher, whom he describes as an "established and respected scholar" with a cheerful and flamboyant disposition. (Hedsel studied under numerous esoteric teachers, eventually becoming a teacher himself.) Hedsel quotes his teacher as saying:

The initiation centres have long recognised that Mankind is in thrall to the Moon—that ordinary men and women are sleeping under the influence of the lunar powers. This is sometimes symbolised through the typical lunar symbols of the serpent—the Egyptian snake, *Apep,* or the alchemical serpent, wrapped around the human form. It is usually portrayed as possessor of the spine of man or woman, yet which belongs to the Moon. This may account for the fact that serpents entwine in the hair of the Moon goddess Hecate.

... The snakes in the hair of Hecate are a sign of the extent to which the serpent still whispers imaginative words into the thinking of Mankind. These imaginative words are pictures derived from the Moon, the realm of imagination.[37]

"Ordinary men and women are sleeping under the influence of the lunar powers" sounds like something Gurdjieff would have said. In fact, the general content of the above quote is largely in agreement with his Fourth Way system. As for the serpent references, these take on an interesting meaning in light of what we know about the "special organ" that Gurdjieff called Kundabuffer.

In Greek mythology, the god Hermes (whose association with the Egyptian lunar deity Thoth has already been explained) is said to have carried a staff called the caduceus, which he received as a gift from his half-brother Apollo. The caduceus is similar to (and therefore often confused with) the staff carried by the Greco-Roman god of medicine, Asclepius. The staff of Asclepius has a single serpent coiled around it and is a symbol of medicine. The caduceus, on the other hand, has two serpents coiled around it, in addition to a pair of wings at the top (in token of Hermes's speed), and is a symbol of commerce. The serpents are coiled in opposite directions, their heads facing each other.

Many would recognize the caduceus as the emblem of the U.S. Army Medical Corps—a mistake since, as we know, the symbol has nothing to do with medicine. Nowadays, largely because of this emblem, the caduceus has adopted medical connotations. Many researchers interpret the caduceus as an ancient representation of the DNA double-helix. It's also been interpreted as a representation of Kundalini energy: "coiled serpent lying at the base of the spine." "The wand is the spine, and the serpents are the Kundalini force, or serpent power, which resides in the Earth and at the base of the spine," to quote Rosemary Ellen Guiley.[38] Now of course, from Gurdjieff's perspective, there is nothing at all positive about Kundalini, which keeps us in a state of hypnotic sleep. Interestingly, we find that Hermes was a dream god and he used his caduceus—this magic wand of his—to put mortals to sleep. If one wanted to be creative about it, one could suggest the true meaning of the caduceus relates to the insertion of reptilian DNA into the human species, resulting in our being placed under the spell of the Moon.

Hedsel was told by his teacher that there is not one Moon but two: the Lighted Moon and the Dark Moon. (Gurdjieff, as we know, also spoke of two moons.) The latter he referred to as "the Moon of snake-infested Hecate,"[39] relating back to the "serpent" that "still whispers imaginative words into the thinking of Mankind." It is revealed in *The Zelator* that the Lighted Moon "dispenses a beneficent influence, while the Dark Moon is evil."[40] The Dark Moon, which is personified as Lilith, lies behind the Lighted Moon, which is personified as the archangel Gabriel; thus, we cannot see the Dark Moon in the same way we cannot see the far side of the Moon, which remains perpetually turned away from us. In order to grasp the nature of the Dark Moon, one must know something about the Qabalistic Tree of Life.

The cornerstone of Western esotericism, the Qabalah (meaning "tradition" in Hebrew) is a profound and complex mystical tradition of Jewish origin. Central to the Qabalah is the Tree of Life, a glyph consisting of ten spheres, called Sephiroth ("numbers" or "categories"), whose specific arrangement forms "a blueprint for understanding all things and relationships in the universe, including the essence of God and the soul of humanity."[41] Each Sephirah is a divine energy, and they are linked together by a total of twenty-two paths. At the top of the Tree is Kether (crown), which might be defined as "pure consciousness." Kether gives rise to Chokmah (wisdom), which in turn gives rise to Binah (understanding). This is followed by Chesed (mercy), then Geburah (strength), Tiphareth (beauty), Netzach (victory), Hod (glory), Yesod (foundation), and finally Malkuth (kingdom).

The Tree is a living, moving system, similar to a river or an electric circuit. It could be compared to a series of ponds positioned downhill from one another; Kether acts as an ever-flowing stream that feeds them. As one pond becomes full, it overflows into the pond below it, which overflows into the next pond—and so on all the way down the line. According to the British occultist Dion Fortune (1890–1946), "each Sephirah contains the potentiality of all that come after it in the scale of down flowing manifestation."[42] Malkuth, positioned at the very bottom of the Tree, is the sphere of the Earth, of matter. Being the final "pond," it receives the emanations, or "waters," of all the Sephiroth above it. It "represents," says Fortune, "the end-results of all the activities of the Tree."[43] Of particular relevance to us in this chapter is the sphere that precedes Malkuth: Yesod, which is ruled by the Moon. We'll discuss the characteristics of Yesod in a moment.

The Tree is made up of three triangles with Malkuth being the only sphere that doesn't form a triangle; it is cut off from the Tree. The upper (supernal) triangle consists of Kether, Chokmah, and Binah; the middle (ethical) triangle consists of Chesed, Geburah, and Tiphareth; and the lower (astral) triangle consists of Netzach, Hod, and Yesod. Each of these three triangles exists in a state of balance, each consisting of a positive, a negative, and a reconciling sphere (which corresponds to Gurdjieff's law of three).

Additionally, the Tree can be split into three vertical rows, or pillars. On the right-hand side is the pillar of Mercy (which is male and positive); on the left-hand side is the pillar of Severity (female and negative); and in the middle is the pillar of Mildness or Equilibrium (neutral). In the case of the lower (astral) triangle, we have Netzach on the right, Hod on the left, and, in the middle, Yesod.

The Tree can be divided horizontally into four separate worlds, or stages of manifestation. At the top, consisting only of Kether, is Atziluth, the divine world of archetypes. Briah, the next world down, consisting of Chokmah and Binah, is the world of creation. Next is Yetzirah, the world of formation. This encompasses the spheres of Chesed, Geburah, Tiphareth, Netzach, Hod, and Yesod. The fourth and final world is Assiah, the world of action, which consists only of Malkuth. It's easy to recognize that Assiah corresponds to the physical world. As for Yetzirah, this corresponds to the astral plane.

The astral is the realm of emotions, of dreams, of illusion, and of spirits, demons, angels, and other entities. It knows no bounds. It can be a place of beauty and endless delight or one's worst nightmare. Because creation flows downward, events that occur in the astral end up manifesting on the physical plane. Occultists speak of different divisions within the astral: lower astral, middle astral, higher astral, and so forth. In terms of density, the lower astral is closer to the physical realm than is the higher astral. (Which is another way of saying that the lower astral is further from Kether than is the higher astral, making it, by comparison, less divine.) It figures that Yesod, being the astral sphere closest to Malkuth, must equate to the lower astral plane.

Some sources erroneously classify Yesod as *the* astral realm. I've also seen Yesod described as the "threshold" to the astral realm: a portal between the physical and spiritual worlds. It is more correct to say Yesod is the lower astral. This view is supported by the occult author Israel Regardie (1907–1985), who describes the domain of Yesod as "that gross region of the metaphysical cosmos containing the cast-off astral remnants of living creatures, the bestial and mental filth sloughed off by human beings in the ascent after death to higher spheres."[44] These "cast-off astral remnants" are known as "astral shells." I'll have more to say about them in a moment.

Occultists also speak of the "etheric plane" and the "etheric body" or "etheric double." The etheric body is said to surround and interpenetrate the physical body, and, being intermediate between the physical and the astral, to serve as a link between the two. According to some, although the astral body exists beyond the realm of the physical, the etheric body is just dense enough to be detected physically, provided the right conditions are present. Plants, animals, and humans—even inanimate objects like stones—are said to possess an etheric double. Whereas the astral is the spiritual body of emotions (the desire body, as it's sometimes called), the etheric is the spiritual body of memory. It "maintains the physical in cellular activity," to quote from *The Zelator*.[45] The etheric is virtually synonymous with Rupert Sheldrake's theory of morphic fields. Sheldrake, a Cambridge-educated biochemist and author, defines a morphic field as "a field within and around a morphic unit [a unit of form or organisation; for example, an atom] which organizes its characteristic structure and pattern of activity."[46]

The etheric plane—the etheric level of reality—is an invisible, energetic structure, or "mesh," around which the physical world maintains its arrangement. The tenuous "material" of the etheric plane is known, not surprisingly, as "etheric substance." Another term for it is "astral light." This Israel Regardie defines as "an omnipresent and all-permeating fluid or medium of extremely subtle matter," which is "diffused throughout all space, interpenetrating and pervading every visible form and object."[47]

Yesod, despite its association with the lower astral plane, is more specifically allied with the etheric plane. To quote Gareth Knight, an expert on the Qabalah: "Yesod is the Sephirah of the etheric plane and so not only is it the powerhouse or machinery* of the physical world it also holds the framework in which the particles of dense matter are enmeshed."[48] Whether one wishes to classify Yesod as the sphere of the etheric or the sphere of the astral is perhaps beside the point, since it's not as though the two realms are considered distinct—as if one were oil and the other water. Rather, it's believed that the two overlap, the astral being an extension of the etheric, which is itself an extension of the physical. Nevertheless, the specifics of the matter needn't bother us for the moment. All we need to recognize is Yesod, the sphere of the Moon, is the most immediate realm from the physical and, what's more, that it supports the physical realm on an invisible, energetic level.

According to occult thought, just as insects and people possess an etheric body, so too do planets and the Earth itself. However, the etheric body of the Earth does not belong to the Earth alone; it also belongs to the Moon. It is a shared etheric body, such that the Earth and the Moon are connected on a very deep level. Interestingly, the British

---

*It's interesting to see Yesod described in such terms, in light of what we know about the Moon's part-artificial nature.

occultist Fortune describes the Moon as the "senior partner" in the Earth-Moon etheric relationship. She elaborates on this by saying that "the Moon is the positive pole of the battery, and the Earth the negative one."[49] From a conventional perspective, when a lightbulb is connected to a battery the current within that circuit flows from the positive to the negative terminal. (In actual fact, the current flows in the other direction.) This aside, Fortune seems to be saying the Moon does the "pushing" and the Earth the "receiving."

Fortune says the Earth is not only subject to ocean tides produced by the Moon but also to etheric tides produced by the Moon—once again because the two celestial bodies share the same etheric double. These etheric tides play an important role in the growth of plants on Earth and in matters related to reproduction. It is when the Moon is full, she says, "lunar activities are at their most active." At new Moon, on the other hand, "etheric energy is at its lowest."[50] Fortune describes the genesis of the Moon as a process whereby it split from the Earth during a period when "evolution was on the cusp between the etheric phase of its development and the phase of dense matter." Thus, the "really important part of [the Moon's] composition is etheric, because it was during the phase of evolution when life was developing the etheric form that the Moon had its heyday."[51]

If what Fortune tells us is true, then no wonder we fail to comprehend the relationship between ourselves and the Moon. We feel the Moon affects us deeply, and it has the power to drive us mad and so on; yet, to our frustration, we remain unable to determine the physical mechanism behind this relationship. Our mistake, according to the occultist, is that we're looking at the matter from a materialistic perspective, whereas what is needed is an etheric perspective. Says Fortune, "In dealing with the rhythms of Luna we are dealing with etheric, not physical, conditions."[52]

The Tree of Life exists in a state of balance and harmony, and whatever embodies these qualities is divine. It figures, then, that what we call "evil" is simply an extreme form of imbalance or disharmony. In the Qabalah, every Sephiroth has its imbalanced, or evil, counterpart—its "shadow side," if you will. These are known as the Qliphoth (meaning "shells" or "husks"), the impure spiritual forces at work in the universe and within each human being. Really, then, there is not one Tree but two: a divine, Sephirothic tree and a nefarious, Qliphothic tree. The Sephirothic and the Qliphothic, without necessarily being opposites, are two sides of the same spiritual coin—like atheism and religion, for example. Kether represents unity, and so we find, not surprisingly, its Qliphothic counterpart Thaumiel (meaning "twins of God") represents duality. Associated with Kether is Metatron, who, in rabbinic literature, is the highest of the angels. Associated with Thaumiel, on the other hand, is Satan, the adversary of God, as well as the Ammonite deity Moloch, to whom children were once sacrificed.

Gabriel is the deity assigned to Yesod, while the Qliphothic counterpart of Yesod is Gamaliel ("recompense of God"). And who do we find haunting the shadowy sphere of Gamaliel? The demon Lilith, of course. If we wish to comprehend the nature of the Dark Moon then we ought to know something about Lilith. The "demon of screeching," as she is sometimes called, Lilith is an important figure in Jewish mythology. References to her appear in a number of rabbinic texts, among them the *Sefer ha-zohar* ("Book of Splendor")—the principal text of the Qabalah—the Babylonian Talmud, the *Alphabet of Ben Sira,* and the Dead Sea Scrolls. Within this body of literature are various depictions of Lilith. Either we are told she was Adam's first wife or that she was the mother of Adam's demonic offspring, following Adam's separation from Eve. While details differ between accounts, they all agree on one thing: Lilith was cruel, rebellious, vindictive, powerful, vampiric, highly seductive, and utterly damned.

According to one account, Lilith and Adam were created by God in the form of conjoined twins. Their relationship as husband and wife was far from harmonious. Lilith was angry and upset that she'd been denied equal status with Adam, which resulted in her leaving him. Hearing of this, God sent three angels to retrieve her. By the time they found

Lilith, she had become a lover to demons and had spawned innumerable demonic offspring. Lilith was told by the angels that if she failed to return to her husband she would be punished harshly by suffering the loss of her offspring at the rate of one hundred per day. Lilith, of course, refused to comply. In retaliation for the loss of her offspring, Lilith took to terrorizing the living, targeting pregnant women and young infants especially, which she kidnaps or strangles. In a variation on this myth, Lilith is created by God in the same manner as Adam; but whereas Adam is formed from dust, she is formed from impure sediment and filth. Refusing to be subservient to Adam and to lie beneath him during intercourse, she left him.

We are told that not only does Lilith target pregnant women and infants but also men who sleep alone, within whom she induces nocturnal emissions. Lilith belongs to that class of demons known as succubi. A succubus is a female demon who has sexual intercourse with sleeping men. (An incubus is the male equivalent of a succubus.) In the Zohar it is written that Lilith "wanders about at night time, vexing the sons of men and causing them to defile themselves."[53] Lilith's demonic offspring are called lilim, and it is said that when she seduces men she also collects their seed, which she uses to produce these vile creatures. Whenever men engage in sexual fantasies and acts of onanism, Lilith is sure to be fluttering close by, ready to collect any wasted seed or misspent sexual energy. Lilith's vampiric attributes are easy to recognize, and it's no wonder modern-day storytellers have made a blood-sucking vampire out of her (and not just any vampire, but the "queen" of them all). An entire book could be written about Lilith, so rich and complex is the mythology surrounding her. It makes perfect sense that she should personify the Dark Moon.

It's worth taking a moment to examine the famous painting of Lilith by John Collier (1850–1934), which does a brilliant job of capturing her complex nature. An Englishman who married into the famous Huxley family, Collier exhibits a thorough knowledge of occult matters. His painting of Lilith, dated at 1887, shows a nude, pail-skinned woman, with long ginger hair that reaches all the way to her rear (see plate 11). She is standing comfortably on the damp, green earth with eyes closed. Coiled lasciviously around her body is a large, dark serpent, the presence of which, we can tell, is a source of great pleasure for her. The two of them—a pale, nubile woman and the phallic, scaly beast—seem to be in a state of profound communion, their heads virtually pressed together. It would not be unreasonable to suggest that the serpent in Collier's painting is meant to be the serpent that tempted Adam and Eve in the Garden of Eden. Indeed, certain Qabalistic texts refer to Lilith as a serpent and associate her with the Eden serpent. What interests me most about Collier's painting is its possible connection with what is referred to in *The Zelator* as the "alchemical serpent, wrapped around the human form."

In the Qabalah, Yesod relates to the sexual organs, while, from an occult perspective, sexual activity has more to do with the spiritual than the physical body. It is said that our physical bodies are sustained by etheric energy and that any sexual act involves the movement of this energy. In the case of sexual practices that are healthy and balanced, the energy of Yesod is "grounded" in Malkuth and the "circuit" completed, as is the natural and harmonious way of things. In the case of unhealthy sexual practices, particularly those that involve onanism, the energy generated in Yesod flows not to Malkuth and so is never grounded. Rather, it lingers in the realm of the lower astral, like so much pollution and garbage, becoming sustenance for the likes of Lilith and other Qliphothic entities. They are called Qliphoth—"shells" or "husks"—because of their empty, vampiric nature. The late Kenneth Grant, who was a student of Crowley, defines them as "seductively potent energies which remain over after manifest entities have been dissolved."[54] The majority of Qliphothic entities, he says, are "habitations of the phantom forms generated by sexual desires and morbid cravings constantly produced by dwellers of Earth."[55] According to the occultist, then, not only do demons exist, but they exist because we feed them. Our "astral garbage" is what keeps them alive—ensouls them, even.

Practitioners of black magic focus on the realm of the demonic—the Qliphothic Tree. To them demons are allies, to be invoked when required. Crowley, for example, claimed to have established contact with a number of demonic entities from whom he received profound esoteric knowledge. Apparently, while staying in Cairo, Egypt, in 1904, Crowley (helped in part by his mediumistic wife, Rose Kelly) came into contact with an entity that called itself Aiwass. Aiwass manifested, he says, in the form of a disembodied voice that emanated from a corner of the room. As for the nature of the voice, he describes it as being "of deep timbre, musical and express. . . ." This was accompanied by a mental picture of Aiwass, consisting of a "tall, dark man in his thirties, well-knit, active and strong, with the face of a savage king, and eyes veiled lest their gaze should destroy what they see."[56]

Over the course of three days, Aiwass dictated to Crowley his famous *The Book of the Law* (1904), which states as its primary precept: "Do what thou wilt shall be the whole of the Law." Crowley was never able to determine whether Aiwass was an entity in his own right—a god, demon, or preterhuman intelligence—or, conversely, a part of himself, his "Holy Guardian Angel." The latter hypothesis appealed to him the most.

One of Crowley's most startling "entity encounters" occurred in 1918, at which time he was living in New York City. Crowley, assisted by a woman named Roddie Minor, performed a series of occult rituals known collectively as the Amalantrah Working, which involved the use of drugs and sex magic. The Working, which employed Minor's mediumistic ability, is said to have created a kind of interdimensional gateway, allowing an alien being, whom Crowley called Lam, to cross over into our universe.

A portrait of Lam drawn by Crowley shows a sinister-looking being with a large, egg-shaped head, no hair whatsoever, a slit for a mouth, a small nose, no ears, and eyes similar to those of an Asian. The entity looks uncannily similar to what is known in the field of ufology as a "grey." It's been suggested, in fact, that the "portal" opened by the Amalantrah Working was key to the emergence of the UFO phenomenon some thirty years later, the portal having grown wider over time. Crowley believed that Lam, which means, in Tibetan, "way" or "path," was the soul of a dead Tibetan lama. Others have suggested that Lam was wholly demonic in nature.

Of course, demonology is a topic that few people take seriously, and it is not my intention to convince the reader that demons, spirits, aliens, and other such beings exist. Whether these beings are real or a product of human imagination is up to the reader to decide. It's important to realize that for the occultist or shaman, nonhuman beings are as much a part of reality as the earth beneath one's feet. Crowley, though we may label him eccentric, even mad, did not doubt the reality of his encounters with Aiwass, Lam, and other beings.

In *The Zelator* a number of references are made to demonic entities, which are described as having more of an internal than external reality. Living inside each of us, we are told, is a "dark double," a kind of negative, parasitic doppelganger. The dark double "so closely resembles its host that it may, when seen as a separate entity, be taken as the host itself." The dark double is further described as extremely intelligent and cunning, yet completely lacking in warmth, joy, creativity, and other important human qualities. It is not human, but "a remnant of a much older stream of human development, and is now something of an interloper in human life. . . . This double is an intruder within the human being it inhabits."[57] Apparently, while pursuing a path of spiritual development, one is certain to encounter and must deal with their dark double. Hedsel recounts what it was like to come face to face with his.

When I first encountered the double I was utterly shocked. It was rather like looking in a mirror and seeing not one's own reflection, but that of a dark monster which aped one's own external appearance and actions. It was a low-grade simian copy of myself. This double spoke with a simulacrum of my own voice, yet it was cold and aloof—totally egotistical, and totally lacking in interest in other human beings. .... I [eventually] began to understand why the esoteric literature could describe it as both natural and unnatural. It is natural in so far as we all have a double indwelling; it is unnatural in so far as it is a leech, a drain on energy, rather than an invited guest. It is natural in so far as it participates in our life; it is unnatural in that it is not in the least interested in our spiritual well-being or personal destiny.[58]

What is described by Hedsel sounds a lot like the reptilian brain—the reptilian side of one's being. Intelligence, cunning, egotism, an absence of warmth, creativity, and joy—these are all reptilian traits. Also, it would not be inaccurate to call the reptilian brain "a remnant of a much older stream of human development," in the sense that it predates the rest of the brain. As for it not being human, this too makes sense. Could the dark double be a metaphor, then, for the reptilian side of our being? Or are we dealing with something more complex?

In order for the initiate to move forward spiritually the dark double must be eradicated, states *The Zelator*. This is achieved through a process of fission in the alchemical sense of the term, whereby the light is liberated from the darkness, or the pure liberated from the impure. One could use the metaphor of obtaining pure drinking water from a saucepan full of undrinkable saltwater. Once heat is applied, the pure water (the light) rises as stream, while the salt (the darkness) remains behind in the bottom of the pan.

As I mentioned earlier, there is a theory among occultists that states that the Moon and the Earth were formerly one body, the former having split from the latter. But, physically separate though the two bodies are, they are spiritually connected in the sense that they share the same etheric body. In *The Zelator* the genesis of the Moon is described in terms of both physical and alchemical fission. Hedsel was told by his American teacher that the separation of the Moon from the Earth was required to "allow life on Earth to continue its spiritual development unimpeded."[59] Had the dense matter that constitutes the Moon remained with the Earth, it would have "weighed down human development too deeply." He adds, "Just as we . . . must slough off the darkness to reach into the light, so the planetary bodies must also involve themselves in a similar fission."[60] This is a far cry from Gurdjieff's view that the Moon came into being as a result of a very unfortunate accident.

Speaking of Gurdjieff, the reader will recall that he likened the Moon to purgatory. Also, as we've seen, there are many myths that equate the Moon with purgatory, usually involving a man or a woman who was banished to the Moon as punishment for their sins. In *The Zelator*, the purgatory-Moon connection is explored in some depth. States Hedsel's teacher, "The skull-face of the Moon . . . is a perpetual memorial to the inexorable consequences of human sin."[61] Here we are being told that the Moon represents humanity's sin. It was this "sin," or darkness, that had to be alchemically shed from the Earth so that life on the planet could "continue its spiritual development unimpeded."

Purgatory, in the Christian sense, is a place of spiritual purification, where, by means of great suffering and hardship, one's sins are cleansed. To make it through purgatory is to earn the right to ascend to heaven. One could imagine purgatory as some kind of ghastly slave labor camp—a kind of Soviet Gulag, perhaps, only many times worse. I find it very curious that the entities Swann encountered during his "psychic visit" to the Moon happened to be laboring away, and, what's more, in the nude. The lack of clothing suggests poverty and discomfort. Were the entities in question not aliens but "sinners"?

In occult lore we find the deeply mysterious notion of the "eighth sphere," according to which the inhabitants of the Moon are neither aliens nor sinners but soul-consuming demons. Mention of the eighth sphere can be found in the literature of Theosophy, the topic first revealed to the general public with the publication of A. P. Sinnett's *Esoteric Buddhism* in 1883. Before then the eighth sphere was one of those occult secrets reserved to the initiated few. The topic also shows up in the work of Rudolf Steiner (1861–1925), the founder of anthroposophy (who, it ought to be acknowledged, originally belonged to the Theosophical Society). Sinnett states in his book that the eighth sphere is the Moon—an error which Steiner, among others, endeavored to correct. Indeed, the term refers not to the Moon itself but to a supposed invisible sphere that is connected to the Moon.

According to Hedsel's teacher, the Moon functions as a "counterweight" to the eighth sphere, which is not so much a sphere but a "vacuum." Being a kind of vacuum, the eighth sphere "sucks things into its own shadowy existence." It is a place inhabited and controlled by demonic "shadow beings" who, in some respects, "are more intelligent than Man, for they are not limited by the power of love, as is Mankind." These evil beings are intent on capturing and populating their world with as many human souls as possible. They have even installed on the Earth special terminals or "soul conduits" by means of which they collect, from the Earth "certain forms of human soul matter." There is nothing at all positive about the eighth sphere and its denizens, whose efforts are "contrary to the evolutionary development which has been planned for the world."[62]

Hedsel's teacher blames mediumship as the primary activity responsible for the generation of this "soul matter," or energy. Thus, when a group of spiritualists on Earth conduct a séance, for example, they are actually "creating fodder for the nourishment of the eighth sphere." Mediumship, we are informed, is a fallacy in the sense "that it cannot penetrate through into the realm of the dead."[63] It puts people in touch not with spirits but with "shadows." By "shadows," Hedsel's teacher is referring to entities of a Qliphothic nature, more specially, astral corpses. It is said that when a person dies they shed their astral body (hence the term "astral corpse") while their "soul" ascends to a higher realm. The astral corpse remains in the astral plane before eventually dissolving altogether. Some astral corpses, in an effort to prolong their existence, turn into "vampires," stealing etheric energy from the living. Although mediums think they're contacting the spirits of the dead, they are in fact summoning astral corpses.*

If there really is an eighth sphere populated by demonic shadow beings then humanity has a lot to be worried about. (As if we don't have enough to worry about already!) The idea of the eighth sphere seems to relate to Gurdjieff's notion of humanity being "food for the Moon." I suspect we're dealing with the same topic presented from two different perspectives. It also has to be asked whether the eighth sphere is equivalent to Gamaliel, the Qliphothic counterpart to Yesod, also known as the Dark Moon. Or does the eighth sphere have no Qabalistic significance? Being but a novice when it comes to esoteric matters, I'm afraid I cannot answer this question.

We are told that the Moon functions as a "counterweight" to the eighth sphere. But what is meant by this? As any dictionary will tell you, a counterweight, or counterbalance, is a "force or influence that offsets or checks an opposing force." Take, for example, a seesaw with two children of similar weight seated on opposite ends. In this instance we could say that "child A" is counterbalanced by "child B." Now, astronomers speak of black holes, cosmic bodies with gravitational fields so intense that nothing, not even light, can escape their pull. Black holes are typically formed from dead massive stars—stars that, having exhausted all of their fuel, have become so unstable that they've gravitationally collapsed inward upon themselves. Because black holes are small and emit no light they cannot be observed directly. In one sense they are invisible. According to one theory, black holes are actually a type of wormhole, a hypothetical shortcut in space-time. Theoretically speaking, then, one could enter a black hole and end up in another

------

*Anyone in doubt of the dangers associated with mediumship and channeling should read Joe Fisher's *The Siren Call of Hungry Ghosts: A Riveting Investigation Into Channeling and Spirit Guides.* In it Fisher describes his involvement in a channeling circle and how he came to believe that the "spirit guides" with whom the group had established communication were liars, manipulators, and essentially malevolent. Sadly, Fisher committed suicide in 2001, at the age of fifty-three, by throwing himself off a cliff near his hometown of Fergus, Ontario, Canada.

universe. Could the eighth sphere, I wonder, be some kind of miniature black hole? A gateway to another reality? If so, it's bound to exhibit considerable gravitational influence; unless, of course, something of great mass was there to offset this influence, to keep it in check. Is that "something" the Moon?

This book was written as an attempt to plumb the depths of the Moon. And while it's true that we've explored some fascinating lunar territory, more questions have been raised than answered. In *The Zelator* the Moon is referred to as "the greatest problem of esoteric lore." The book also states that the Moon "is not at all what it appears to be."[64] I agree with both statements but would amend the former to include science as well as esoteric lore. In terms of scientific knowledge, little is known about the Moon, especially as concerns the "bigger picture" of the matter. That the Moon is "Earth's sole natural satellite" is stated as a basic fact, yet even this is open to dispute. First of all, it would seem that the Moon isn't wholly natural; it has numerous artificial characteristics. Second, it could just as convincingly be argued that the Moon is Earth's twin planet as Earth's satellite.

In fact, writing this book has left me with the very strong feeling that the Moon is Earth's twin planet. I feel that by perceiving the Moon in this way, rather than as Earth's insignificant satellite, our understanding of it will move forward in leaps and bounds. Isaac Asimov, whom I've quoted twice already in relation to the Moon, argued in favor of the Moon being Earth's twin planet, making the two a double-planet system as opposed to a planet-satellite system. Using a mathematical concept he called the "tug-of-war" value, he demonstrated that the Sun's hold on the Moon is twice as strong as the Earth's hold on the Moon. This stands in contrast to every other satellite in the solar system, which is more strongly attracted to its primary than it is to the Sun. Asimov concluded, "We might look upon the Moon, then, as neither a true satellite of the Earth nor a captured one, but as a planet in its own right, moving about the Sun in careful step with the Earth."[65]

The giant impact hypothesis clearly hasn't solved the puzzle of the

Moon's origin. In late 2011, I got in touch via e-mail with Clive R. Neal, a professor of Earth Sciences at the University of Notre Dame, whose primary field of research is the origin and evolution of the Moon. I wish to make it clear that Neal is in no way associated with any of the ideas presented in this book, least of all with the spaceship moon theory. He is a purely conventional scientist with a sound reputation. In response to my question "Of all the Moon's mysteries, which do you think is the biggest and why?" he wrote, "If the Moon formed by a giant impact of a Mars-sized planet into Earth, why are volatiles present in the lunar samples? All volatiles would have been driven off. We need to rethink that hypothesis!"[66] We need to do more than "rethink" it; we need to dispose of it.

One of my motivations for writing this book was to encourage people to look at the Moon in a new and radical way. Original and radical thinking is necessary if we are to succeed in getting to the bottom of some of the Moon's biggest mysteries. I wish to caution, however, that this approach will bear little fruit if we neglect to keep our feet firmly planted on the ground, to balance the radical with the rational. There has occurred in recent years a revival of interest in the spaceship moon theory, resulting in much lively discussion on the Internet as to whether the Moon is truly a spaceship that was brought here from somewhere far away. This is encouraging to see. Yet it worries me that many have embraced this theory *in its entirety,* presumably having failed to take into account the evidence that argues against it.

There is much of value in the spaceship moon theory, while the idea of the Moon as a spaceship is a powerful and alluring metaphor. Again, this theory is accurate only in so far as it states that the Moon began its life as a perfectly natural world but was "hijacked" and artificially modified to an extensive degree. Yes the Moon has numerous artificial characteristics. But is it literally a spaceship? Remember, the Moon and Earth have a shared ancestry. The Moon did not arrive from somewhere else but has accompanied the Earth all along. It is therefore unlikely to be a spaceship.

My own interpretation of the Moon's origin, which I don't expect the reader to accept, is a combination of the coaccretion and spaceship moon theories. I've decided to call this the "hybrid double planet theory." As explained earlier, the coaccretion theory and the idea of the Moon as Earth's twin planet go hand in hand (which is why the coaccretion theory is also known as the double planet theory). The beauty of the coaccretion theory is that it puts the Moon and Earth on equal footing. It views the Moon as Earth's "sister." As for the term "hybrid" in the title of this theory, this relates to the Moon possessing both natural and artificial characteristics.

The coaccretion theory explains why the Moon and Earth have identical oxygen isotope signatures and also why the two are approximately the same age. On the other hand it fails to explain: (1) why the Moon and Earth are so different in terms of density and chemical composition; (2) the high angular momentum of the Earth-Moon system; and (3) the Moon's orbital inclination. These weaknesses vanish if we accept, as per the spaceship moon theory, that the Moon underwent an extensive amount of artificial modification early in its history; in other words, that it is a hybrid rather than exclusively natural world. It is not unreasonable to suggest that this remarkable feat of "planetary engineering" included shifting the Moon to its current orbital position.

It's impossible to present in this book a complete explanation of the hybrid double planet theory, as it is very much a work in progress. It is my hope that this theory, incomplete though it is, will stimulate discussion among those who are open to exploring the riddle of the Moon's origin in an unconventional manner. Perhaps one or more intelligent individuals will bring this theory to fruition.

In his book *Ancient Aliens on the Moon*, Mike Bara rejects, as I do, both the giant impact and spaceship moon theories. His thoughts regarding the origin of the Moon are not far removed from my own: "[The Moon] appears to be either formed from the Earth itself or very nearby at the same time as the Earth. . . . Fanciful stories of it being a celestial body from another part of the solar system (or the galaxy) that was 'driven' here and placed into orbit are most likely wrong. That does not however preclude it from being *inhabited or modified much later in its evolutionary process*" [my emphasis].[67]

Though I hope Bara will forgive the comparison, *Ancient Aliens on the Moon* could be seen as a modern version of George Leonard's *Somebody Else Is on the Moon*. In it Bara analyzes numerous NASA photos to reveal what he believes to be undeniable evidence of ancient alien ruins on the surface of the Moon. We looked at some of this photographic evidence earlier—for example, "the shard" and "the tower." While I do find some of this evidence impressive, it leaves me, for the most part, underwhelmed. It's likely that many of these so-called anomalies can be accounted for in mundane terms. The findings presented in this book strongly suggest that the Moon's most compelling anomalies are to be found not on its surface but beneath its surface. It's a shame that gifted researchers like Bara and his colleague Richard Hoagland chose to focus almost exclusively on the former.

Many months after I'd completed the first draft of this book, some fascinating findings began to emerge concerning NASA's Gravity Recovery and Interior Laboratory (GRAIL) mission. Launched on September 10, 2011, and completed late the following year, GRAIL involved sending two small spacecraft to the Moon, GRAIL A (named Ebb) and GRAIL B (Flo), to map in unprecedented detail its gravitational field and hence shed some light on its interior structure. The two spacecraft orbited the Moon in tandem, some 30 miles above the surface and around 125 miles apart from each other. The mapping process consisted of monitoring variations in distance between the two spacecraft as they made their way around the Moon.

The data obtained by GRAIL has enabled scientists to construct an extremely detailed map of the Moon's gravitational field. Several versions of this map have been posted online by NASA. In the standard version of the map, areas of lower than average gravity are shown in blue, whereas areas of higher than average gravity are shown in red. Generally the red parts of the map (higher than average gravity) correspond to regions of high elevation, for example, mountains; while the blue parts of the map (lower than average gravity) correspond to regions of low elevation, for example, craters. In other words, the colors match up with topographical features. But here's the interesting thing: some of the colors match up with features hidden beneath the lunar surface and are unrelated to topography. The Moon's mascons, for example, stand out as bright red bull's eyes.

As I stated earlier, the Moon's mascons correspond to impact basins and are explained in terms of mantle material bulging upward to compensate for missing crustal material; mantle material, being higher in density than crustal material, exerts a greater gravitational pull than crustal material. A similar yet opposite process occurs in the case of mountains (high ones especially), whereby the crust pushes down into the underlying mantle. The result is a deficiency of mantle material and hence a lower gravitational pull. These two processes come under the topic of what geologists call "isostasy."

A gravitational map with the influence of surface topography removed is known as a Bouguer anomaly map. This map shows variations in crustal thickness as well as variations in crust or mantle density. The Bouguer anomaly version of the gravity map of the Moon is even more interesting than the standard version, because it provides a superior glimpse into the Moon's interior. As with the standard version, areas of lower than average gravity are shown in blue, whereas areas of higher than average gravity are shown in red. Once again, the Moon's mascons stand out as bright red bull's eyes, though in this map they're even more pronounced. In general, where there is red (areas of high gravity) the crust is thin; where there is blue (areas of low gravity) the crust is thick. As expected considering that the crust on the far side of the Moon is a good deal thicker than the crust on the near side, the former is dominated by blue and the latter, red. (See plate 12.)

Using data obtained by GRAIL, planetary geophysicist Mark Wieczorek of the Paris Institute of Earth Physics and colleagues calculated that the average bulk density of the lunar highland crust is 159 pounds per cubic foot, which is 12 percent lower than was previously assumed. (To offer a comparison, granite, a low-density rock, is between 165 and 172 pounds per cubic foot.) To explain the oddly low density of the lunar highland crust, Wieczorek deduced that the crust possesses an average porosity "of 12% to depths of at least a few kilometers." He further reasoned that porosity in some regions could be as high as 21 percent and may extend "to depths of tens of kilometers, and perhaps into the uppermost mantle."[68]

Wieczorek's findings have been interpreted by scientists as an indication that the lunar crust is extremely fractured and pulverized due to the Moon having sustained countless devastating impacts early in its history. To quote NASA: "The 12 percent average porosity of the highland crust is a consequence of fractures generated by billions of years of impact cratering."[69] I reject this explanation and argue instead that the void space in the lunar crust (and possibly uppermost mantle) represents nothing less than vast hollow regions of an artificial nature. It figures too that the Moon "rings like a bell" on account of these hollow regions.

There is a great deal more I'd like to discuss with regard to the findings obtained by GRAIL, but I'm afraid we're nearing the end of our lunar journey.

In confronting the Moon we are also forced to confront the dark and hidden side of ourselves—our "demons" and our "sins." Call it the dark double, if you like. This is unpleasant and frightening, but, if we wish to awaken, necessary. In the film *Moon*, Sam is forced to come to terms with the fact that he's a clone; his entire reality is a fabrication; and he himself is not "real," even his memories those of the original Sam Bell. Although this is hugely unsettling for him, it's also what leads to his awakening. In the end he manages to make peace with his clone—his "double"— and finds freedom within. Each and every one of us is like Sam. Some of us have already stumbled upon our "clone," others haven't; we all will eventually. The truth has a way of catching up with us.

It's significant that Hedsel describes the dark double as a "leech, a drain on energy." He even refers to it as "parasitic." Now, if there's one major theme running through this book its parasitism: the non-mutual relationship between a parasite and its host. Gurdjieff taught we are "food for the Moon"; the Moon is a parasite and humanity is the host. Sam, prior to his "awakening," is nothing more than a slave to his employers, his job being to oversee the mining of helium-3 (to be used back on Earth for fusion *energy* production). Only after he confronts his clone or "double," thereby discovering the terrible truth of his existence, does he stop cooperating with his own enslavement and exploitation. He ceases to be a "parasitic host" and starts to take control of his own destiny. I believe it's up to us to do the same.

The Moon, internally and externally speaking, has more to teach us about ourselves than we could ever imagine. We need to learn her secrets as best we can, for it would seem that the very fate of our "souls" depends on it.

Charles Duke, the tenth man to walk on the Moon, is one of several Apollo astronauts to have undergone a change in spiritual outlook as a result of what they experienced in outer space, having returned from his *Apollo 16* mission with a renewed interest in Christianity. In December of 1971, while on a geology training exercise in Hawaii in preparation for the mission, Duke experienced a dream so strange and remarkable that he remembers it clearly to this day. At the time he had the dream he was in a state of flu-induced delirium, his temperature extremely high. "This was really significant to me because I rarely dream, and this was one of the most vivid dreams I'd ever had," he says.[70] Duke described this dream during an interview in 2000. It is with Duke's dream that I wish to express my final thoughts on the Moon and in so doing conclude this book.

In my dream John [Young] and I were driving the Rover up to North Ray Crater, and as we came over one of the ridges there was a set of tracks out in front of us going east-west. And so we got excited, told Houston, and they said, "Well, follow the tracks." You could tell by the tread marks that the vehicle was going to the east. So we turned right, to the east. And towards the eastern end of the valley we come upon this car that looked very similar to the Lunar Rover and there were two figures in it. So we radioed that we'd found this car, and we start to describe it. So I ran over to the passenger side and I pulled up the visor and I was looking at myself. But it wasn't like a nightmare—it wasn't like you were dead; whoever it was looked like me but it wasn't me. The other fellow looked like John. We decided that the bodies weigh too much, so let's just take a little piece of this, a little piece of that. And so we got all the way back to Earth and they analysed the data, and it was hundreds of thousands of years old. And that was such a vivid dream that for a long time I didn't tell my wife about it, until actually I got back. But I remember when we got to lunar orbit and we could look down and first see our landing site. I looked down to see if I could see a set of tracks, going east-to-west . . .[71]

NOTES

INTRODUCTION

1. Sagan and Shklovski, *Intelligent Life in the Universe*, 373.
2. Knight and Butler, *Who Built the Moon?* 56.
3. Geoffrey Macnab, "Moon Rising: Two New Lunar Movies Are Taking Viewers Back Into Orbit," The Independent, October 23, 2008, www.independent .co.uk/arts-entertainment/films/features/moon-rising-two-new-lunar-movies-are-taking-viewers-back-into-orbit-969621.html (accessed May 31, 2013).
4. Nick Redfern. E-mail to Louis Proud, April 8, 2011.

CHAPTER I. THE MOON WE KNOW

1. Encyclopedia Britannica, s.v. "illusion."
2. Knight and Butler, *Who Built the Moon?* 4.
3. Ibid., 5.
4. Moore, *Guide to the Moon,* 135.
5. Eric M. Jones, "One Small Step," NASA: Apollo 11 Lunar Surface Journal, 1995, www.hq.nasa.gov/alsj/a11/a11.step.html (accessed May 31, 2013).
6. Ibid.
7. Moore, *Patrick Moore on the Moon.*
8. Ibid.
9. "Ask an Astrophysicist: Human Body in a Vacuum," NASA's Imagine the Universe, 1997, http://imagine.gsfc.nasa.gov/docs/ask_astro/answers/970603 .html (accessed May 31, 2013).

10. Ouspensky, *In Search of the Miraculous*, 77.

11. Moore, *Patrick Moore on the Moon*.

CHAPTER 2. A CLOSER LOOK AT THE MOON

1. Lang, *The Cambridge Guide to the Solar System*, 2nd ed., 190.

2. Ibid., 190–91.

3. Moore, *Guide to the Moon*, 103.

4. Ibid., 171.

5. Galilei and Finocchiaro, *The Essential Galileo*, 52.

6. Lyell, *Principles of Geology*, 2, 222.

7. Brunner, *Moon: A Brief History*.

8. Ibid.

9. "The Barringer Meteorite Crater," The Barringer Crater Company, www
.barringercrater.com/about/history_1.php (accessed May 31, 2013).

10. See www.meteorcrater.com (accessed July 23, 2013).

11. Baldwin, *The Face of the Moon*.

12. Mackenzie, *The Big Splat, or How Our Moon Came to Be*, 121.

13. Moore, *Patrick Moore on the Moon*.

14. Lang, *The Cambridge Guide to the Solar System*, 184.

15. Moore, *Patrick Moore on the Moon*.

16. Tudor Vieru, "LCROSS Finds Water on the Moon," Softpedia, November 14, 2009, http://news.softpedia.com/news/LCROSS-Finds-Water-on-the-Moon-127011.shtml (accessed May 31, 2013).

17. Tariq Malik, "Tons of Water Ice Found on the Moon's North Pole," www
.space.com/7987-tons-water-ice-moon-north-pole.html (accessed May 31, 2013).

18. Mike Wall, "Mining the Moon's Water: Q & A with Shackleton Energy's Bill Stone," *SPACE.com*, January 13, 2011, www.space.com/10619-mining-moon-water-bill-stone-110114.html (accessed May 31, 2013).

19. Encyclopedia Britannica, s.v. "moon," 2011.

20. Courtland, "Found."

21. "ISRO Finds Cave on the Moon; Can Be Used as an Outpost," The Economic Times, http://economictimes.indiatimes.com/isro-finds-cave-on-moon-can-be-used-as-an-outpost/articleshow/7562900.cms (accessed June 1, 2013).

22. Trudy E. Bell, "Bizarre Lunar Orbits," NASA, 2006, http://science.nasa .gov/science-news/science-at-nasa/2006/06nov_loworbit (accessed June 1, 2013).

23. Chaikin, *A Man on the Moon*, 110.

24. Ibid., 126.

25. Nola Taylor Redd, "Rare Volcanoes Discovered on Far Side of the Moon," SPACE.com, July 25, 2011, www.space.com/12419-moon-side-rare-volcanoes .html (accessed June 1, 2013).

26. Diana Lutz, "Unique Volcanic Complex Discovered on Moon's Far Side," http://news.wustl.edu/news/Pages/22512.aspx (accessed June 1, 2013).

CHAPTER 3. ORIGIN UNKNOWN

1. James M. Foard, "The Untold Story of Charles Darwin," www.thedarwin-papers.com/oldsite/number1/Darwinpapers1Htm.htm (accessed June 1, 2013).

2. Bowlby, *Charles Darwin: A New Life*.

3. Darwin, *The Autobiography of Charles Darwin*.

4. Ibid.

5. Workman and Reader, *Evolutionary Psychology*, 4.

6. Encyclopedia Britannica, s.v. "Darwin, Charles."

7. Phipps, *Darwin's Religious Odyssey*.

8. Shapiro, *The Yale Book of Quotations*.

9. "Babes of the Future: Major Leonard Darwin Tells True Purpose of Eugenics," NYTimes.com, http://query.nytimes.com/mem/archive-free/pdf?res=FB081 EFA3B5813738DDDA80A94DA415B828DF1D3 (accessed June 1, 2013).

10. Mackey and Haywood, *Encyclopedia of Freemasonry Part 3*.

11. Galton, *Memories of My Life*.

12. Mackenzie, *The Big Splat, or How Our Moon Came to Be*.

13. Darwin, *The Next Million Years*, 184–85.

14. DeLury, "Sir George Howard Darwin."

15. Darwin. *The Botanic Garden*, 28.

16. Mackenzie, *The Big Splat, or How Our Moon Came to Be*, 80.

17. Ibid., 86.

18. Webb, *Brief Biography and Popular Account of the Unparalleled Discoveries of T. J. J. See*.

19. Mackenzie, *The Big Splat*, 91.

20. "Calls Moon Planet Captured by Earth," *New York Times*, June 26, 1909.

21. Mackenzie, *The Big Splat*, 96.

22. Asimov, *Of Time and Space and Other Things*, 83.

23. Wilson, *Our Mysterious Spaceship Moon*.

24. Mackenzie, *The Big Splat*, 136.

25. Wilson, *Secrets of Our Spaceship Moon*, 119.

26. Ibid., 119.

27. Mackenzie, *The Big Splat*, 149–50.

28. Ibid., 183.

29. Lang, *The Cambridge Guide to the Solar System*, 200.

30. Mike Wall, "Moon's Interior Wet as Earth's, Rocks Indicate," SPACE.com, May 26, 2011, www.space.com/11797-moon-interior-wet-lunar-origins.html (accessed June 1, 2013).

31. Ferris Jabr, "Apollo 17 Moon Rocks Are Surprisingly Wet," *New Scientist*, May 26, 2011, www.newscientist.com/article/dn20511-apollo-17-moon-rocks-are-surprisingly-wet.html (accessed June 1, 2013).

32. Nield, "Moonwalk."

33. Melosh, "An Isotopic Crisis for the Giant Impact Origin of the Moon?"

34. Ruzicka, Snyder, and Taylor, "Giant Impact and Fission Hypotheses for the Origin of the Moon."

35. Knight and Butler, *Who Built the Moon?* 56.

CHAPTER 4. THE MYSTERY DEEPENS

1. Encyclopedia Britannica, s.v. "Apollo 13."

2. Lattimer, *All We Did Was Fly to the Moon*, 77.

3. Lewis, *The Voyages of Apollo: The Exploration of the Moon*, 167.

4. Wilson, *Secrets of Our Spaceship Moon*, 104.

5. Wilson, *Our Mysterious Spaceship Moon*, 99.

6. Lewis, *The Voyages of Apollo*, 123.

7. Ibid., 124.

8. Wilkins, *Our Moon*, 120.

9. Ibid., 120.

10. Wilson, *Secrets of Our Spaceship Moon*, 95.

11. Ibid., 97.

12. Ibid.

13. Sagan and Shklovski, *Intelligent Life in the Universe*, 373.

14. "Buzz Aldrin Stokes the Mystery of the Monolith on Mars," Mail Online, www .dailymail.co.uk/sciencetech/article-1204254/Has-mystery-Mars-Monolith-solved.html (accessed June 1, 2013).

15. Dirk Vander Ploeg, "Scientist Claims Mars Moon Phobos Is Hollow!" UFO Digest, 2011, www.ufodigest.com/article/scientist-claims-mars-moon-phobos-hollow (accessed June 1, 2013).

16. Ibid.

17. Jamie Oberg, "Russia's Dark Horse Plan to Get to Mars," Discover Magazine .com, June 2009, http://discovermagazine.com/2009/jun/21-russias-dark-horse-plan-to-get-to-mars/article_view?b_start:int=1&-C= (accessed June 1, 2013).

18. Lang, *The Cambridge Guide to the Solar System*, 177.

19. Wilson, *Secrets of Our Spaceship Moon*, 106.

20. Mary Caperton Morton, "Moonquake Mystery Deepens," Earth Magazine, August 19, 2009, www.earthmagazine.org/article/moonquake-mystery-deepens (accessed June 1, 2013).

21. Ibid.

22. Clive R. Neal, "On the Lookout for Quakes on the Moon," Geotimes, July 2006, www.geotimes.org/july06/feature_MoonQuakes.html (accessed June 1, 2013).

23. Trudy E. Bell, "Moonquakes," NASA, 2006, http://science.nasa.gov/ science-news/science-at-nasa/2006/15mar_moonquakes (accessed June 1, 2013).

24. Kate Fink, "Strange Quarks Attack," SciTini, October 11, 2006, www.bu.edu/ phpbin/news-cms/news/?dept=1127&id=41016&template=226 (accessed June 1, 2013).

25. Ibid.

26. Knight and Butler, *Who Built the Moon?* 72.

27. Wilson, *Secrets of Our Spaceship Moon*, 107.

28. Ibid.

29. Ibid.

30. Ibid., 25.

31. Farouk El-Baz. E-mail to Louis Proud, December 13, 2011.

32. Louis Proud. E-mail to Farouk El-Baz, December 13, 2011.

33. Farouk El-Baz. E-mail to Louis Proud, December 13, 2011.

34. Vasin and Shcherbakov, "Is the Moon the Creation of Intelligence?"

35. Ibid.

36. Ibid.

37. Ibid.

38. Ibid.

39. Lang, *The Cambridge Guide to the Solar System,* 178.

40. Vasin and Shcherbakov, "Is the Moon the Creation of Intelligence?"

41. Ibid.

42. Zdenek Kopal, *The Moon in the Post-Apollo Era.*

43. Diana Lutz, "Unique Volcanic Complex Discovered on Moon's far Side," Washington University in St. Louis, July 24, 2011, http://news.wustl.edu/news/Pages/22512.aspx (accessed June 1, 2013).

44. Ibid.

45. "Moon Rock Analysis Uncovers 'Unearthly Pair,'" The Deseret News, January 13, 1972, http://news.google.com/newspapers?nid=336&dat=19720113&id=k tZjAAAAIBAJ&sjid=e38DAAAAIBAJ&pg=7278,2362668 (accessed June 1, 2013).

46. "Mimas: Overview," NASA, http://solarsystem.nasa.gov/planets/profile .cfm?Object=Mimas (accessed June 1, 2013).

47. Wilson, *Our Mysterious Spaceship Moon,* 104–5.

48. Wilson, *Secrets of Our Spaceship Moon,* 138.

49. Ibid., 139.

50. Lewis, *The Voyages of Apollo,* 78.

51. Kelly Beatty, "New Insights on Lunar Swirls," Sky & Telescope, March 24, 2011, www.skyandtelescope.com/community/skyblog/newsblog/118592329 .html (accessed June 1, 2013).

52. Encyclopedia Britannica, s.v. "Roche limit."

CHAPTER 5. OUR RELATIONSHIP WITH THE MOON

1. Nadia Macleod, "Moon Cycles and Women," 1999–2011, www.menstruation .com.au/periodpages/mooncycles.html (accessed June 1, 2013).

2. Cashford, *The Moon.*

3. Gooch, *The Dream Culture of the Neanderthals,* 38.

4. Ibid., 40.

5. James Owen, "Orangutans May Be Closest Human Relatives, Not Chimps," National Geographic News, June 23, 2009, http://news.nationalgeographic.com/news/2009/06/090623-humans-chimps-related.html (accessed June 1, 2013).

6. James, *The Cult of the Mother Goddess.*

7. Cashford, *The Moon,* 12.

8. David Whitehouse, "Prehistoric Moon Map Unearthed," BBC News, April 22, 1999, http://news.bbc.co.uk/2/hi/science/nature/325290.stm (accessed June 1, 2013).

9. Ibid.

10. Stooke, "Neolithic Lunar Maps at Knowth and Baltinglass, Ireland."

11. David Whitehouse, "Prehistoric Moon Map Unearthed," BBC News, April 22, 1999, http://news.bbc.co.uk/2/hi/science/nature/325290.stm (accessed June 1, 2013).

12. Krupp, *Beyond the Blue Horizon,* 69.

13. Horne, *The Code of Hammurabi.*

14. Ibid.

15. Cashford, *The Moon,* 147.

16. Bruce, *The Mark of the Scots,* 183.

17. Razwy and Ali, *The Qur'an Translation,* 126.

18. Leonard Shlain, "Dr. Leonard Shlain Discusses His Book 'Alphabet Versus the Goddess,'" Bodhi Tree Bookstore, 1970–2011, www.bodhitree.com/lectures/Shlain.html (accessed June 1, 2013).

19. Ibid.

20. Hawke, *Praise to the Moon,* 55.

21. Cashford, *The Moon,* 14–15.

22. Graves. *The Greek Myths,* 466.

23. Hawke, *Praise to the Moon,* 10–11.

24. Cashford, *The Moon,* 71.

25. Ibid.

26. Ibid.

27. Shakespeare, *A Midsummer Night's Dream,* scene II.

28. Henderson, *Notes on the Folklore of the Northern Counties of England and the Borders.*

29. Cashford, *The Moon,* 168.

30. Ibid.

CHAPTER 6. THE MOON MATRIX

1. Ewing, *Insanity*.
2. Ibid.
3. Terry, *The Ultimate Evil*.
4. Hawke, *Praise to the Moon*, 47–48.
5. Guiley, *Moonscapes*.
6. Ibid.
7. Parley, *Peter Parley's Almanac for Old and Young*, 80.
8. Oliven, "Moonlight and Nervous Disorders."
9. Brunner, *Moon*.
10. Lieber, *The Lunar Effect*, 16.
11. Ibid., 22–23.
12. Ibid., 26.
13. Ibid., 27.
14. Wilson, *From Atlantis to the Sphinx*, 96.
15. Rotton and Kelly, "Much Ado About the Full Moon."
16. Kelly, Rotton, and Culver, "The Moon Was Full and Nothing Happened."
17. Radin, *The Conscious Universe*, 179.
18. See www.deanradin.com/NewWeb/bio.html (accessed June 1, 2013).
19. "The Amazing Grunion," California Department of Fish and Game, www.dfg.ca.gov/marine/grunion.asp (accessed June 1, 2013).
20. Watson, *Supernature*, 29.
21. Tony Phillips, "Earth's Magnetic Field Does Strange Things to the Moon," NASA, 2008, http://science.nasa.gov/science-news/science-at-nasa/2008/17apr_magnetotail (accessed June 1, 2013).
22. Andrew Fazekas, "Full Moons Get Electrified by Earth's Magnetic 'Tail,'" National Geographic News, November 18, 2010,http://news.nationalgeographic.com/news/2010/11/101118-science-space-full-moon-electric-charge (accessed June 1, 2013).
23. Tony Phillips, "Earth's Magnetic Field Does Strange Things to the Moon," NASA, 2008, http://science.nasa.gov/science-news/science-at-nasa/2008/17apr_magnetotail (accessed June 1, 2013).
24. Lieber, *The Lunar Effect*, 112.
25. Stolov and Cameron, "Variations of Geomagnetic Activity with Lunar Phase."
26. Watson, *Supernature*, 30.
27. Maia Weinstock, "A-maze-ing Mole Rats," Discover Magazine, April 21, 2004,

http://discovermagazine.com/2004/apr/a-maze-ing-mole-rats (accessed June 1, 2013).

28. Pickrell, "Rat Radar."

29. Richard Gray, "Monarch Butterflies Use Internal Compass to Find Their Way," The Telegraph, January 27, 2010, www.telegraph.co.uk/earth/wildlife/7855868/Monarch-butterflies-use-internal-compass-to-find-their-way.html (accessed June 1, 2013).

30. Jason Palmer, "Human Eye Protein Senses Earth's Magnetism," BBC News, June 21, 2011, www.bbc.co.uk/news/science-environment-13809144 (accessed June 1, 2013).

31. Ed Yong, "Humans Have a Magnetic Sensor in Our Eyes, But Can We Detect Magnetic Fields?" Discover Magazine, June 21, 2011, http://blogs.discovermagazine.com/notrocketscience/2011/06/21/humans-have-a-magnetic-sensor-in-our-eyes-but-can-we-see-magnetic-fields/#.UdeTTay2bcg (accessed June 1, 2013).

32. Levitt, Electromagnetic Fields, 73.

33. Baker, "A Sense of Magnetism."

34. Yong, "Humans Have a Magnetic Sensor in Our Eyes."

35. Berk, Dodd, and Henry, "Do Ambient Electromagnetic Fields Affect Behavior?"

36. Ibid.

37. Catherine Brahic, "Does the Earth's Magnetic Field Cause Suicides?" New Scientist, April 24, 2008, www.newscientist.com/article/dn13769-does-the-earths-magnetic-field-cause-suicides.html (accessed June 1, 2013).

38. Martensen, The Brain Takes Shape.

39. Encyclopedia Britannica, s.v. "Circadian Rhythm."

40. Brahic, "Does the Earth's Magnetic Field Cause Suicides?"

41. Nishimura and Fukushima, "Why Animals Respond to the Full Moon."

42. Ibid.

43. Radin, The Conscious Universe, 179.

44. Ibid., 184.

45. Ibid., 186.

46. Ibid.

47. Icke, In the Light of Experience.

48. Icke, Human Race Get Off Your Knees, 21.

49. Ibid.

50. Ibid., 23.

51. Ibid., 25.

52. Ibid., 28.

53. Ibid., 30.

54. Ibid., 35.

55. Ibid.

56. Ibid., 36.

57. Icke, *In the Light of Experience*.

58. Icke, *Human Race Get Off Your Knees*, 37.

59. Ibid., 36.

60. Ibid.

61. Icke, *In the Light of Experience*.

62. Icke, "David Icke on Wogan Now & Then," October 21, 2007, www.youtube .com/watch?v=KqTAIX2WlTQ (accessed July 6, 2013).

63. Icke, *Remember Who You Are*.

64. Natalie Clarke, "He Once Claimed He's the Son of God and the World is Run by Alien Lizards, but the Story of David Icke's Marriage Breakdown Is Almost as Weird," Mail Online, January 9, 2012, www.dailymail.co.uk/femail/ article-2083287/David-Ickes-marriage-breakdown-He-claimed-hes-Son-God-world-run-alien-lizards-story-marriage-breakdown-weird.html (accessed June 1, 2013).

65. Icke, *Remember Who You Are*.

66. Icke, *Human Race Get Off Your Knees*.

67. Ibid., 310.

68. Ibid.

69. Ibid.

70. Icke, *Children of the Matrix*.

71. Encyclopedia Britannica, s.v. "Naga."

72. The Bible: Authorized King James Version with Apocrypha.

73. Smith, *The Evolution of the Dragon*, 156.

74. "How Did Human Emotions Evolve?" The Human Brain: An Owner's Manual, 2001, http://library.thinkquest.org/C0114820/emotional/origins .php3 (accessed June, 2, 2013).

75. Icke, *Remember Who You Are*.

76. Ibid.

77. Icke, *Human Race Get Off Your Knees*, 286.

78. Ibid., 287.

79. Ibid., 316.

80. Ibid.

81. Ibid., 363.

82. Icke, *Remember Who You Are.*

83. Icke, *Human Race Get Off Your Knees,* 414.

84. Icke, *Remember Who You Are.*

85. Ibid.

CHAPTER 7. FOOD FOR THE MOON

1. "Powerful and Magnetic Personality," Gurdjieff and the Fourth Way: A Critical Appraisal, http://gurdjiefffourthway.org/pdf/magnetic.pdf (accessed June 2, 2013), 1.

2. Wilson, *The Occult.*

3. Wilson, *Dreaming to Some Purpose.*

4. Bennett, *Witness,* 244.

5. Gurdjieff, *Meetings with Remarkable Men.*

6. Ibid.

7. Ibid.

8. Ibid.

9. Ibid.

10. Needleman, "G. I. Gurdjieff and His School."

11. Gurdjieff, *Meetings with Remarkable Men.*

12. Ibid.

13. Ibid.

14. Ibid.

15. Ibid.

16. Ouspensky, *In Search of the Miraculous,* 7.

17. Ibid., 10.

18. Ibid., 11.

19. Ibid.

20. John Pentland, *The Encyclopedia of Religion,* s.v. "P. D. Ouspensky" (New York: Macmillan, 1987), Vol. 11, 143.

21. Ouspensky, *In Search of the Miraculous,* 18.

22. Ibid., 21.

23. Ibid., 44.

24. Ibid., 50.

25. Ibid., 23–24.

26. Ibid., 24.

27. Ibid., 85.

28. Mirabello, *The Odin Brotherhood.*

29. Ouspensky, *In Search of the Miraculous,* 57.

30. Ibid.

31. Ibid.

32. Ibid.

33. Ibid.

34. Ibid., 86.

35. Ibid., 83.

36. Ibid.

37. Ibid., 85.

38. Ibid.

39. Ibid.

40. The Bible: Authorized King James Version with Apocrypha.

41. Ouspensky, *In Search of the Miraculous,* 138.

42. Ibid.

43. Ibid.

44. Ibid.

45. Ibid., 85.

46. Ibid., 85–86.

47. Ibid., 47.

48. Ouspensky, *A Further Record.*

49. Ibid.

50. Ouspensky, *In Search of the Miraculous,* 219.

51. Ibid., 145.

52. Ibid., 219.

53. Ibid., 220.

54. Gurdjieff, *Views From the Real World,* 151.

55. Ouspensky, *In Search of the Miraculous,* 155.

56. Gurdjieff, *Beelzebub's Tales to His Grandson,* 3.

57. Zigrosser, "Gurdjieff–Salzmann–Orage."

58. Gurdjieff, *Beelzebub's Tales to His Grandson,* 82.

59. Ibid., 61.

60. Ibid., 62–63.

61. Myers, "Gurdjieff, the Moon and Organic Life."

62. Charles Q. Choi, "Earth Had Two Moons That Crashed to Form One, Study Suggests," SPACE.com, August 3, 2011, www.space.com/12529-earth-2-moons-collision-moon-formation.html (accessed June 2, 2013).

63. Shirley, *Gurdjieff,* 287.

64. Shirley, John. E-mail to Louis Proud, April 20, 2011.

65. Shirley, John. E-mail to Louis Proud, April 22, 2011.

66. Gurdjieff, *Beelzebub's Tales to His Grandson,* 63.

67. William Patterson. Letter to Louis Proud, July 4, 2011.

CHAPTER 8. LIFE ON THE MOON

1. "William Herschel Quotes," Thinkexist.com, http://thinkexist.com/quotes/william_herschel (accessed June 2, 2013).

2. Bartholomew and Radford, *The Martians Have Landed!* 79–83.

3. Ibid.

4. Clark, "The Men Who Once Lived in the Moon."

5. Goodman, *The Sun and the Moon.*

6. Middlehurst, Burley, Moore, and Welther, *Chronological Catalog of Reported Lunar Events (NASA TR R-227).*

7. Ibid.

8. Ibid.

9. Ibid.

10. Trudy E. Bell and Dr. Tony Phillips, "Moon Storms," NASA, 2005, http://science.nasa.gov/science-news/science-at-nasa/2005/07dec_moonstorms (accessed June 2, 2013).

11. Moore, *Patrick Moore on the Moon.*

12. Marrs, *Alien Agenda,* 21.

13. Clark, "The Men Who Once Lived in the Moon."

14. *Encyclopedia Britannica,* s.v. "Unidentified Flying Object (UFO)."

15. Marrs, *Alien Agenda.*

16. Wilkins, *Flying Saucers on the Attack.*

17. Keyhoe, *The Flying Saucer Conspiracy.*

18. Jessup, *The Expanding Case for the UFO,* 110–94.

19. Ibid., 110–94.

20. Ibid.

21. Chaikin, *A Man on the Moon*, 110.

22. Wilson, *Secrets of Our Spaceship Moon*, 261.

23. Leonard, *Somebody Else Is on the Moon*, 23.

24. Ibid.

25. Ibid., 25.

26. Ibid., 29.

27. Ibid., 56.

28. Ibid., 62.

29. Oberg, *UFOs & Outer Space Mysteries*, 88.

30. Ibid., 92.

31. Leonard, *Somebody Else Is on the Moon*, 208.

32. Ibid., 131.

33. Bara, *Ancient Aliens on the Moon*, 39–40.

34. Farouk El-Baz. E-mail to Louis Proud, July 19, 2013.

35. Hoagland and Bara, *Dark Mission*, 186.

36. Vallee, *Messengers of Deception*, 203.

37. Joseph P. Farrell, "The Idea That Will Not Go Away: Bases on the Moon," Giza Death Star, May 16, 2011, http://gizadeathstar.com/2011/05/the-idea-that-will-not-go-away-bases-on-the-moon (accessed June 2, 2013).

38. Steckling, *We Discovered Alien Bases on the Moon*.

39. Hoagland and Bara, *Dark Mission*, 187.

40. Ibid., 194.

41. Bruce Cornet, "Interpretation of Anomalous Structures on the Moon," ufo-bbs.com, May 15, 1994, www.ufo-bbs.com/txt3/2398.htm (accessed July 7, 2013).

42. Ibid.

43. Hoagland, and Bara, *Dark Mission*.

44. Ibid., 244.

45. Ibid., 210.

46. George Noory (radio show host on Coast to Coast AM) in discussion with David Livingston, Ken Johnston, and Steve Troy, February 12, 2003, www.coasttocoastam.com/show/2003/02/12 (accessed June 3, 2013).

47. Ibid.

48. Ibid.

49. Good, *Need to Know,* 186.

50. Davies and Wagner, "Searching for Alien Artefacts on the Moon."

51. Ibid.

52. Thomas O'Toole, "Six Mysterious Statuesque Shadows Photographed on Moon by Orbiter," Washington Post, November 23, 1966, www.astrosurf.com/lunascan/2cusp.htm (accessed July 6, 2013).

53. Jury, "Regular Geometric Patterns Formed by Moon 'Spires.'"

54. Ibid.

55. Sanderson, "Mysterious 'Monuments' on the Moon."

56. "Scientist Sees a Pattern in Moon 'Spires,'" Los Angeles Times, February 1, 1967, www.astrosurf.com/lunascan/4cusp.htm (accessed July 7, 2013).

57. Carlotto, "3D Analysis of the 'Blair Cuspids' and Surrounding Terrain."

58. See "The Disclosure Project," 2010, www.disclosureproject.org (accessed June 2, 2013).

59. Greer, and Loder, *Disclosure Project Briefing Document,* 293–94.

60. Ibid., 293–94.

61. Ibid., 292–93.

62. Ibid.

63. Ibid.

64. "Project Horizon Report: Volume I, Summary and Supporting Considerations," history.army.mil, United States Army, June 9, 1959, www.history.army.mil/faq/horizon/Horizon_V2.pdf (accessed June 3, 2013).

65. Richard Dolan, "Musings on a Secret Space Program," Para-News, June 6, 2010, http://richardthomasblogger.blogspot.com.au/2010/06/guest-article-by-richard-dolan.html (accessed June 3, 2013).

66. Wilson, *Our Mysterious Spaceship Moon,* 48. Wilson fails to provide the title of Otto Binder's article, but states that it appeared in *Saga* UFO Special #3.

67. Ibid.

68. Flindt and Binder, *Mankind.*

69. Ibid.

70. Chatelain, *Our Cosmic Ancestors,* 25.

71. Ibid., 24.

72. Ibid., 18.

73. Ibid., 24.

74. Ibid.

75. Ibid., 25.

76. Strieber, *Transformation*, 240.

77. Chatelain, *Our Cosmic Ancestors*, 25.

CHAPTER 9. THE LUNAR SPHERE

1. Morehouse, *Remote Viewing*.

2. Marrs, *Psi Spies*, 34.

3. Ibid.

4. Ibid., 35.

5. Ibid., 36.

6. Ibid., 38.

7. Ibid.

8. Ibid.

9. Gruber, *Psychic Wars*, 28.

10. Ibid.

11. Marrs, *Psi Spies*, 39.

12. Gruber, *Psychic Wars*, 15.

13. Marrs, *Psi Spies*, 166–67.

14. Ibid., 167.

15. Swann, *Penetration*.

16. Ibid., 10.

17. Ibid., 13.

18. Ibid.

19. Ibid., 14.

20. Ibid., 19.

21. Ibid., 24.

22. Ibid., 25.

23. Ibid., 27.

24. Ibid., 30.

25. Ibid.

26. Ibid., 31.

27. Ibid., 36.

28. Ibid., 44–45.

29. Ibid., 49.

30. Ibid., 20.

31. Ibid., 66.

32. Ibid., 112–13.

33. Hedsel, *The Zelator*, 21.

34. Ibid., 71.

35. Ibid., 11.

36. Ibid., 183.

37. Ibid., 197.

38. Guiley, *The Encyclopedia of Magic and Alchemy*, 51.

39. Hedsel, *The Zelator*, 203.

40. Ibid., 310.

41. Regardie, *The Tree of Life*, 482.

42. Fortune, *The Mystical Qabalah*.

43. Ibid.

44. Regardie, *The Tree of Life*, 77.

45. Hedsel, *The Zelator*, 21.

46. See www.sheldrake.org/Resources/glossary (accessed June 3, 2013).

47. Regardie, *The Tree of Life*, 71.

48. Knight, *A Practical Guide to Qabalistic Symbolism*, 175.

49. Fortune, *The Mystical Qabalah*.

50. Ibid.

51. Ibid.

52. Ibid.

53. Sperling and Simon, *The Zohar, Volume 1*, 83.

54. Grant, *Hecate's Fountain*.

55. Grant, *Nightside of Eden*.

56. Guiley, *TheEncyclopedia of Demons and Demonology*, 46.

57. Hedsel, *The Zelator*, 47.

58. Ibid., 48.

59. Ibid., 240.

60. Ibid., 241.

61. Ibid., 202.

62. Ibid., 432–33.

63. Ibid.

64. Ibid., 242.

65. Asimov, *Of Time and Space and Other Things*, 84.

66. Clive R. Neal. E-mail to Louis Proud, December 21, 2011.

67. Bara, *Ancient Aliens on the Moon*, 16.

68. Wieczorek et al., "The Crust of the Moon as Seen by GRAIL."

69. "Closer Look at Lunar Highland Crust," NASA, www.nasa.gov/mission_pages/grail/multimedia/pia16588.html (accessed July13, 2013).

70. Duke and Duke, *Moonwalker*, 187.

71. "Stephen Baxter Interviews Charlie Duke," homepage.mac.com/sjbradshaw/baxterium/bax_duke_iv.html.

BIBLIOGRAPHY

Asimov, Isaac. *Of Time and Space and Other Things.* New York: Doubleday, 1965.

Baker, Robin R. "A Sense of Magnetism," *New Scientist* 87, no. 1219 (September 18, 1980).

Baldwin, Ralph B. *The Face of the Moon.* Chicago: University of Chicago Press, 1949.

Bara, Mike. *Ancient Aliens on the Moon.* Kempton, IL: Adventures Unlimited Press, 2012.

Bartholomew, Robert E., and Benjamin Radford. *The Martians Have Landed!: A History of Media-Driven Panics and Hoaxes.* Jefferson, NC: McFarland & Company Inc., 2012.

Bennett, John G. *The Autobiography of John G. Bennett.* Tucson, AZ: Omen Press, 1974.

Berk, Michael, Seetal Dodd, and Margaret Henry. "Do Ambient Electromagnetic Fields Affect Behavior? A Demonstration of the Relationship between Geomagnetic Storm Activity and Suicide." *Bioelectromagnetics* 27 (2006): 151–55.

Bruce, Duncan A. *The Mark of the Scots: Their Astonishing Contributions to History, Science, Democracy, Literature and the Arts.* New York: Kensington Publishing Corp., 1996.

Bowlby, John. *Charles Darwin: A New Life.* New York: W. W. Norton & Company Inc., 1991.

Browne, Janet. *Charles Darwin: Voyaging.* New York: Alfred A. Knopf Inc., 1995.

Brueton, Diana. *The Moon: Myth, Magic and Fact.* New York: Barnes & Noble Books, 1998.

Brunner, Bernd. *Moon: A Brief History*. New Haven, CT: Yale University Press, 2010.

Carlotto, Mark J. "3D Analysis of the 'Blair Cuspids' and Surrounding Terrain." *New Frontiers in Science* 1, no. 2 (2002): 27–42.

Cashford, Jules. *The Moon: Myth and Image*. London: Cassell Illustrated, 2002.

Chaikin, Andrew. *A Man on the Moon: The Voyages of the Apollo Astronauts*. New York: Penguin Books USA Inc., 1994.

Chatelain, Maurice. *Our Cosmic Ancestors*. Sedona, Arizona: Temple Golden Publications, 1987.

Clark, Jerome. "The Men Who Once Lived in the Moon." *Fate* 61, no. 5 (May 2008): 10–18.

Courtland, Rachel. "Found: First 'Skylight' on the Moon." *New Scientist*. October 22, 2009.

Darwin, Charles. *The Autobiography of Charles Darwin*. London: Bibliolis Books Ltd., 1887.

Darwin, Charles Galton. *The Next Million Years*. London: Praeger, 1973.

Darwin, Erasmus. *The Botanic Garden: A Poem in Two Parts*. London: Jones & Company, 1825.

Davies, P. C. W., and R. V. Wagner. "Searching for Alien Artefacts on the Moon." *Acta Astronautica* 89 (November 23, 2011): 261–65.

DeLury, Alfred T. "Sir George Howard Darwin." *Journal of the Royal Astronomical Society of Canada* 7 (1913): 114–19.

Duke, Charlie, and Dotty Duke. *Moonwalker*. Rose Petal Press, 2011.

Ewing, Charles Patrick. *Insanity: Murder, Madness, and the Law*. New York: Oxford University Press, 2008.

Flindt, Max H., and Otto O. Binder. *Mankind: Child of the Stars*. Greenwich, CT: Fawcett Publications Inc., 1974.

Fortune, Dion. *The Mystical Qabalah*. London: Society of Inner Light, 1935.

Galilei, Galileo, and Maurice A. Finocchiaro. *The Essential Galileo*. Indianapolis, IN: Hackett Publishing Company Inc., 2008.

Galton, Sir Francis. *Memories of My Life*. New York: E. P. Dutton, 1909.

Gooch, Stan. *The Dream Culture of the Neanderthals: Guardians of the Ancient Wisdom*. London: Wildwood House Ltd., 1979.

Good, Timothy. *Need to Know: UFOs, the Military, and Intelligence*. New York: Pegasus Books LLC, 2007.

Goodman, Matthew. *The Sun and the Moon: The Remarkable True Account of*

Hoaxers, Showmen, Dueling Journalists, and Lunar Man-Bats in Nineteenth-Century New York. New York: Basic Books, 2008.

Grant, Kenneth. *Hecate's Fountain.* London: Skoob Books Publishing, 1992.

———. *Nightside of Eden.* London: Fredrick Muller Ltd., 1977.

Graves, Robert. *The Greek Myths.* New York: Penguin Books, 1955.

Greer, Steven M., and Theodore C. Loder. *Disclosure Project Briefing Document.* Crozet, VA: The Disclosure Project, April 2001.

Gruber, Elmar R. *Psychic Wars: Parapsychology in Espionage—and Beyond.* London: Blandford, 1999.

Guiley, Rosemary Ellen. *Moonscapes: A Celebration of Lunar Astronomy, Magic, Legend and Lore.* New York: Prentice Hall Press, 1991.

Guiley, Rosemary Ellen. *The Encyclopedia of Demons and Demonology.* New York: Infobase Publishing Inc., 2009.

———. *The Encyclopedia of Magic and Alchemy.* New York: Infobase Publishing Inc., 2006.

Gurdjieff, G. I. *Beelzebub's Tales to His Grandson.* New York: E. P. Dutton & Co. Inc., 1964.

———. *Meetings with Remarkable Men.* London: Routledge & Kegan Paul Ltd., 1963.

———. *Views From the Real World: Early Talks of Gurdjieff.* London: E. P. Dutton & Co. Inc., 1975.

Harding, M. Esther. *Woman's Mysteries: Ancient & Modern.* Boston, MA: Shambhala Publications, Inc., 1971.

Hawke, Elen. *Praise to the Moon: Magic & Myth of the Lunar Cycle.* St. Paul, MN: Llewellyn Publications, 2002.

Hedsel, Mark. *The Zelator: A Modern Initiate Explores the Ancient Mysteries.* York Beach, ME: Samuel Weiser, Inc., 1998.

Henderson, William. *Notes on the Folklore of the Northern Counties of England and the Borders.* London: Longmans, Green, and Co., 1866.

Hoagland, Richard C., and Mike Bara. *Dark Mission: The Secret History of NASA.* Port Townsend, WA: Feral House, 2007.

Horne, Charles F. *The Code of Hammurabi.* Hong Kong: Forgotten Books, 2007.

Icke, David. *Children of the Matrix.* Wildwood, MO: Bridge of Love Publications USA, 2001.

———. *Human Race Get Off Your Knees: The Lion Sleeps No More.* Ryde, Isle of Wight: David Icke Books Ltd., 2010.

——. *In the Light of Experience: The Autobiography of David Icke*. London: Warner Books, 1993.

——. *Remember Who You Are*. Ryde, Isle of Wight: David Icke Books Ltd., 2012.

James, E. O. *The Cult of the Mother Goddess: an Archaeological and Documentary Study*. London: Thames and Hudson, 1959.

Jessup, M. K. *The Expanding Case for the UFO*. New York: The Citadel Press, 1957.

Jury, William. "Regular Geometric Patterns Formed by Moon 'Spires,'" 26, no. 13, March 30, 1967.

Kelly, I. W., J. Rotton, and R. Culver. "The Moon was Full and Nothing Happened: A Review of Studies on the Moon and Human Behavior and Lunar Beliefs." *The Sceptical Inquirer* 10 (1985–86): 129–43.

Kephas, Aeolus. *Homo Serpiens: An Occult History of DNA from Eden to Armageddon*. Kempton, IL: Adventures Unlimited Press, 2009.

Keyhoe, Donald E. *The Flying Saucer Conspiracy*. New York: Henry Holt and Company Inc., 1955.

Knight, Christopher, and Alan Butler. *Who Built the Moon?* London: Watkins Publishing, 2005.

Knight, Gareth. *A Practical Guide to Qabalistic Symbolism*. Boston, MA: Red Wheel/Weiser, 2001.

Kopel, Zdenek. *The Moon in the Post-Apollo Era*. Dordrecht/Boston: Reidel, 1974.

Krupp, E. C. *Beyond the Blue Horizon: Myths and Legends of the Sun, Moon, Stars, and Planets*. New York: HarperCollins, 1991.

Lachman, Gary. *In Search of P.D. Ouspensky: The Genius in the Shadow of Gurdjieff*. Wheaton, IL: Quest Books, 2004.

Lang, Kenneth R. *The Cambridge Guide to the Solar System*. 2nd ed. New York: Cambridge University Press, 2011.

Lattimer, Dick. *All We Did Was Fly to the Moon*. Alachua, FL: Whispering Eagle Press, 1983.

Leonard, George. *Somebody Else Is on the Moon*. New York: Pocket Books, 1976.

Levitt, B. Blake. *Electromagnetic Fields: A Consumer's Guide to the Issues and How to Protect Ourselves*. San Diego: Harcourt Brace & Company, 1995.

Lewis, Richard S. *The Voyages of Apollo: The Exploration of the Moon*. New York: Quandrangle, 1974.

Lieber, Arnold L. *The Lunar Effect: Biological Tides and Human Emotions*. New York: Anchor Press/Doubleday, 1978.

Lyell, Charles. *Principles of Geology 2*. London: John Murray, 1832.

Mackenzie, Dana. *The Big Splat, or How Our Moon Came to Be*. Hoboken, NJ: John Wiley & Sons Inc., 2003.

Mackey, Albert G., and H. L. Haywood. *Encyclopedia of Freemasonry Part*. Whitefish: MT: Kessinger Publishing, 2003.

Marrs, Jim. *Alien Agenda: Investigating the Extraterrestrial Presence Among Us*. New York: HarperCollins, 1997.

Marrs, Jim. *Psi Spies*. Phoenix, AZ: AlienZoo Publishing, 2000.

Martensen, Robert L. *The Brain Takes Shape: An Early History*. New York: Oxford University Press Inc., 2004.

McCrae, Niall. *The Moon and Madness*. Exeter: Imprint Academic, 2011.

Melosh, H. J. "An Isotopic Crisis for the Giant Impact Origin of the Moon?" 72nd Annual Meteoritical Society Meeting, 2009.

Middlehurst, Barbara M., Jaylee M. Burley, Patrick Moore, and Barbara L. Welther. *Chronological Catalog of Reported Lunar Events (NASA TR R-227)*. Washington, DC: NASA, July 1968.

Mirabello, Mark. *The Odin Brotherhood*. Oxford: Mandrake of Oxford, 1992.

Moore, Patrick. *Guide to the Moon*. Cambridge: Lutterworth Press, 1976.

———. *Patrick Moore on the Moon*. London: Cassell, 2001.

Morehouse, David. *Remote Viewing: The Complete User's Manual for Coordinate Remote Viewing*. Boulder, CO: Sounds True Inc., 2008.

Myers, Richard. "Gurdjieff, the Moon and Organic Life." *The Gurdjieff Journal* 11, issue 1.

Needleman, Jacob. "G. I. Gurdjieff and His School." *Gurdjieff International Review* 3 (1999).

New York Times. "Calls Moon Planet Captured by Earth," June 26, 1909.

———. "Scientist Sees Pattern in Moon 'Spires.'" February 1, 1967.

Nield, Ted. "Moonwalk." *Geoscientist* 19, no. 9 (September 2009): 8.

Nishimura, T., and M. Fukushima. "Why Animals Respond to the Full Moon: Magnetic Hypothesis." *Bioscience Hypotheses* (2009): 399–401.

Oberg, James E. *UFOs & Outer Space Mysteries: A Sympathetic Skeptic's Report*. Virginia Beach, VA: Donning Company, 1982.

Oliven, J. F. "Moonlight and Nervous Disorders: A Historical Study," *American Journal of Psychiatry* 99 (1943): 579–84.

O'Toole, Thomas. "Six Mysterious Statuesque Shadows Photographed on Moon by Orbiter." *Washington Post*, November 23, 1966.

Ouspensky, P. D. *A Further Record: Extracts From Meetings 1928–1945*. London: Arkana, 1986.

———. *In Search of the Miraculous: Fragments of An Unknown Teaching*. Orlando: Harcourt Inc., 1949.

Parley, Peter. *Peter Parley's Almanac for Old and Young*. Philadelphia: Desilver, Thomas, & Company, 1836.

Phipps, William E. *Darwin's Religious Odyssey*. Harrisburg, PA: Trinity Press International, 2002.

Pickrell, John. "Rat Radar: Rodent Uses Natural 'GPS.'" *National Geographic*. January 29, 2004.

Radin, Dean. *The Conscious Universe: The Scientific Truth of Psychic Phenomena*. New York: HarperCollins, 1997.

Razwy, Sayed A. A., and Ali Abdullah Yusuf. *The Qur'an Translation*. New York: Tahrike Tarsile Qar'an, Inc., 2007.

Regardie, Israel. *The Tree of Life: An Illustrated Study in Magic*. Woodbury, MN: Llewellyn Publications, 2001.

Rotton, J., and I. W. Kelly. "Much Ado about the Full Moon: A Meta-Analysis of Lunar-Lunacy Research." *Psychological Bulletin* 97 (March 1985): 286–306.

Ruzicka, A., Snyder, G.A., and L. A. Taylor. "Giant Impact and Fission Hypotheses for the Origin of the Moon: A Critical Review of Some Geochemical Evidence." *International Geology Review* 40 (1998): 851–64.

Sanderson, Ivan T. "Mysterious 'Monuments' on the Moon." *Argosy* 371, no. 2, August 1970.

Sagan, Carl, and Ivan Shklovski. *Intelligent Life in the Universe*. San Francisco: Holden Day, 1966.

Shakespeare, William. *A Midsummer Night's Dream*. Hong Kong: Forgotten Books, 2012.

Shapiro, Fred R. *The Yale Book of Quotations*. New Haven, CT: Yale University Press, 2006.

Sherrill, Thomas J. "A Career of Controversy: The Anomaly of T. J. J. See." *Science History Publications Ltd.* (1999): 25–50.

Shirley, John. *Gurdjieff: An Introduction to His Life and Ideas*. New York: Tarcher/Penguin, 2004.

Smith, G. Elliott. *The Evolution of the Dragon*. Philadelphia, PA: Albert Saifer, 1918.

Sperling, Harry, and Maurice Simon. *The Zohar, Volume 1*. New York: Soncino Press, 1984.

Steckling, Fred. *We Discovered Alien Bases on the Moon*. Los Angeles: GAF Publishing, 1981.

Stolov, Harold L., and A. G. W. Cameron. "Variations of Geomagnetic Activity with Lunar Phase." *Journal of Geophysical Research* 69, no. 23 (1964): 4975–82.

Stooke, P. J. "Neolithic Lunar Maps at Knowth and Baltinglass, Ireland." *Journal for the History of Astronomy* (February 1994): 39–55.

Strieber, Whitley. *Transformation: The Breakthrough*. New York: Beech Tree Books, William Morrow & Co. Inc., 1988.

Swann, Ingo. *Penetration: The Question of Extraterrestrial and Human Telepathy*. Rapid City, SD: Ingo Swann Books, 1998.

Terry, Maury. *The Ultimate Evil: An Investigation into America's Most Dangerous Satanic Cult*. New York: Doubleday, 1987.

Vallee, Jacques. *Messengers of Deception: UFO Contacts and Cults*. Berkeley, CA: And/Ord Press, 1978.

Vasin, Mikhail, and Alexander Scherbackov. "Is the Moon the Creation of Intelligence?" *Sputnik* July 1970.

Watson, Lyall. *Supernature: A Natural History of the Supernatural*. London: Hodder and Stoughton Ltd., 1973.

Webb, W. L. *Brief Biography and Popular Account of the Unparalleled Discoveries of T. J. J See*. Lynn, MA: Thos. P. Nichols & Son Co., 1913.

Wilkins, H. P. *Our Moon*. London: Fredrick Muller, Ltd., 1954.

Wilkins, Harold T. *Flying Saucers on the Attack*. New York: Henry Holt and Company Inc., 1955.

Wilson, Colin. *Atlantis to the Sphinx: Recovering the Lost Wisdom of the Ancient World*. London: Virgin Books Ltd., 1996.

———. *Dreaming to Some Purpose*. London: Century, 2004.

———. *The Occult*. London: Grafton Books, 1979.

Wilson, Don. *Our Mysterious Spaceship Moon*. London: Sphere Books Ltd., 1976.

———. *Secrets of Our Spaceship Moon*. London: Sphere Books Ltd., 1980.

Wieczorek, Mark A., et al., "The Crust of the Moon as Seen by GRAIL." *Science* 339, no. 6120 (February 8, 2013): 671–75.

Workman, Lance, and Will Reader. *Evolutionary Psychology: An Introduction*. Cambridge: Cambridge University Press, 2004.

Zigrosser, Carl. "Gurdjieff—Salzmann—Orage." *Gurdjieff International Review* 8 (2004).

INDEX

Tunguska meteorite strike, 29
Twain, Mark, 12
Twardowski, Mr., 176–77
two moons theories, 272, 274–75, 278, 335

UFOs
    Apollo 13 mission sabotaged by, 315–16
    Disclosure Project, 308–11
    images altered by NASA, 308–9
    images burned by NASA, 309
    modern era of, 286–87
    Moon connection with, 287–89
    seen by Apollo astronauts, 312–16
    simplistic view of, 316–17
    See also transient lunar phenomena (TLP)
uranium on the Moon, 127, 128, 129
Urey, Harold, 90, 107, 133
Utchat of Thoth, 72

Vallee, Jacques, 296
Vasin, Mikhail, 119, 120, 129
Venus, 21–22, 39
Venus of Laussel, 144, 150, 151, plate 7
Verne, Jules, 275
virtual-reality universe, 233, 235–36
volcanic activity
    Compton-Belkovich volcanic complex, 126–27
    maria formed by, 15, 34, 124–25
    melt inclusions from, 97–99
    in spaceship moon theory, 124–25
Volvox, 199–200

Wagner, Robert, 303–4
Wallace, Alfred Russel, 67
Ward, William, 94
water
    ice on the Moon, 48–53
    Moon beliefs and, 172–74
    in Moon's interior, 97, 98–99, 114
Watt, James, 65
Webb, William, 87
Wedgwood, Josiah, 63–64, 68, 72
Weir, Peter, 7
Wells, H. G., 8–9, 272
Whitehurst, John, 65
Wieczorek, Mark, 353–54
Wilkins, Harold T., 287
Wilkins, Hugh Percy, 106–7, 284–85
Wilson, Colin, 192, 242–43
Wilson, Don, 116–17, 120, 123, 132–33, 139
Wilson, Donald K., 25–26
Withering, William, 65
Wittcomb, Samuel, 291
Wolf, Karl, 309–10

Yazidis, 245–46
Yelov, Abram, 246, 247
Yesod, 337, 338, 340, 342
Young, John, 31

Zelator, The, 332–34, 335, 337, 342, 344–47, 349
Zero Point Field, 27
Ziggurat of Ur, 156, plate 8
zirconium in lunar maria, 93, 122